Barron's Regents Exams and Answers
Earth Science

DAVID BEREY
Educational Consultant

Former member of
Science Department
Roslyn High School
Roslyn, L.I., N.Y.

EDWARD J. DENECKE, JR.
Staff Development Specialist

Science Technical Assistance Center
Whitestone, N.Y.

BARRON'S

Barron's Educational Series, Inc.

All inquiries should be addressed to:
Barron's Educational Series, Inc.
250 Wireless Boulevard
Hauppauge, New York 11788
http://www.barronseduc.com

ISBN 0-8120-3165-2
ISSN 1069-2959

PRINTED IN THE UNITED STATES OF AMERICA
9 8 7 6 5 4 3 2 1

Contents

Regents Examinations, Answers, and Self-Analysis Charts   69

Preface

TO THE STUDENT:

This book has been designed to help you check your mastery of basic earth science concepts. It contains the following special features which should help you in your study:

(1) *Wrong answers explained.* Have you ever answered a question on a test incorrectly and wondered why your answer was wrong? Since you surely have, you should find the explanations for wrong choices helpful in clearing up misconceptions. In addition you will find that reading explanations for questions answered correctly will help to further strengthen your background.

(2) *Earth Science Reference Tables.* These tables are provided to students for use while taking the New York State Earth Science Regents examination. Some questions require the use of these tables and this is usually indicated in the question. For other questions you may find information in the tables helpful. It is therefore a good idea to become generally familiar with what is contained in the tables.

(3) *Topic Outline matrix.* The topic outlines from the New York State Earth Science syllabus and the New York State Program Modification in Earth Science have been cross referenced to questions from past examinations. This should help you to locate questions to review particular concepts. See "HOW TO USE THE TOPIC OUTLINE" for additional instructions.

The author is indebted to Mr. George Matthias, Croton-Harmon H.S., Croton, New York, who reviewed early drafts and made valuable suggestions.

D.B.

How to Use This Book

FORMAT OF THE EARTH SCIENCE REGENTS EXAMINATION

The examination consists of a performance portion containing five tasks to measure laboratory skills and a written portion containing multiple-choice questions.

THE PERFORMANCE TEST

Since laboratory experiences are an essential part of a science course, a portion of the Earth Science Regents Examination is devoted to assessing laboratory skills. Tasks have been identified from laboratory experiments that you will have performed during the school year. These tasks, which represent skills that you are expected to have mastered, change only slightly, if at all, from year to year.

The performance portion of the examination is administered separately from the written portion, normally two weeks earlier. Arrangements for administering the performance exam are made at each school in accordance with guidelines set by the New York State Education Department.

The scoring for each task is based upon accuracy. Values within a certain range are granted the full two points allotted to each task. Values within a slightly larger, less precise range earn one point. It is possible to accumulate a maximum of 10 points on the performance portion of the regular examination and 15 points on the performance portion of the program modification edition.

Additional information regarding the performance test, including an indication of the six tasks to be completed, will be provided by your teacher when this portion of the examination is given. The following is an outline set of the six tasks that have been included in past examinations. The time allowed for completing the tasks at each station is 6 minutes.

Note: The following description represents that information the State Education Department has stated may be shared with students before taking the performance part of the examination. You should be familiar with the skills being assessed because you have used them in laboratory activities throughout the year. However, you will not be allowed to practice the entire test or any of the individual stations before this performance component is administered.

Station 1 . . *Identification*

Using a mineral identification kit and key, the student will determine the characteristics of two mineral samples and identify each sample by name.

Station 2 . . *Classification*

Using rock identification charts, the student will classify two rock samples as igneous, sedimentary, or metamorphic and state the reason for each classification, in one or more complete sentences.

Station 3 . . *Angular Measurement*

Using a plastic hemisphere that models the apparent path of the sun, an external protractor, a ruler, and masking tape, the student will locate the position of the sun at a given time and measure the distance between that position and a fixed point.

Station 4 . . *Mass–Density*

Using a single-pan, triple-beam decigram balance, a mineral density chart, and a calculator, the student will find the density, determine the mass, and calculate the volume of a given mineral sample.

Station 5 . . *Settling Time*

Using a column of fluid, three sizes of plastic particles of the same density, a stopwatch, and a calculator, the student will determine the average settling time for each of the three sizes of particles.

Station 6 . . *Graphing*

Using data obtained from Station 5, the student will construct a line graph of average settling time vs. particle diameter and will determine the settling time for another given particle diameter.

THE WRITTEN TEST—REGULAR EXAMINATION

The written portion of the regular examination represents 90 points of the total score and has two parts. Part I contains 55 questions taken from the 14 topics in the Earth Science Syllabus. Each question in this part must be answered. Part II contains ten groups of five questions each. You are to select seven of these groups and answer all five questions in each group. Most groups contain questions that relate to only one or two of the topics in the syllabus. You should initially select the seven groups of questions to answer on the basis of the strength of your knowledge of the various topics in the course of study. If time permits, you may attempt to answer the questions in any of the other three groups. You may decide to substitute one or more of these groups for your original choices if you feel more confident of your answers. In that case be sure to erase completely your answers to the groups you originally chose.

Although the questions on each written test are unique, they are similar to questions on past Regents examinations. By studying questions that have been asked in the past, therefore, you can strengthen your skills and knowledge in preparation for the examination you will take.

THE WRITTEN TEST—PROGRAM MODIFICATION EDITION

The written portion of the program modification edition represents 75 points of the total score and has three parts.

Part I contains 40 questions taken from the 9 core topics in the Earth Science Program Modification syllabus. Each question in this part must be answered.

Part II contains six groups of ten questions each. You are to select *one* of these groups and answer all ten questions in that group. Each group contains questions that relate to only one of the optional/extended topics in the program modification syllabus. You should initially select the group of questions to answer on the basis of the strength of your knowledge of the various topics in the course of study. If time permits, you may attempt to answer the questions in any of the other groups. You may decide to substitute one or more of these groups for your original choice if you feel more confident of your answers. In that case be sure to erase completely your answers to the group you originally chose.

Part III contains a varying number of free-response questions, which may be multi-part, but whose value totals 25 points. The questions in Part III may cover any content in the 9 core units. Free-response questions will test your graphing, mapping, mathematical, and reasoning skills in a more comprehensive way than can be accomplished with a multiple-choice question. Review free-response questions from previous examinations to familiarize yourself with this type of question. This part of the test is a recent addition and each examination may contain questions unlike those in previous examinations. Keep these points in mind:

- Write in complete sentences whenever you are asked to answer in your own words.
- Show all work, and use correct units in mathematical problems.
- Be familiar with map scales, legends, constructing isolines, and drawing profiles.

Although the questions on each written test are unique, they are similar to questions on past Regents examinations. By studying questions that have been asked in the past, therefore, you can strengthen your skills and knowledge in preparation for the examination you will take.

THE LOCAL COMPONENT—PROGRAM MODIFI-CATION EDITION

The local component is a long-term project approved by your teacher. It is worth 10 points toward your examination score. Scientists are rarely able to perform an experiment and obtain meaningful results within a time frame of one or two class periods. Therefore, local component projects require an extended period of time. In performing these projects, you should have the opportunity to observe change, gather data, record information, and make conclusions about the world around you. All projects should be approved by the teacher before they are conducted. Although library research is an important aspect of nearly any scientific inquiry, students are strongly encouraged to make the gathering and analysis of their own observations and data the primary emphasis of their long-term projects.

Tips for Local Component Projects:
- Keep the focus narrow. Do something small, but do it very well. Stress quality rather than quantity. A report need not be long to be excellent.

- The project should involve doing science, not just writing about science. Try to gather original documentation and/or data.
- New technology will allow the use of equipment and media that were not available in the past. Word processors, computers, and still or video cameras can add to the value of your report.
- Get an early start. When contacting people or writing for information, allow plenty of time for responses. Some projects will require data collected over a number of months.

CONTENTS OF THIS BOOK

THE TOPIC OUTLINE AND THE EARTH SCIENCE REFERENCE TABLES

The first part of this book consists of two special features, the Topic Outline and the Earth Science Reference Tables. For detailed instructions on how to use the *Topic Outline,* see page 25. The *Earth Science Reference Tables* in this book are the same ones that are provided to students for use while taking the Regents examination. You should be familiar with these tables at the time of the exam since you will have used them during the school year. Some questions require the use of these tables, and this fact is usually indicated in the question. For other questions you may find information in the tables helpful. It is therefore a good idea to become generally familiar with what is contained in the tables.

REGENTS EXAMINATIONS AND ANSWERS

In the second, much longer part of the book there are six actual Earth Science Regents examinations (three regular and three modified) with Answer Keys. Every correct answer is fully explained. Also, to help clear up misconceptions that may lead students astray, in many cases explanations for wrong answers are provided as well.

Test-Taking Tips

The following pages contain several tips to help you achieve a good grade on the Earth Science Regents exam. They are divided into GENERAL HELPFUL TIPS and SPECIFIC HELPFUL TIPS.

GENERAL HELPFUL TIPS

TIP 1

Be Confident and Prepared

SUGGESTIONS
- Review previous tests.
- Use a clock or watch, and take previous exams at home under examination conditions, (i.e., don't have the radio or television on.)
- Get a review book. (The preferred book is Barron's *Let's Review: Earth Science*.)
- Talk over the answers to questions on these tests with someone else, such as another student in your class or someone at home.
- Finish all your homework assignments.
- Look over classroom exams that your teacher gave during the term.
- Take class notes carefully.
- Practice good study habits.
- Know that there are answers for every question.
- Be aware that the people who made up the Regents exam want you to pass.

- Remember that thousands of students over the last few years have taken and passed an Earth Science Regents. You can pass too!
- On the night prior to the exam day: lay out all the things you will need, such as clothing, pens, and admission cards.
- Go to bed early; eat wisely.
- Bring at least two pens to the exam room.
- Bring your favorite good luck charm/jewelry to the exam.
- Once you are in the exam room, arrange things, get comfortable, be relaxed, attend to personal needs (the bathroom).
- Keep your eyes on your own paper; do not let them wander over to anyone else's paper.
- Be polite in making any reasonable requests of the exam room proctor, such as changing your seat or having window shades raised or lowered.

TIP 2

Read Test Instructions and Questions Carefully

SUGGESTIONS

- Be familiar with the test directions ahead of time.
- Decide upon the task(s) that you have to complete.
- Know how the test will be graded.
- Know which question or questions are worth the most points.
- Give only the information that is requested.
- Where a choice of questions exists, read all of them and answer *only* the number requested.
- Underline important words and phrases.
- Ask for assistance from the exam room proctor if you do not understand the directions.

TIP 3

Budget Your Test Time in a Balanced Manner

SUGGESTIONS

- Bring a watch or clock to the test.
- Know how much time is allowed.
- Arrive on time; leave your home earlier than usual.
- Prepare a time schedule and try to stick to it. Remember that Regents exams are longer than classroom tests, so you will need to pace yourself accordingly.
- Answer the easier questions first.
- Devote more time to the harder questions and to those worth more credit.
- Don't get "hung up" on a question that is proving to be very difficult; go on to another question and return later to the difficult one.
- Ask the exam room proctor for permission to go to the lavatory, if necessary, or if only to "take a break" from sitting in the room.
- Plan to stay in the room for the entire three hours. If you finish early, read over your work—there may be some things that you omitted or that you may wish to add. You also may wish to refine your grammar, spelling, and penmanship.

TIP 4

Be "Kind" to the Exam Grader/Evaluator

SUGGESTIONS

- Assume that you are the teacher grading/evaluating your test paper.
- Answer questions in an orderly sequence.
- Write legibly.
- Answer Part III questions with complete sentences.
- Proofread your answers prior to submitting your exam paper. Have you answered all the Part I and (if appropriate) Part III questions and the required number of Part II questions?

TIP 5

Use Your Reasoning Skills

SUGGESTIONS
- Answer *all* questions.
- Relate (connect) the question to anything that you studied, wrote in your notebook, or heard your teacher say in class.
- Relate (connect) the question to any film you saw in class, any project you did, or to anything you may have learned from newspapers, magazines, or television.
- Decide whether your answers would be approved by your teacher.
- Look over the entire test to see whether one part of it can help you answer another part.
- Be cautious when changing an answer. Try to remember why you selected the first answer to be sure that the new answer is better.

TIP 6

Don't Be Afraid to Guess

SUGGESTIONS
- In general, go with your first answer choice.
- Eliminate obvious incorrect choices.
- If you are still unsure of an answer, make an educated guess.
- There is no penalty for guessing; therefore, answer ALL questions. An omitted answer gets no credit.

Let's now review the six GENERAL HELPFUL TIPS for short-answer questions:

SUMMARY OF TIPS
1. Be confident and prepared.
2. Read test instructions and questions carefully.
3. Budget your test time in a balanced manner.
4. Be "kind" to the exam grader/evaluator.
5. Use your reasoning skills.
6. Don't be afraid to guess.

SPECIFIC HELPFUL TIPS FOR THE SHORT-ANSWER (MULTIPLE-CHOICE) QUESTIONS

TIP 1

Answer the Easy Questions First

The best reason for using this hint is that it can build up your confidence. It also enables you to use your time more efficiently. You should answer these questions first, while skipping over and circling the numbers of the more difficult questions. You can always return to these later during the exam. Easy questions usually contain short sentences and few words. The answer can often be arrived at quickly from the information presented.

EXAMPLE

Which object best represents a true scale model of the shape of the Earth?

1 a Ping-Pong ball	3 an egg
2 a football	4 a pear
	(from June 1997 Regents)

The correct answer, choice 1, is obvious if you know the true shape of the Earth because only a Ping-Pong ball has a completely round shape.

TIP 2

Consider What the Questioner Wants You to Do, and Underline Key Words

Remember that the people who made up the Regents questions had specific tasks they want you to accomplish. These tasks can be understood if you read instructions carefully, underline key words, and put yourself "in their shoes." Try to figure out what they would want a student to do with a given question. Determine for yourself exactly what is being tested. Underlining helps you focus on the key ideas in the question.

EXAMPLE

Oxygen is the most abundant element by volume in the
Earth's

1 inner core
2 crust

3 hydrosphere
4 troposphere
(from June 1997 Regents)

The key words in the stem that need to be underlined are "Oxygen"
and "volume." Charts in the *Earth Science Reference Tables* contain
information about the composition of each of these zones. Oxygen is
most abundant by volume only in the crust, choice 2. The other
choices have different elements that are most abundant. If you had
based your answer on "mass" instead of "volume" in this case you
would still be correct but in other cases your answer might have been
wrong.

TIP 3

Look for Clues Among Choices in the Question, as Well as in Other Questions

---

By reading questions and choices carefully, you may often find words
and phrases that provide clues to an answer. This hint is important
because it assists you in making links and connections between various
questions.

EXAMPLE

Hot springs on the ocean floor near the mid-ocean ridges
provide evidence that

1 climate change has melted huge glaciers
2 marine fossils have been uplifted to high elevations
3 meteor craters are found beneath the oceans
4 convection currents exist in the asthenosphere
(from June 1997 Regents)

Choice 4 is correct. The "clue" in the question is in the nature of hot
springs and that they are caused by heat. Your knowledge of Earth
Science should tell you that convection currents involve the transfer of
heat energy so they could be the cause of hot springs. None of the
other choices can directly result in the heating of water.

TIP 4

Examine All Possibilities
Beware of Tricky and Tempting Foils (Decoys)

By remembering this tip, you will be careful to survey all possible responses before making a selection. A given choice, or two, may initially appear to be the correct answer. This often happens with complex concepts that include two or more related ideas.

EXAMPLE

Why do the locations of sunrise and sunset vary in a cyclical pattern throughout the year?

1 The Earth's orbit around the Sun is an ellipse.
2 The Sun's orbit around the Earth is an ellipse.
3 The Sun rotates on an inclined axis while revolving around the Earth.
4 The Earth rotates on an inclined axis while revolving around the Sun.

(from June 1997 Regents)

Choice 4 is correct because the changes in the locations of sunrise and sunset are caused by the tilt in the Earth's axis. Choice 1 is a true statement but it is not the cause of changes in sunrise and sunset. When reading answers, be careful not to forget the question or you may accidentally choose the first true statement you encounter.

TIP 5

Always Select the More Broad, Encompassing Choice

This tip is most helpful when two or more choices are correct, but you conclude that one choice is broader (or more encompassing) than the other. Indeed, one choice may actually include the other or others.

EXAMPLE

A conglomerate contains pebbles of shale, sandstone and granite. Based on this information which inference about the pebbles in the conglomerate is most accurate?

1 They were eroded by slow-moving water.
2 They came from other conglomerates.
3 They are all the same age.
4 They had various origins.

(from June 1997 Regents)

Choice 4 is the best answer because it provides for the broadest possibilities. The conglomerate could have come from the weathering of other conglomerates, but this is far less likely. You can sometimes eliminate a choice because it is too restrictive and therefore less likely. This applies to choice 3, which suggests that the pebbles are all the same age. Although this is possible, it is less likely to be true without additional information.

TIP 6

Use a Process of Elimination

This tip provides a very good way of arriving at an answer. It is particularly useful when you face a difficult question and are unsure of the best response. Also, it increases your chances of coming up with the correct answer and (1) assists in discarding *unacceptable* choices, and (2) narrows the possible *acceptable* choices. (You may wish to physically cross out on the question page the choices that you decide are incorrect.)

EXAMPLE

By which process does water vapor leave the atmosphere and form dew?

1 condensation	3 convection
2 transpiration	4 precipitation

(from June 1997 Regents)

The correct choice is 1. Suppose you had forgotten the term condensation. You might still answer the question correctly by eliminating the other choices. You could eliminate transpiration because it refers to the process by which plants release moisture, convection because it refers to a method of heat transfer, and precipitation because it does not cause dew to form. This leaves only condensation as a possible choice.

TIP 7

Detect Differences Among the Choices Presented

You should be careful in picking out foils and disclaimers, something we mentioned in Tip 4. In a similar manner, this tip helps you in choosing among general and specific answers. This is especially important when you have reduced your selections to two choices.

EXAMPLE

On a clear April morning near Rochester, New York, which surface will absorb the most insolation per square meter?

1 a calm lake
2 a snowdrift

3 a white-sand beach
4 a freshly plowed
   farm field

(from June 1997 Regents)

Choice 4 is correct because insolation is most readily absorbed when a surface is dark colored and irregular. It is easy to eliminate choices 2 and 3 because the brightness of these surfaces suggests that they reflect back much of the insolation. You might be inclined to select choice 1 because the insolation does readily pass through the surface of the water. The key, however, is the word "absorb," which describes what actually happens when the insolation reaches the surface of the farm field.

TIP 8

Don't Choose an Answer That Is Correct in Itself but Incorrect as It Relates to the Question

It is crucial to keep this tip in mind when evaluating a question that has attractive choices. They will appear attractive because each has an element of truth; however, you must decide which one of these has the greatest relationship to the question itself.

EXAMPLE

The best evidence that the Earth rotates on its axis comes from observations of

1 seasonal changes of constellations
2 the apparent motion of a Foucault pendulum
3 the changing altitude of the Sun at noon
4 the changing altitude of Polaris

(from January 1997 Regents)

Choice 2 is correct. Choices 1 and 3 are also true statements, but they relate to the revolution of the Earth around the Sun rather than the rotation of the Earth around its axis. The rotation and the revolution of the Earth are closely related concepts so it is easy to pick an incorrect answer if the question is not interpreted correctly.

TIP 9

Look for "Giveaways" and "Freebies"

There are times when a test question practically gives away the answer. You can determine such rare moments by focusing on obvious words, prefixes, grammatical construction, and other revealing tips.

EXAMPLE

The primary purpose of a classification system is to enable people to

1 make measurements that are very accurate
2 eliminate inaccurate inferences
3 organize observations in a meaningful way
4 extend their powers of observation

(from January 1997 Regents)

Choice 3 is the correct answer. The "giveaway" here is the word "organize." If you know that a classification system is a way of grouping things, then it is easy to relate the word "organize" to a classification system.

TIP 10

Think Out the Answer Before Looking at the Possible Choices

This tip helps to stimulate memory recall. As you read a question, your "intellectual radar" may pick up something that will jog loose a key thought, concept, or fact in your mind.

EXAMPLE

The theory of continental drift suggests that the

1 continents moved due to changes in the Earth's orbital velocity
2 continents moved due to the Coriolis effect, caused by the Earth's rotation
3 present-day continents of South America and Africa are moving toward each other
4 present-day continents of South America and Africa once fit together like puzzle parts

(from June 1996 Regents)

Choice 4 is the correct answer. When answering complex questions, it often helps to think out the correct answer first before looking at the choices. Then try to match the choices to the general answer you have in mind. Knowing that the continents were once part of one large landmass that broke apart and that this is explained by the theory of continental drift can lead you to selecting choice 4.

TIP 11

Make Informed and Educated Guesses

If you are not sure of an answer, don't be afraid to guess. Any answer is better than no answer—you have nothing to lose, as there is no penalty for guessing. Remember that the correct answer is there, somewhere, right on the exam page, waiting for you to find it. If you've eliminated one or more options, then your chances of picking the right answer increases from one out of four to one out of three, etc. A word of caution—guessing should be used only as a last resort. Do not go into the Regents exam room expecting to pass by guessing your way through the questions. There is no substitute for careful, diligent, exam preparation long before the exam date itself.

EXAMPLE

A fine-grained igneous rock was probably formed by

1 weathering and erosion
2 great heat and pressure that did not produce melting
3 rapid cooling of molten material
4 burial and cementation of sediment

(from June 1996 Regents)

Choice 3 is the correct answer. You could eliminate choices 1 and 4 because they relate to sedimentary rocks. Suppose, however, that you couldn't remember between igneous and metamorphic rocks which is which. At least, at this point, you have reduced your chances of guessing correctly to one out of two. Of course, in this case you could obtain additional help by consulting the *Earth Science Reference Tables* to help you distinguish between igneous and metamorphic rocks.

Let's now review the eleven SPECIFIC HELPFUL TIPS for short-answer questions:

SUMMARY OF TIPS

1. Answer the easy questions first.

2. Consider what the questioner wants you to do, and underline key words.

3. Look for clues among choices in the question, as well as in other questions.

4. Examine all possibilities. Beware of tricky and tempting foils (decoys).

5. Always select the more broad, encompassing choice.

6. Use a process of elimination.

7. Detect differences among the choices presented.

8. Don't choose an answer that is correct in itself but incorrect as it relates to the question.

9. Look for "giveaways" and "freebies."

10. Think out the answer before looking at the possible choices.

11. Make informed and educated guesses.

In addition to these SPECIFIC HELPFUL TIPS, here are five more bonus ones:

1. Read each question twice.
2. Generally, try to go with your first inclination.
3. Avoid looking for patterns.
4. Be aware that universals (i.e., *always, never, only*) should usually be disregarded as possible correct choices.
5. Remember that sometimes you may find helpful information in the *Earth Science Reference Tables* even when the question does not specifically refer you to them. It is therefore helpful to be familiar with what is contained in the reference tables prior to the exam.

HELPFUL TIPS FOR THE PART III QUESTIONS

Part III is a constructed-response section that counts for 25 points. All students must answer every question in this part of the examination. Some questions may require the use of the *Earth Science Reference Tables*.

TIP 1

Know What a Constructed-Response Question Is

Constructed response means that you must provide, in writing, your own response to a question, rather than choosing your answer from a list of possible choices. There are two types of constructed-response questions: open response and free response. In an open-response question there is only one correct answer, whereas a free-response question has many alternative answers.

Example of an **open-response** type question.

Base your answers to questions 109 and 111 on the diagram below and on your knowledge of Earth Science. The diagram represents the apparent path of the Sun on the dates indicated for an observer in New York State. The diagram also shows the angle of Polaris above the horizon.

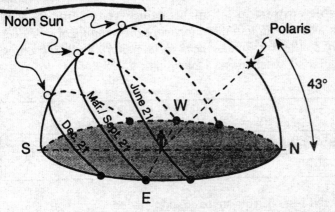

Noon Sun

Polaris

43°

Mar./Sept. 21

Dec. 21

June 21

W

S

N

E

109 State the latitude of the location represented by the diagram to the *nearest degree*, including the latitude direction in your answer. [2]

Question 109 is an open-response question; it has only one possible answer—43° N.

Example of a **free-response** type question.

108 The climate of most locations near the Equator is warm and moist.

a Explain why the climate is warm. [1]

_____

_____

_____

b Explain why the climate is moist. [1]

_____

_____

_____

Although a correct response must be scientifically correct, there are many alternative ways of wording your answers to this question, so it is a free-response question.

Acceptable answers to (a) might include statements such as: *"Locations near the Equator receive more direct sunlight."* or *"The intensity of sunlight is greater near the Equator."* Both responses, although different, would receive full credit.

TIP 2

Understand Why These Kinds of Questions Are on the Examination

---

Constructed-response questions are designed to:
- assess your *understanding* of earth science concepts rather than straight recall of factual information
- require you to demonstrate that you can pull all the pieces together to solve a problem
- give the teacher insight into how you think by requiring you to defend your answer in writing or show all of your work
- assess knowledge and skills that cannot easily be assessed using a multiple-choice format

---

TIP 3

Know What Will Be Expected of You for This Part of the Examination

---

In this part of the examination, you will be expected to demonstrate in a variety of ways that you have certain knowledge, skills, and abilities. You may be asked to construct a graph or draw a diagram. You may be asked to solve a mathematical problem and show all of your work. You may be asked to provide explanations for physical phenomena, or write an essay on a topic in earth science.

Some of the skills and abilities that may be assessed in Part III of the examination are:

Science Inquiry Skills
- Use data to construct a reasonable explanation.
- Communicate, critique, and analyze work.
- Develop descriptions, explanations, predictions, and models using evidence.
- Think critically and logically to make the relationships between evidence and explanations.
- Recognize and analyze alternative explanations and predictions.
- Use mathematics to ask questions, to gather, organize, and present data, and to structure explanations.
- Formulate and revise scientific explanations and models using logic and evidence.
- Communicate and defend a scientific argument.

Earth Science Skills and Abilities
- Find Polaris using the Big Dipper.
- Relate the altitude of Polaris to the observer's latitude
- Apply Eratosthene's method for determining the Earth's circumference.
- Use latitude and longitude to locate positions on a map.
- Construct and interpret isoline maps, such as topographic or weather maps.
- Calculate gradient on an isoline map.
- Relate crystal size to cooling rate.
- Classify rocks and minerals by their characteristics.
- Use the rock identification charts in the *Earth Science Reference Tables*.
- Analyze a seismogram to locate the epicenter, *P*- and *S*-wave travel times, and origin time of the earthquake.
- Relate observations of earth phenomena to plate tectonics
- Explain differences in weathering rates of earth materials.
- Recognize bedding patterns and relate them to the mode of deposition.
- Recognize glacial features and relate them to specific erosional or depositional processes.
- Use the principle of superposition to establish relative age of rock layers and features.
- Correlate rock outcrops.
- Find the age of a sample given the proportions of original isotope and decay product.

- Interpret weather station models and synoptic weather maps.
- Given a synoptic weather map, predict movements of weather systems and predict the weather at selected localities.
- Discuss the various forms of electromagnetic radiation shown in the *Earth Science Reference Tables*.
- Interpret a heating curve for water.
- Relate the angle and duration of insolation to surface temperatures.
- Explain the greenhouse effect.
- Use water budget graphs to analyze climate.
- Interpret diagrams of the Earth's orbit.
- Illustrate the relative positions of the noon sun throughout the year in New York State.
- Translate among celestial observations, two-dimensional and three-dimensional models of Earth's motions.
- Calculate the eccentricity of an orbit.
- Explain phases of the moon.
- Compare and contrast the geocentric and heliocentric models of the universe.
- Compare and contrast conditions on Earth with those on nearby planets.
- Apply the Doppler Effect to determine motion of celestial objects.
- Be able to discuss current issues such as global warming, ozone depletion, deforestation, recycling, El Niño, and the effects of population growth on natural processes.

TIP 4

Work to Improve Your Score on These Types of Questions

---

1. **Determine what the question is asking you to do. Underline key words in the question.**

    Most questions of this type have an action word that identifies what you must do to earn points. Below, the action words in a question you looked at earlier are bold and underlined. Other key words are underlined.

State the latitude of the location to the nearest degree. **Include** the latitude direction in your answer. [2]

2. **Check to see what points are being awarded for.**

    The points are listed in brackets after each question. If more than one point is being awarded, you can usually make a good guess about what each point will be given for if you read the question again carefully. In the example above, you have probably guessed that one point is given for stating the latitude to the nearest degree, and one point is given for including the correct direction. The answer was 43° North. If you wrote 43°, you would only get one point. To get full credit, you must state both 43° and North.

3. **Always include units in your answers and check to make sure the form of your answer matches the form asked for in the question.**

    Consider this example:

Air Temperature	21°C
Barometric Pressure	993.1 mb
Wind Direction	From the East
Windspeed	25 knots

State the underlined barometric pressure in its proper form, <u>as used on a station model.</u>

What you are asked to do is state the barometric pressure. But if you write 993.1 mb, you would be marked incorrect because the form asked for was "as used on a station model." On a station model, the decimal point is dropped and the last three digits only are listed. Your answer would have to be given as 931 because that is what would appear on a station model.

4. **When asked to answer in the form of a statement or essay, use complete sentences and organize your answer carefully.**

    Complete sentences begin with a capital letter, have a subject and a verb, and end with a period. Very often, a point is awarded solely for answering in a complete sentence—even if the answer is wrong!! So read over your statements or essays carefully.

5. **Examine the answer sheet for Part III for clues to how the question is to be answered.**

   The answer sheet for Part III will often give you a format for answering the question. For example, look at the question and answer sheet for that question below.

115 Using one or more complete sentences, state one
    reason that ultravoilet rays are dangerous. [2]

_____

_____

_____

If the answer sheet provides several blank lines, you are probably being asked to write a short essay. If it has a blank graph, you are probably being asked to construct a graph. Be sure to include all elements asked for in the question. For example, in the graph below you must not only plot points, but plot them with specific shapes, label axes, and choose appropriate scales for each axis.

101 Plot the data for rock sample *A* for the 20 minutes
    of the investigation. Surround each point with a
    small triangle and connect the points. [1]

   Example: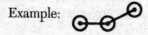

102 Plot the data for rock sample *B* for the 20 minutes
    of the investigation. Surround each point with a
    small triangle and connect the points. [1]

   Example: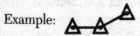

103 Plot the data for rock sample *C* for the 20 minutes
    of the investigation. Surround each point with a
    small triangle and connect the points. [1]

   Example: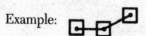

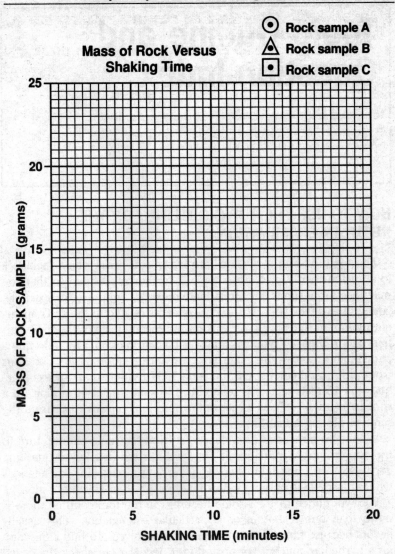

Mass of Rock Versus Shaking Time

Rock sample A
Rock sample B
Rock sample C

MASS OF ROCK SAMPLE (grams)

SHAKING TIME (minutes)

6. **Review questions from Part III of previous exams.**
   Use the Answers Explained section of your book to review how Part III questions are answered and how point values are awarded. Focus more on the types of questions being asked, rather than on the specific questions.

Topic Outline and Question Index

HOW TO USE THE TOPIC OUTLINE FOR THE REGULAR EXAMINATION

The topic outline is divided into five major areas which include a total of fourteen topics. Each topic is further subdivided so that you may locate questions for individual concepts. Suppose, for example, you wish to check your mastery of concepts related to the sun's apparent motion in the sky. If you look on the second page of the topic outline under Topic IV, you will find "A-1.4 Sun motion." By following a parallel line across the page you can locate appropriate questions from past examinations. On the June 1997 examination, for example, question 15 tests concepts related to the sun's apparent motion. In a similar manner you can locate other questions on this topic from the same or past examinations.

When you have identified questions from the topic outline, turn to the appropriate examination in the book for the first keyed question. Try to answer it and then check your response, reviewing explanations where desirable. Continue in this manner for other keyed questions.

You will notice that for some headings in the topic outline there is more than one test item on a particular examination. This usually occurs because more than one concept is involved. To find a question on a particular concept you may find it necessary to check the keyed questions on several examinations.

HOW TO USE THE TOPIC OUTLINE FOR THE PROGRAM MODIFICATION EDITION

The topic outline is divided into 9 core Units and 6 optional/extended Topics. The core Units are numbered from 1 to 9; the optional/extended Topics are lettered from A to F. Each unit or topic is further subdivided so that you may locate questions for individual concepts. Suppose, for example, you wish to check your mastery of concepts related to tectonic plate boundaries. If you look in the topic outline under Unit 3, you will find "5. Diverg, conv & transf boundaries." By following a parallel line across the page you can locate appropriate questions from past examinations. On the 1997 examination, for example, question 8 tests concepts related to the plate boundaries. In a similar manner you can locate questions on this topic from other past examinations.

When you have identified questions from the topic outline, turn to the appropriate examination in the book for the first keyed question. Try to answer it and then check your response, reviewing explanations where desirable. Continue in this manner for other keyed questions.

You will notice that for some headings in the topic outline there is more than one test item on a particular examination. This usually occurs because more than one concept is involved. To find a question on a particular concept you may find it necessary to check the keyed questions on several examinations.

TOPIC OUTLINE* AND QUESTION INDEX— REGULAR EXAMINATION

*Earth Science Syllabus. New York State Department, The University of the State of New York, 1970

	EXAMINATION QUESTIONS		
	January 1997	June 1997	January 1998
B-2 Position description			
B-2.1 Vector-scalar properties			
B-2.2 Fields	62, 65, 80	6, 61–65	
TOPIC IV—EARTH MOTIONS			
A. Celestial observations			
A-1 Motion of objects in the sky			
A-1.1 Star paths			
A-1.2 Planetary motions		103	82
A-1.3 Satellite motion		10	11
A-1.4 Sun motion	10	15, 67, 80	10, 17
B. Terrestrial observations			
B-1			
B-1.1 Foucault pendulum	11		
B-1.2 Coriolis effect		11	
C. Time			
C-1 Frames of reference for time			
C-1.1 Earth motions		66	
D. Solar system models			
D-1 Geocentric and heliocentric models			
D-1.1 Geocentric model			
D-1.2 Heliocentric model			
D-2 Simple celestial model			
D-2.1 Geometry of orbits	9, 12	8	8, 83, 84
D-2.2 Force and energy transformations	56–58, 60	9	5, 61–65, 81, 85
Area 3 The earth's energy budgets			
TOPIC V—ENERGY IN EARTH PROCESSES			
A. Electromagnetic energy and energy transfer			
A-1 Electromagnetic energy			
A-1.1 Properties			12, 14, 102
A-1.2 Solar energy			
A-1.3 Earth energy			
A-2 Energy transfer			
A-2.1 Conduction	13	12	13
A-2.2 Convection			
A-2.3 Radiation			
B. Energy transformation			
B-1 Transformation in Earth processes			
B-1.1 Latent heat	14	93	
B-1.2 Movement of matter			99
B-1.3 Wavelength absorption and radiation		14, 16	
B-1.4 Friction			
C. Energy relationship in Earth processes			
C-1 Conservation of energy			
C-1.1 Closed system	71–75	13, 92, 94, 95	79
TOPIC VI—INSOLATION AND THE EARTH'S SURFACE			
A. Insolation at the Earth's surface			
A-1 Insolation factors			
A-1.1 Angle	16		15, 76
A-1.2 Duration	15		18
A-1.3 Absorption			77
A-1.4 Reflection			
A-1.5 Scattering			
A-1.6 Energy conversion			
B. Terrestrial radiation			
B-1 Radiation factors			

	January 1997	June 1997	January 1998
B-1.3 Moisture storage			
B-1.4 Moisture utilization			
B-1.5 Moisture deficit	24	28	
B-1.6 Moisture recharge			
B-1.7 Moisture surplus			
B-2 Streams			
B-2.1 Stream discharge and the water budget			26
B-3 Climates and the local water budget			
B-3.1 Climatic regions	81, 82		
C-Climate pattern factors			66–70
C-1 Factors			
C-1.1 Latitude		78, 79	
C-1.2 Elevation			
C-1.3 Large bodies of water and ocean currents	84		25
C-1.4 Mountain barriers	85		27
C-1.5 Wind belts	19		9
C-1.6 Storm tracks			22
Area 4 The rock cycle			
TOPIC IX—THE EROSIONAL PROCESS			
A. Weathering			
A-1 Evidence of weathering			
A-1.1 Weathering process		2	
A-1.2 Weathering rates	23		100
A-1.3 Soil formation		29	29
A-1.4 Soil solution			
B. Erosion			
B-1 Evidence of erosion			
B-1.1 Displaced sediments			
B-1.2 Properties of transported materials			
B-2 Factors affecting transportation			
B-2.1 Gravity			
B-2.2 Water erosion	27, 32, 89, 90	30, 81–83, 104	31, 32, 71–75
B-2.3 Wind and ice erosion		26	
B-2.4 Effect of erosional agents			51
B-2.5 Effect of man			
B-2.6 Predominant agent		39	
TOPIC X—THE DEPOSITIONAL PROCESS			
A. Deposition			
A-1 Factors			
A-1.1 Size		24	86
A-1.2 Shape		31	59
A-1.3 Density	31		
A-1.4 Velocity	86–88	84, 85	30
B. Erosional-depositional system			
B-1 Characteristics			
B-1.1 Erosional-depositional change			
B-1.2 Dominant process			
B-1.3 Erosional-depositional interface			
B-1.4 Dynamic equilibrium			
B-1.5 Energy relationships			
TOPIC XI—THE FORMATION OF ROCKS			
A. Rocks and sediments			
A-1 Comparative properties			
A-1.1 Similarities	34	56	
A-1.2 Differences			

	EXAMINATION QUESTIONS		
	January 1997	June 1997	January 1998
Area 5 The history of the earth			
TOPIC XIII—INTERPRETING GEOLOGICAL HISTORY			
A. Geologic events			
A-1 Sequence of geologic events			
A-1.1 Chronology of layers	45, 55	89	97
A-1.2 Igneous intrusions and extrusions	67	50	92, 94
A-1.3 Faults, joints, and folds	98, 99		44
A-1.4 Internal characteristics			
B. Correlation techniques			
B-1 Correlations			
B-1.1 Continuity			
B-1.2 Similarity of rock			
B-1.3 Fossil evidence	47, 66	46, 48	45
B-1.4 Volcanic time markers		48	
B-1.5 Anomalies to correlation			
C. Determining geologic ages			
C-1 Rock record			
C-1.1 Fossil evidence	46	45	
C-1.2 Scale of geologic time	44		
C-1.3 Erosional record	42	90	98
C-1.4 Geologic history of an area	97	51, 86, 87	48, 54
C-2 Radioactive decay			
C-2.1 Decay rates	48	47	49
C-2.2 Half-lives		44	50
C-2.3 Decay product ratios	49		
D. The fossil record			
D-1 Ancient life			
D-1.1 Variety of life forms	54		46, 47
D-1.2 Evolutionary development		88	
TOPIC XIV—LANDSCAPE DEVELOPMENT AND ENVIRONMENTAL CHANGE			
A. Landscape characteristics			
A-1 Quantitative observations			
A-1.1 Hillslopes			
A-1.2 Stream patterns			
A-1.3 Social associations			
A-2 Relationship of characteristics			
A-2.1 Landscape regions	50, 52, 53, 63	54, 96, 97, 99	52, 53, 55
B. Landscape development			
B-1 Environmental factors			
B-1.1 Uplifting and leveling force	96, 100	49, 55, 98	
B-1.2 Climate		52	
B-1.3 Bedrock		33	
B-1.4 Time			
B-1.5 Dynamic equilibrium			
B-1.6 Man		5	

TOPIC OUTLINE AND QUESTION INDEX— PROGRAM MODIFICATION EDITION

	EXAMINATION QUESTIONS		
	June 1995	June 1996	June 1997
UNIT 1: EARTH DIMENSIONS			
A - SHAPE			
1. Observation of stars to find shape, esp. North Star & Big Dipper	34	35, 109	31
2. Earth is slightly oblate; photographs from space	2		
3. Earth appears perfectly circular			
4. Smoothness			
B - SIZE			
1. Determination using Polaris		1	
2. Polar & equatorial diameters			
C - PARTS			
1. Atmosphere is thin			
2. Hydrosphere is thin & intermediate in density			20
3. Lithosphere is rock & soil			
D - MAPS			
1. Latitude & longitude; poles & equator as references	113	36	1
2. Field maps & isolines	11	101, 102	101
3. Topo maps & contour lines		101–103	2, 109, 110
4. Change with time; human influence			
5. Calculation of gradient	3	104	102
UNIT 2: MINERALS & ROCKS			
A - EARTH COMPOSITION			
1. Mineral resources			
2. Rocks are composed of minerals	109		4
3. Identification & classification; color, crystal forms, cleavage, streak, hardness, luster, density	1		
4. Few are common mica, magnetite			
5. One or more minerals		4	
6. Classification by origin	4		
B - IGNEOUS ROCKS			
1. Crystallize from magma		·	
2. Cooling rate & crystal size; glass, obsidian, granite	8		6
C - SEDIMENTARY ROCKS			
1. Compressed, cemented layers; texture: shale, sandstone, conglomerate	5		
2. Rounded grains & layers; grains & cement	5, 6		
3. Evaporite & organic rocks; rock salt & coal			
4. Form at or near surface			
5. May contain fossils			
D - METAMORPHIC ROCKS			
1. Crystallization w/o melting; high temperature & pressure	7	5	3
2. Foliation, banding, high density		3	
E - CHANGES			
1. Rock cycle	40		5
TOPIC A: ROCKS, MINERALS AND RESOURCES			
A - MINERALS			
1. Grouped by composition; acids & carbonates			46
2. Silicon/oxygen tetrahedra	47	50	
3. Arrangement & bonding	46	43	49
4. Oxygen & silicon are most common silicates	45	41	50
B - IGNEOUS ROCKS			
1. Texture, composition, density & color used to identify	48	46	41

	EXAMINATION QUESTIONS		
	June 1995	June 1996	June 1997
2. Intrusive & extrusive texture	42	44	
3. Felsic continents, mafic oceans		48	
C - SEDIMENTARY ROCKS			
1. Fragmental, chemical or organic; horizontal layers		42	
2. Fragmental: grain size classification	43	49	45
3. Organic & chem: composition & texture			44
4. Fossils & environment; land/oceanic environment			
D - METAMORPHIC			
1. Inferring parent rocks			42, 43
2. Contact metamorphism transition zones	44	45	
3. Regional metamorp. & mtns.; pressure induced changes	41		
4. Grouped by comp & texture including foliation & banding		47	47
5. Continuum of change			
E - CONSERVATION			
1. Vital fossil fuels & plastics			
2. Distribution of wealth & politics	49, 50		
UNIT 3: THE DYNAMIC CRUST			
A - EARTHQUAKES			
1. Zones of activity	112		
2. Wave speed & rock density			7
3. Richter & Mercalli scales	54		
4. Earthquake & volcano hazards			
B - DYNAMIC LITHOSPHERE			
1. Folds, faults, displaced fossils	11, 114		106
2. Changing bench marks			
3. Fit of the continents; plate tectonics		6	103–105
4. Major & minor plates			51
5. Diverg, conv & transf boundaries		7	8
TOPIC B: EARTHQUAKES AND EARTH'S INTERIOR			
A - EARTHQUAKES			
1. P & S-waves	52, 56		
2. P-waves are faster	56, 58	57, 59	
3. Origin time & distance	57, 59	56, 58	57, 59, 60
4. Epicenter: 3 or more stations	60		58
B - EARTH'S INTERIOR			
1. S-waves: solids only			
2. Liquid core of the Earth			
3. Density & temp increase w/depth			54
4. Seismic & meteorite evidence of layered Earth			
5. Meteorites, seismic & lab experiments suggest Fe & Ni core			
C - PLATE MOVEMENTS			
1. Rock & fossil correlations	10, 53		
2. Fossil climates	51		
3. Oceans are younger			
4. Magnetic stripes		53	
5. Heat flow: convection, hot spots		51	55
6. Force to move plates		51	
7. Rifting, subduction, faults	55	54, 55	52
D - PROPERTIES OF CRUST			
1. Continents thicker; low density		60	53
2. Ocean bottoms are basaltic; continents are like granite		52	
UNIT 4: SURFACE PROCESSES & LANDSCAPES			
A - WEATHERING			
1. Physical & chemical	106	12	11

	EXAMINATION QUESTIONS		
	June 1995	June 1996	June 1997
2. Exposure to biosphere, hydrosphere & atmosphere			
3. Climate factor	9		
4. Particle size & surface area	104		
5. Mineral composition factor	22, 107	10	
B - WEATHERING PRODUCTS			
1. Soils: minerals & organic			
2. Weathering & biological processes			
3. Horizons, leaching; many NY soils have no C horizon			12
4. Human influence			
C - EROSION			
1. Most sediments are transported; different mineral composition	12		
2. Agents driven by gravity, wind, water & ice			9, 10
3. Running water is most common			
4. Stream volume & velocity		4	
5. Particle size & stream velocity		14	112
6. Position of fastest flow; erosion & deposition			111
7. Sediments carried by solution, suspension, bouncing & rolling	13		
D - DEPOSITION			
1. Size, shape & density		15	
2. Sorting & depositional features			
3. Deltas, sandbars, graded bedding			
4. Gravity & ice deposits unsorted		13	
E - LANDFORMS			
1. Climate, rocks & structures; dry climates cause angular forms		8	107
2. Glaciers & NYS landforms: U-valleys, closed depressions, hills	17, 75		13
3. Drainage patterns: radial, rectangular & dendritic			
TOPIC C: OCEANOGRAPHY			
A - WHY IMPORTANT			
1. Bottom resources	61		
2. Materials in solution	63	69	67–70
3. Margins & bottom features; profiles & gradients		61	61, 62
4. Change through time; deposition, slides & earthquakes	64		
5. Magnetic reversals at ridges; conveyor motion	66	62, 63	63
6. Moving hot spots			
7. Terrestrial & organic sediments	65		
B - OCEAN WAVES & CURRENTS			
1. Winds & density differences; planetary winds, colder go deeper		65	64
2. (Coriolis effect), curves right in Northern Hemisphere	68	68	
3. Distributes uneven energy; water's high specific heat	70		66
4. Surface currents wind generated (Fetch) & wave height	67	64	
5. Tsunami from earthquakes		70	
6. Oceans distribute pollutants	62		64
C - COASTAL PROCESSES			
1. Waves collapse to make breakers			
2. Long shore currents; effects of piers & jettys	69	66	
3. Nature of beach sediments			65
4. Erosion & deposition		67	
TOPIC D: GLACIAL GEOLOGY			
A - WHAT ARE GLACIERS?			
1. Snowfall exceeds melt; polar & high altitude		78	74
2. Flow downhill & outward	71	79	72, 80
3. Push, carry & drag till	77	72	76

	EXAMINATION QUESTIONS		
	June 1995	June 1996	June 1997
4. Mixed sizes, erratics	74		
5. Abrasion by rock on rock; grooves & striations	73	74	75
6. At least a mile thick in NY; covered Catskills & Adirondacks			
7. Traceable erratics	80		
8. Trap dust and air samples			71
B - GLACIAL HISTORY OF NYS			
1. Superimposed till layers			
2. Ecological & climate changes	78		77
3. NYS soils are till			
4. Sea level changes; land fossils off Long Island	78	76	78
C - GLACIAL LANDSCAPES			
1. Erosional features	76	77	
2. Depositional features: esp. Long Island and Finger Lakes	72, 77	73, 80	79
3. Bedrock features: polish, striations & grooves even in highest mtns. of NYS			
4. Outwash shows layering		71	
5. Sand and gravel resources		75	
UNIT 5: EARTH'S HISTORY			
A - GEOLOGICAL SEQUENCE			
1. (Law of superposition)	16, 21, 110	30	15
2. Igneous intrusions/extrusions			
3. Faults & folds are younger		16	
B - CORRELATION			
1. "Walking" the outcrop			
2. Similar rocks & fossils; changes in environment (facies)			14
3. Index fossils: widespread but limited in time		17	
4. Volcanic ash & rapid deposits			
C - GEOLOGIC HISTORY			
1. Time scale from fossils	15	19	19
2. Movements of continents & mtn. building have been dated	14	18, 22	56, 108
3. Human existence is very brief			
4. Buried erosion surfaces	16, 108		
5. (Uniformitarianism)			
D - ABSOLUTE AGES			
1. Unstable radioactive nuclei			
2. Half life is predictable		20, 21	17
3. (Decay product ratios)			
4. When to use C-14, others	79		
E - EVOLUTION			
1. Evidence of different life forms			
2. Most life forms now extinct			
3. Fossil evidence of evolution			
4. Fossil evidence of past environments	23		16
UNIT 6: METEOROLOGY			
A - DESCRIPTION & MEASUREMENT			
1. Weather is condition of atmosphere			
2. Daily temperature cycle			
3. Dewpoint measures moisture; use sling psychrometer	18		22, 23, 113
4. Definition of relative humidity			
5. Air pressure from weight of air measured with a barometer	19		
6. Wind has magnitude & direction, named by direction they are from; wind vane & anemometer			114
B - RELATIONS AMONG VARIABLES			
1. Temperature and vapor capacity			

	June 1995	June 1996	June 1997
EXAMINATION QUESTIONS			
3. Precipitation can evaporate, infiltrate or flow to streams; Earth is a closed system			
B - SOLAR ENERGY			
1. Sun is Earth's major source			
2. Intensity & angle		27	27
3. Angle & time of day, season			
4. Day length & season; temperature & length of daylight			
5. Sunlight is absorbed & reflected			24
6. (Albedo) of land & water	26		
7. Earth re-radiates heat energy; change in wave length			40
8. (Greenhouse effect); CO_2, sea level changes			
C - CLIMATE FACTORS			
1. Definition of climate; use of water budget graphs		108	
2. Precipitation & temperature		108	
3. Effects of altitude & latitude		105, 106	
4. Water bodies modify climates; (high specific heat)		107	26
5. Prevailing winds & ocean currents			
6. Mountain barriers: climates on windward & leeward sides		29	
UNIT 8: THE EARTH IN SPACE			
A - CELESTIAL OBSERVATIONS			
1. Sun path is an arc			
2. Sun moves east to west; in NYS, noon sun is never at (zenith)	30		
3. Stars (etc.) seem to rotate 15°/hr. around Polaris	31		31, 33
4. Definition of constellations			
5. (Geocentric theory) & observations			
6. (Heliocentric theory) also works			
B - REVOLUTION WITH TILT			
1. Sun's path w/season & latitude		33	
2. Sun higher in summer (not 90°)			
3. Noon positions at other locations			
4. Changing positions of sunrise & set	32		34
5. Day length w/season & latitude		31	32
6. Seasons due to tilt & revolution			
7. Tilt & revolution cause seasons		32, 94	
C - GRAVITY			
1. Elliptical orbits			
2. Satellite speed & distance	36–38, 91		35
3. Definition of eccentricity	36		37
4. Earth orbit is nearly circular			
5. Period of moon orbit			118
6. Cause of moon phases	33		116
7. Causes of tides			
8. Cause/predictability of eclipses		34	117
9. Gravity w/mass & distance			38
D - EARTH IN THE UNIVERSE			
1. Speed of light to measure space			
2. Sun is a typical star, but close			
3. Sun orbits Milky Way; great interstellar distances			
4. Solar system in arm of Spiral Galaxy; Milky Way visible in night sky	39		
5. Universe includes many galaxies			38
TOPIC F: ASTRONOMY EXTENSIONS			
A - HELIOCENTRIC & GEOC. MODELS			
1. Geocentric model: Earth is stationary			
2. Geocentric model accounts for most observations	100		97
3. Geocentric model must include complex planetary orbits	95		94

	EXAMINATION QUESTIONS		
	June 1995	June 1996	June 1997
4. Heliocentric is simpler		96	
B - TESTING BOTH MODELS			
1. Foucalt pendulum supports heliocentric model		92	98
2. Coriolis effect supports heliocentric model			
C - EARTH & NEARBY PLANETS			
1. Temperature is determined by distance from the sun	97		91
2. Only Earth has liquid water	92	95	99
3. Inner: rocky; outer: mostly gas	93	93	
4. Only Earth has free oxygen			
5. Atmospheres of Mars & Venus are mostly carbon dioxide		97	95, 96
6. Moon and Mercury have no atm. Lunar rocks show no weathering	94		
7. Impact structures are common on Mercury, Moon & Mars			93
8. Exploration for human benefit			
D - EVOLUTION OF UNIVERSE			
1. Red shift increases w/distance	99		100
2. Universe probably expanding		98	
3. Expansion accounts for 15–20 billion years	98	100	
4. Began as big bang			
5. Future depends upon mass expansion, implosion & oscillation models	96		
UNIT 9: ENVIRONMENTAL AWARENESS			
A - INTERDEPENDENT SYSTEMS			
1. Interrelations of systems	35		
2. Technology accelerates change			
3. Definition of pollution: ground, water or air			
4. Industry & population density			39
5. Increasing human population disease control & food production	35		
B - BALANCE WITH NATURE			
1. Limited resources; use of natural cycles			
2. Conservation & planning			
3. Depletion and substitutions			
4. Need for research, global warming & ozone depletion		115	
5. Need for research & education			
PROBLEM SOLVING SKILLS			
1. Applying mathematics	105		
2. Classifying			
3. Communicating			
4. Creating models	101–103	110, 111	
5. Developing vocabulary			
6. Formulating hypotheses			
7. Generalizing			
8. Identifying variables			
9. Inferring		111, 115	
10. Interpreting data	101–103	2, 5, 91, 99	48, 61, 68, 73, 92, 119, 120
11. Making decisions			
12. Manipulating ideas		115	
13. Manipulating materials			
14. Measuring			
15. Observing			
16. Predicting		86, 111	
17. Questioning			
18. Recording data		112–114	
19. Replicating			
20. Using cues			

Earth Science Reference Tables and Charts

This edition of the Earth Science Reference Tables was first used in the classroom during the 1993–94 school year. The first examination for which these tables were used was the June 1994 Regents Examination in Earth Science.

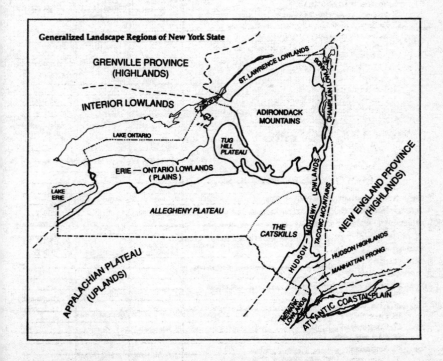

Generalized Landscape Regions of New York State

Generalized Bedrock Geology of New York State

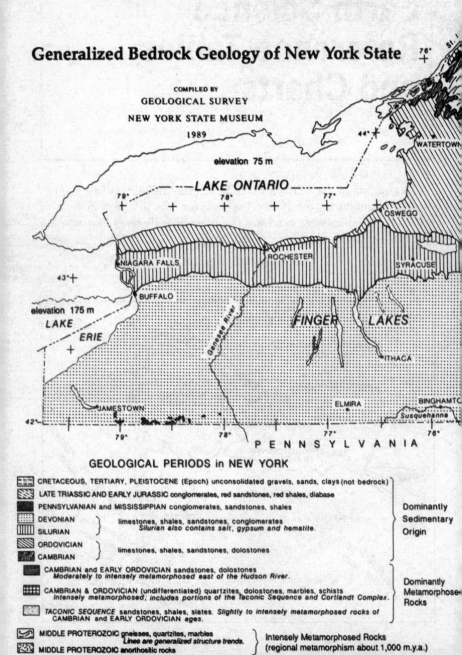

COMPILED BY
GEOLOGICAL SURVEY
NEW YORK STATE MUSEUM
1989

elevation 75 m

LAKE ONTARIO

WATERTOWN

OSWEGO

NIAGARA FALLS   ROCHESTER   SYRACUSE

BUFFALO

elevation 175 m

LAKE
ERIE

Genesee River

FINGER LAKES

ITHACA

JAMESTOWN   ELMIRA   BINGHAMTON

Susquehanna

P E N N S Y L V A N I A

GEOLOGICAL PERIODS in NEW YORK

CRETACEOUS, TERTIARY, PLEISTOCENE (Epoch) unconsolidated gravels, sands, clays (not bedrock)

LATE TRIASSIC AND EARLY JURASSIC conglomerates, red sandstones, red shales, diabase

PENNSYLVANIAN and MISSISSIPPIAN conglomerates, sandstones, shales — Dominantly Sedimentary Origin

DEVONIAN } limestones, shales, sandstones, conglomerates
SILURIAN } *Silurian also contains salt, gypsum and hematite.*

ORDOVICIAN }
CAMBRIAN } limestones, shales, sandstones, dolostones

CAMBRIAN and EARLY ORDOVICIAN sandstones, dolostones
Moderately to intensely metamorphosed east of the Hudson River.

CAMBRIAN & ORDOVICIAN (undifferentiated) quartzites, dolostones, marbles, schists
intensely metamorphosed; includes portions of the Taconic Sequence and Cortlandt Complex. — Dominantly Metamorphosed Rocks

TACONIC SEQUENCE sandstones, shales, slates. *Slightly to intensely metamorphosed rocks of CAMBRIAN and EARLY ORDOVICIAN ages.*

MIDDLE PROTEROZOIC gneisses, quartzites, marbles
Lines are generalized structure trends.   Intensely Metamorphosed Rocks
MIDDLE PROTEROZOIC anorthositic rocks   (regional metamorphism about 1,000 m.y.a.)

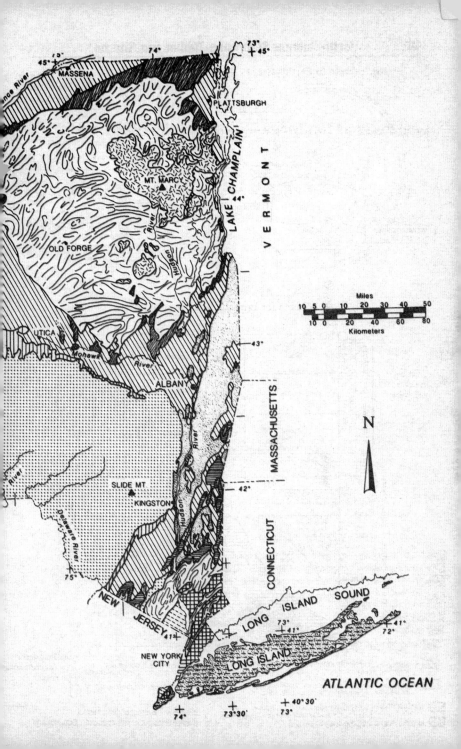

Rock Cycle in Earth's Crust

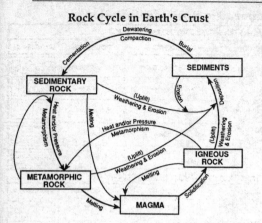

Relationship of Transported Particle Size to Water Velocity*

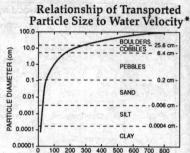

*This generalized graph shows the water velocity needed to maintain, but not start movement. Variations occur due to differences in particle density and shape.

Scheme for Igneous Rock Identification

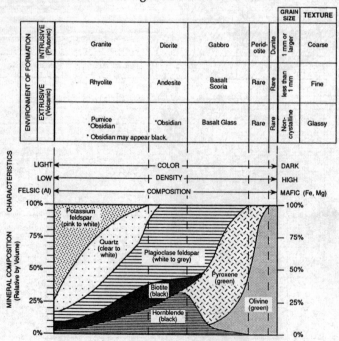

Note: The intrusive rocks can also occur as exceptionally coarse-grained rock, Pegmatite.

Scheme for Sedimentary Rock Identification

INORGANIC LAND-DERIVED SEDIMENTARY ROCKS					
TEXTURE	GRAIN SIZE	COMPOSITION	COMMENTS	ROCK NAME	MAP SYMBOL
Clastic (fragmental)	Mixed, silt to boulders (larger than 0.001 cm)	Mostly quartz, feldspar, and clay minerals; May contain fragments of other rocks and minerals	Rounded fragments	Conglomerate	
			Angular fragments	Breccia	
	Sand (0.006 to 0.2 cm)		Fine to coarse	Sandstone	
	Silt (0.0004 to 0.006 cm)		Very fine grain	Siltstone	
	Clay (less than 0.0006 cm)		Compact; may split easily	Shale	

CHEMICALLY AND/OR ORGANICALLY FORMED SEDIMENTARY ROCKS					
TEXTURE	GRAIN SIZE	COMPOSITION	COMMENTS	ROCK NAME	MAP SYMBOL
Nonclastic	Coarse to fine	Calcite	Crystals from chemical precipitates and evaporites	Chemical Limestone	
	Varied	Halite		Rock Salt	
	Varied	Gypsum		Rock Gypsum	
	Varied	Dolomite		Dolostone	
	Microscopic to coarse	Calcite	Cemented shells, shell fragments, and skeletal remains	Fossil Limestone	
	Varied	Carbon	Black and nonporous	Bituminous Coal	

Scheme for Metamorphic Rock Identification

TEXTURE		GRAIN SIZE	COMPOSITION	TYPE OF METAMORPHISM	COMMENTS	ROCK NAME	MAP SYMBOL
FOLIATED	Slaty	Fine	CHLORITE MICA QUARTZ FELDSPAR AMPHIBOLE GARNET PYROXENE	Regional	Low-grade metamorphism of shale	Slate	
	Schistose	Medium to coarse			Medium-grade metamorphism; Mica crystals visible from metamorphism of feldspars and clay minerals	Schist	
	Gneissic	Coarse		(Heat and pressure increase with depth, folding, and faulting)	High-grade metamorphism; Mica has changed to feldspar	Gneiss	
NONFOLIATED		Fine	Carbonaceous		Metamorphism of plant remains and bituminous coal	Anthracite Coal	
		Coarse	Depends on conglomerate composition		Pebbles may be distorted or stretched; Often breaks through pebbles	Meta-conglomerate	
		Fine to coarse	Quartz	Thermal (including contact) or Regional	Metamorphism of sandstone	Quartzite	
			Calcite, Dolomite		Metamorphism of limestone or dolostone	Marble	
		Fine	Quartz, Plagioclase	Contact	Metamorphism of various rocks by contact with magma or lava	Hornfels	

GEOLOGIC HISTORY OF NEW

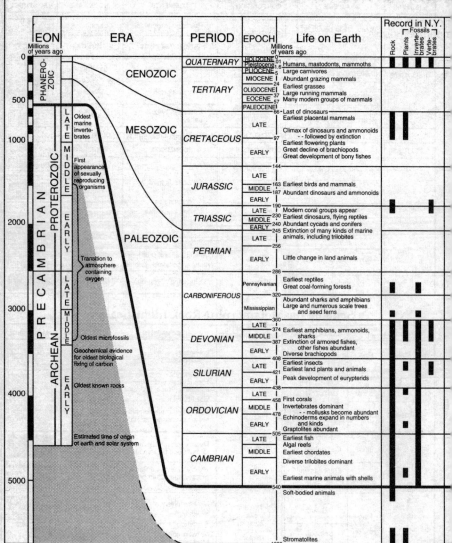

YORK STATE AT A GLANCE

Important Fossils of New York	Tectonic Events Affecting Northeast North America	Important Geologic Events in New York	Inferred Position of Earth's Landmasses
CONDOR MASTODONT FIG-LIKE LEAF	Passive Margin / Rifting	Advance and retreat of last continental ice Uplift of Adirondack region Sandstones and shales underlying Long Island and Staten Island deposited on margin of Atlantic Ocean Development of passive continental margin Kimberlite and lamprophere dikes Atlantic Ocean continues to widen Initial opening of Atlantic Ocean	TERTIARY 59 Million years ago CRETACEOUS 119 Million years ago
COELOPHYSIS		Intrusion of Palisades Sill Rifting Massive erosion of Paleozoic rocks	
CLAM	Transform Collision	Appalachian (Alleghanian) Orogeny caused by collision of North America and Africa along transform margin	TRIASSIC 232 Million years ago
AMMONOID BRACHIOPOD NAPLES TREE PLACODERM FISH EURYPTERID CORAL HEAD GRAPTOLITE	Transform Collision	Catskill Delta forms Erosion of Acadian Mountains Acadian Orogeny caused by collision of North America and Avalon and closing of remaining part of Iapetus Ocean Evaporite basins; salt and gypsum deposited Erosion of Taconic Mountains; Queenston Delta forms Taconian Orogeny caused by closing of western part of Iapetus Ocean and collision between North America and volcanic island arc	PENNSYLVANIAN 306 Million years ago DEVONIAN/MISSISSIPPIAN 363 Million years ago
TRILOBITE	Subduction / Continental Collision	Iapetus passive margin forms	ORDOVICIAN 458 Million years ago
STROMATOLITES	Rifting / Passive Margin	Rifting and initial opening of Iapetus Ocean Erosion of Grenville Mountains Grenville Orogeny: Ancestral Adirondack Mtns. and Hudson Highlands formed Subduction and volcanism Sedimentation, volcanism	

Inferred Properties of Earth's Interior

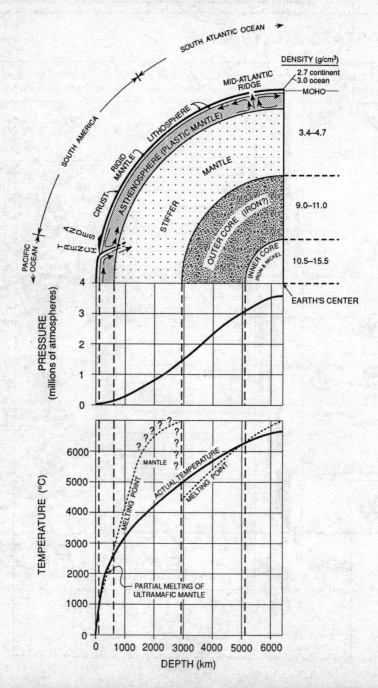

Tectonic Plates

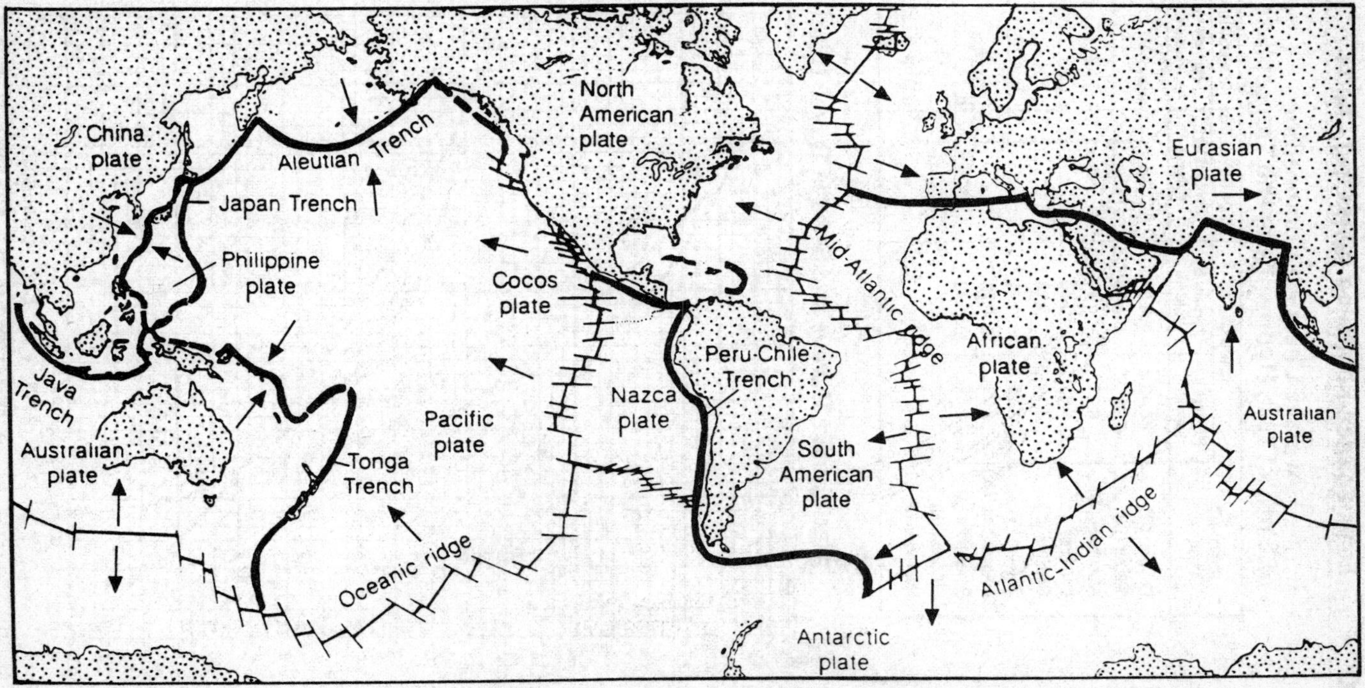

Eurasian plate

Australian plate

African plate

Atlantic-Indian ridge

Mid-Atlantic ridge

South American plate

Antarctic plate

Peru-Chile Trench

North American plate

Nazca plate

Cocos plate

Pacific plate

Oceanic ridge

Aleutian Trench

Japan Trench

Tonga Trench

Philippine plate

China plate

Australian plate

Java Trench

Average Chemical Composition of Earth's Crust, Hydrosphere, and Troposphere

ELEMENT (symbol)	CRUST		HYDROSPHERE	TROPOSPHERE
	Percent by Mass	Percent by Volume	Percent by Volume	Percent by Volume
Oxygen (O)	46.40	94.04	33	21
Silicon (Si)	28.15	0.88		
Aluminum (Al)	8.23	0.48		
Iron (Fe)	5.63	0.49		
Calcium (Ca)	4.15	1.18		
Sodium (Na)	2.36	1.11		
Magnesium (Mg)	2.33	0.33		
Potassium (K)	2.09	1.42		
Nitrogen (N)				78
Hydrogen (H)			66	

Earthquake P-wave and S-wave Travel Time

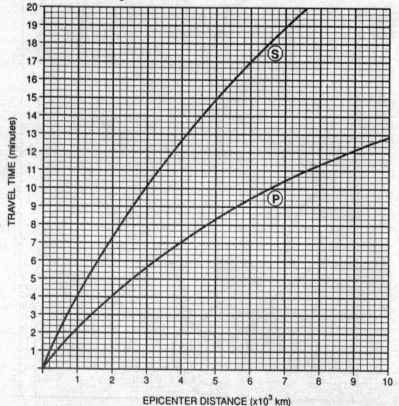

TRAVEL TIME (minutes)

EPICENTER DISTANCE ($\times 10^3$ km)

Dewpoint Temperatures

Dry-Bulb Temperature (°C)	Difference Between Wet-Bulb and Dry-Bulb Temperatures (C°)														
	1	2	3	4	5	6	7	8	9	10	11	12	13	14	15
−20	−33														
−18	−28														
−16	−24														
−14	−21	−36													
−12	−18	−28													
−10	−14	−22													
−8	−12	−18	−29												
−6	−10	−14	−22												
−4	−7	−12	−17	−29											
−2	−5	−8	−13	−20											
0	−3	−6	−9	−15	−24										
2	−1	−3	−6	−11	−17										
4	1	−1	−4	−7	−11	−19									
6	4	1	−1	−4	−7	−13	−21								
8	6	3	1	−2	−5	−9	−14								
10	8	6	4	1	−2	−5	−9	−14	−28						
12	10	8	6	4	1	−2	−5	−9	−16						
14	12	11	9	6	4	1	−2	−5	−10	−17					
16	14	13	11	9	7	4	1	−1	−6	−10	−17				
18	16	15	13	11	9	7	4	2	−2	−5	−10	−19			
20	19	17	15	14	12	10	7	4	2	−2	−5	−10	−19		
22	21	19	17	16	14	12	10	8	5	3	−1	−5	−10	−19	
24	23	21	20	18	16	14	12	10	8	6	2	−1	−5	−10	−18
26	25	23	22	20	18	17	15	13	11	9	6	3	0	−4	−9
28	27	25	24	22	21	19	17	16	14	11	9	7	4	1	−3
30	29	27	26	24	23	21	19	18	16	14	12	10	8	5	1

Relative Humidity (%)

Dry-Bulb Temperature (°C)	Difference Between Wet-Bulb and Dry-Bulb Temperatures (C°)														
	1	2	3	4	5	6	7	8	9	10	11	12	13	14	15
−20	28														
−18	40														
−16	48	0													
−14	55	11													
−12	61	23													
−10	66	33	0												
−8	71	41	13												
−6	73	48	20	0											
−4	77	54	32	11											
−2	79	58	37	20	1										
0	81	63	45	28	11										
2	83	67	51	36	20	6									
4	85	70	56	42	27	14									
6	86	72	59	46	35	22	10	0							
8	87	74	62	51	39	28	17	6							
10	88	76	65	54	43	33	24	13	4						
12	88	78	67	57	48	38	28	19	10	2					
14	89	79	69	60	50	41	33	25	16	8	1				
16	90	80	71	62	54	45	37	29	21	14	7	1			
18	91	81	72	64	56	48	40	33	26	19	12	6	0		
20	91	82	74	66	58	51	44	36	30	23	17	11	5	0	
22	92	83	75	68	60	53	46	40	33	27	21	15	10	4	0
24	92	84	76	69	62	55	49	42	36	30	25	20	14	9	4
26	92	85	77	70	64	57	51	45	39	34	28	23	18	13	9
28	93	86	78	71	65	59	53	47	42	36	31	26	21	17	12
30	93	86	79	72	66	61	55	49	44	39	34	29	25	20	16

Selected Properties of Earth's Atmosphere

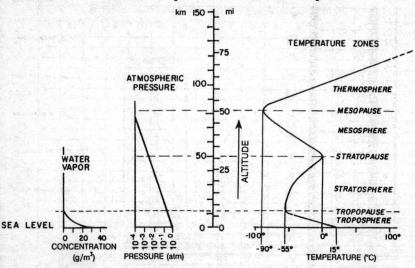

Planetary Wind and Moisture Belts in the Troposphere

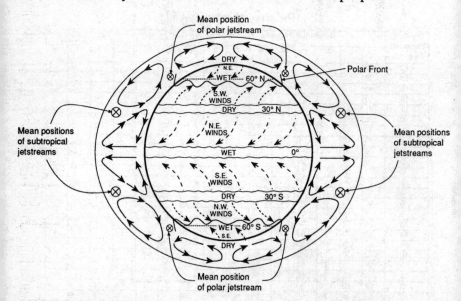

The drawing shows the locations of the belts near the time of an equinox. The locations shift somewhat with the changing latitude of the Sun's vertical ray. In the Northern Hemisphere the belts shift northward in summer and southward in winter.

Temperature

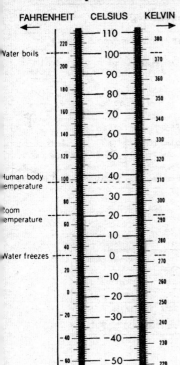

FAHRENHEIT ← CELSIUS KELVIN →

Water boils
Human body temperature
Room temperature
Water freezes

Lapse Rate

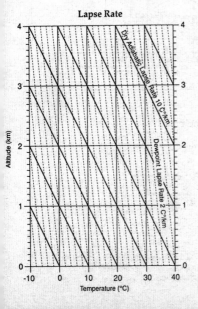

Pressure

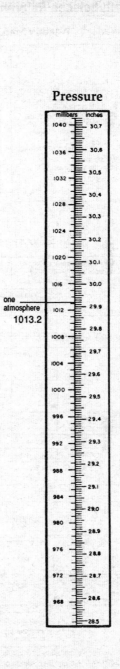

millibars   inches

one atmosphere
1013.2

cm

Weather Map Information

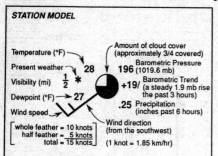

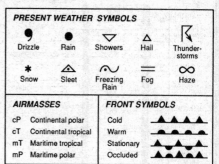

Solar System Data

Planet	Mean Distance from Sun (millions of km)	Period of Revolution	Period of Rotation	Eccentricity of Orbit	Equatorial Diameter (km)	Density (g/cm³)
MERCURY	57.9	88 days	59 days	0.206	4,880	5.4
VENUS	108.2	224.7 days	243 days	0.007	12,104	5.2
EARTH	149.6	365.26 days	23 hours 56 min 4 sec	0.017	12,756	5.5
MARS	227.9	687 days	24 hours 37 min 23 sec	0.093	6,787	3.9
JUPITER	778.3	11.86 years	9 hours 50 min 30 sec	0.048	142,800	1.3
SATURN	1,427	29.46 years	10 hours 14 min	0.056	120,000	0.7
URANUS	2,869	84.0 years	11 hours	0.047	51,800	1.2
NEPTUNE	4,496	164.8 years	16 hours	0.009	49,500	1.7
PLUTO	5,900	247.7 years	6 days 9 hours	0.250	2,300	2.0

Electromagnetic Spectrum

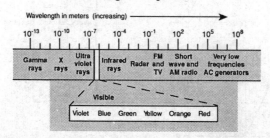

Equations and Proportions

Equations

Percent deviation from accepted value	$\text{deviation (\%)} = \dfrac{\text{difference from accepted value}}{\text{accepted value}} \times 100$		C_p = specific heat

Percent deviation from accepted value
$$\text{deviation (\%)} = \frac{\text{difference from accepted value}}{\text{accepted value}} \times 100$$

Eccentricity of an ellipse
$$\text{eccentricity} = \frac{\text{distance between foci}}{\text{length of major axis}}$$

Gradient
$$\text{gradient} = \frac{\text{change in field value}}{\text{change in distance}}$$

Rate of change
$$\text{rate of change} = \frac{\text{change in field value}}{\text{change in time}}$$

Circumference of a circle $\quad C = 2\pi r$

Eratosthenes' method to determine Earth's circumference
$$\frac{\angle a}{360°} = \frac{s}{C}$$

Volume of a rectangular solid $\quad V = \ell w h$

Density of a substance $\quad D = \dfrac{m}{V}$

Latent heat
$$\begin{cases} \text{solid} \longleftrightarrow \text{liquid} \quad Q = mH_f \\ \text{liquid} \longleftrightarrow \text{gas} \quad Q = mH_v \end{cases}$$

Heat energy lost or gained $\quad Q = m\,\Delta T C_p$

C_p = specific heat
C = circumference
d = distance
D = density
F = force
h = height
H_f = heat of fusion
H_v = heat of vaporization
$\angle a$ = shadow angle
ℓ = length
s = distance on surface
m = mass
Q = amount of heat
r = radius
ΔT = change in temperature
V = volume
w = width
Note: $\pi \approx 3.14$

Proportions

EURYPTERID
New York State Fossil

Kepler's harmonic law of planetary motion
$\quad (\text{period of revolution})^2 \propto (\text{mean radius of orbit})^3$

Universal law of gravitation
$$\text{force} \propto \frac{\text{mass}_1 \times \text{mass}_2}{(\text{distance between their centers})^2} \quad \left(F \propto \frac{m_1\,m_2}{d^2} \right)$$

Physical Constants

Properties of Water

Latent heat of fusion (H_f)............ 80 cal/g
Latent heat of vaporization (H_v) .... 540 cal/g
Density (D) at 3.98°C ....1.00 g/mL

Specific Heats of Common Materials

MATERIAL		SPECIFIC HEAT (C_p) (cal/g·C°)
Water	solid	0.5
	liquid	1.0
	gas	0.5
Dry air		0.24
Basalt		0.20
Granite		0.19
Iron		0.11
Copper		0.09
Lead		0.03

Radioactive Decay Data

RADIOACTIVE ISOTOPE	DISINTEGRATION	HALF-LIFE (years)
Carbon-14	$C^{14} \rightarrow N^{14}$	5.7×10^3
Potassium-40	$K^{40} \nearrow Ar^{40} \searrow Ca^{40}$	1.3×10^9
Uranium-238	$U^{238} \rightarrow Pb^{206}$	4.5×10^9
Rubidium-87	$Rb^{87} \rightarrow Sr^{87}$	4.9×10^{10}

Astronomy Measurements

MEASUREMENT	EARTH	SUN	MOON
Mass (m)	5.98×10^{24} kg	1.99×10^{30} kg	7.35×10^{22} kg
Radius (r)	6.37×10^3 km	6.96×10^5 km	1.74×10^3 km
Average density (D)	5.52 g/cm³	1.42 g/cm³	3.34 g/cm³

DET 516 (10-93–375,000)
3-028581

Coelophysis
fossil footprints

Surface Ocean Currents

WARM CURRENTS
COOL CURRENTS

Glossary of Earth Science Terms

abrasion the wearing away of rock material that results when one rock particle strikes another. In stream abrasion, particles carried by the stream may strike other particles, causing pieces to break off.

actual evapotranspiration the amount of water vapor that is actually returned to the atmosphere by evaporation and transpiration at a particular location and time. *See also* **potential evapotranspiration**.

adiabatic change the change in temperature of a gas caused by expansion or compression. When a gas expands, for example, it cools and its temperature drops.

air mass a large area in the atmosphere in which the temperature and moisture conditions are similar.

altitude the angle between the line of sight to a star and the horizon; also, the elevation above sea level.

angle of insolation the angle at which the Sun's rays strike the Earth's surface.

anticline a series of folded rock layers that bends upward near the center. *See also* **syncline**.

apparent diameter the size that an object appears to an observer. When objects are close by, they appear larger than when they are farther away.

atmosphere the envelope of air that encircles the Earth. It has been divided into zones based largely on temperature differences.

axis an imaginary line around which an object rotates. The Earth's axis extends from the North Pole to the South Pole.

azimuth for a star, the angle along the horizon between the north pole and the point where the line of sight to a star intersects the horizon.

barometer an instrument for measuring atmospheric pressure.

barometric pressure the amount of force exerted by the air per square inch or centimeter at a particular location.

bedrock the rock layer nearest the Earth's surface, lying directly below any soil layers.

boiling point the temperature at which a liquid changes to a vapor.

capillarity (capillary action) the rising of water against gravity, as when water rises above the water table in soil because of the attraction between the water and the soil.

cementation the process by which sediments are bonded together by material dissolved in water when the water evaporates.

chemical weathering weathering that occurs because of the chemical reaction between material dissolved in water and local rock material. *See also* **physical weathering**.

classification the process of organizing objects on the basis of similarities in their properties.

cleavage the splitting of a mineral in distinctive directions caused by the arrangement of the atoms in a mineral. For example, the mineral mica cleaves in layers because the atoms in the crystal structure of mica are arranged in layers.

climate the average weather conditions, in terms of temperature and moisture, of an area over a long period of time.

clouds masses of water droplets suspended in the air. They form when air is cooled and moisture condenses.

cold front the boundary between two air masses of different temperatures, the point where the colder air mass moves under and pushes up the warmer air mass. *See also* **warm front**.

compaction the process that results when buried sediments are subjected to pressure, which packs them together. Together with cementation, this process causes sediments to be converted to rock.

condensation the process whereby, when moisture in the atmosphere is cooled, it changes from a vapor to a liquid.

conduction the method of heat transfer in solids in which faster moving molecules strike other molecules, causing them to speed up.

continental air mass an air mass that forms over land and therefore is relatively dry.

continental drift a hypothetical slow movement of the continents, which forms the basis of the theory that the present continents were once part of one large landmass that broke up. Since that time, the continents having been drifting apart.

continental glacier a large sheet of ice that covered much of a continent during a period of the Earth's past geologic history.

contour interval the differences in elevation between contour lines on a contour or topographic map.

contour line a line on a topographic map that connects points having the same elevation.

contour map *See* **topographic map**.

convection the method of energy transfer in fluids in which the fluid expands when it is heated and rises. When the fluid cools, it contracts and sinks.

convection cell the circular pattern of movement in a fluid caused by the rising and sinking of the fluid due to differences in density caused by differences in temperature.

core the innermost zone of the Earth's interior. A solid inner core is surrounded by a molten outer core.

coriolis effect an apparent force, due to the rotation of the Earth, that deflects winds toward the right in the Northern Hemisphere and toward the left in the Southern Hemisphere.

correlation the matching of rock layers at different locations, based on composition, thickness, and in some cases fossil content, for the purpose of establishing that they represent the same rock layer.

crust the outermost solid layer of the Earth, extending across the continents and under the oceans. It is thicker under the continents than under the oceans.

density the mass per unit volume of a substance.

deposition the process by which material carried by running water, glaciers, or wind settles out when the velocity of the carrier slows down.

desert an area when the total annual precipitation is far less than the amount of evapotranspiration. Even areas over the oceans may be classified as deserts.

dew moisture that condenses at the Earth's surface when moist air touches a cool area.

dewpoint temperature the temperature at which the air becomes saturated and excess moisture begins to condense.

dike a rock layer that forms when molten rock material flows through breaks in rock layers and then cools and hardens.

discharge the total volume of water flowing in a river or stream per unit of time.

duration of insolation the number of hours that the Sun's rays strike the Earth's surface over a 24-hour period; the number of hours of daylight.

earthquake the large-scale and rapid motion of rock layers that occurs when pressure is released.

electromagnetic spectrum the range of wavelengths of energy released by the Sun. A small range of wavelengths represents visible light. Other bands within the spectrum include ultraviolet radiation, cosmic rays, infrared radiation, and radio waves.

elevation the height above sea level.

ellipse a curve that has two centers, or foci. The sum of the distance between either focus and any point on the ellipse and the distance between the other focus and that point is a constant. The shape of the Earth's orbit around the Sun is an ellipse with the Sun at one focus.

energy sink a mass such as an ice cube that is cooler than its surroundings. Energy flows into a sink.

energy source a mass such as a pot of boiling water that is warmer than its surroundings. Energy flows out from a source.

epicenter the location on the Earth's surface directly above the point of origin or focus of an earthquake.

equator an imaginary circle around the Earth lying halfway between the poles and dividing the Earth's surface into the Northern and Southern hemispheres.

equinox the two times a year when the Sun is directly overhead at the equator at noon. At these times, about March 21 and September 23, the number of hours of day and of night is the same.

erosion the process by which weathered rock material is carried away by agents such as running water, ice, wind, and gravity.

evaporation the process by which water is converted from a liquid to a vapor.

evapotranspiration the combined processes of evaporation and transpiration, representing the method by which water vapor returns to the atmosphere. *See also* **actual evapotranspiration; potential evapotranspiration**.

extinct referring to a plant or animal species that lived in the past but is no longer found alive on the Earth.

fault movement within rock layers where pressure has caused the layers to break. The layers on one side of the break move up, while the opposite layers move down.

felsic referring to igneous rocks composed of minerals with a high content of aluminum. These rocks tend to be lighter in color than others.

focus the point of origin within the Earth's crust or mantle of an earthquake.

formation (rock) a sequence of rock layers that cover a large area and were formed over a period of time.

fossil the preserved remains or traces of an animal or plant that lived in the past.

fossil record groups of fossils found in a rock layer that are used to interpret how and when the rock layer was formed and what the environment was like at that time.

fracture as term used to describe the irregular way in which some minerals break.

freezing point the temperature at which a liquid changes to a solid.

front the boundary between two different air masses. *See also* **cold front; warm front**.

frost moisture that condenses directly from a vapor to a solid when moist air touches a cold surface. When the air temperature is below freezing, frost forms instead of dew.

geosyncline a large basin into which sediments have been deposited over a long period of time. The sediments that were deposited earlier continue to sink as more sediments are added.

gradient the slope of the land or of a river or stream.

greenhouse effect the process by which longer wavelength radiation emitted from the Earth's surface is absorbed by carbon dioxide in the atmosphere, causing a rise in air temperature.

half-life the amount of time it takes for half the mass of a radioactive substance to decay. For example, the half-life of carbon-14 is 5.6×10^3 years.

heat of fusion the amount of energy required to convert one gram of ice at 0°C to water at 0°C.

heat of vaporization the amount of energy required to convert one gram of water at 100°C to vapor at 100°C.

high-pressure center a location on a weather map where winds blow outward, in a clockwise direction, away from the center. This occurs because the air pressure is lower at the center than over the surrounding area. *See also* **low-pressure center**.

humidity the amount of moisture in the air.

hydrosphere the outer zone of the Earth, which consists of the oceans and seas.

hypothesis the attempt to explain a scientific phenomenon on the basis of observations and other relevant information.

igneous referring to rocks that form when molten rock material cools and solidifies.

impermeable referring to a layer of material through which water cannot pass. Most soil layers are permeable, while most rock layers are not.

index fossil a fossil that is found over widespread areas and that formed from an organism that existed for a relatively short geologic period of time.

inference an interpretation or explanation of a natural phenomenon based on observations.

infiltration the process by which water at the Earth's surface penetrates and filters down through porous soil and rock layers.

insolation the radiation reaching the Earth's surface from the Sun. This term is a contraction of "incoming solar radiation."

intrusion the forcible entry of solidified molten rock material between rock layers or into breaks within a rock layer.

isobar a line on a weather map that connects points having the same air or barometric pressure.

land breeze a wind blowing from over the land to over a lake or ocean. It occurs when the water temperature is warmer than the land temperature. *See also* **sea breeze**.

latent heat the amount of heat required to change a substance from one phase to another, such as converting water to ice.

latitude imaginary circles around the Earth parallel to the equator and between the North and South poles.

lava molten rock material that reaches the Earth's surface through volcanoes.

lithosphere the solid outer portion of the Earth.

longitude imaginary circles around the Earth that pass through the North and South poles.

low-pressure center a location on a weather map where winds blow inward, in a counterclockwise direction, into a center. This occurs because the air pressure is lower at the center than in the surrounding area. *See also* **high-pressure center**.

lunar eclipse the phenomenon that occurs when the Earth passes between the Moon and the Sun, causing light from the Sun to be blocked from reaching the surface of the Moon. *See also* **solar eclipse**.

luster the type of shine exhibited by a mineral, based on the way its surface reflects light. Examples are metallic, glassy, and dull lusters.

mafic referring to igneous rocks composed of minerals with a high content of magnesium and iron. They tend to be darker in color than other rocks.

magma molten rock material beneath the Earth's surface.

magnetic north (pole) the point near the geographic North Pole toward which the needle on a compass points.

mantle the zone within the Earth that lies between the crust and the outer core.

maritime air mass an air mass that forms over water and therefore has a relatively high moisture content.

mass the measure of the amount of matter contained in a sample.

meander a curve in a river or stream.

meridian a line of longitude measured in degrees. The prime meridian (0°) passes through Greenwich, England.

metamorphic referring to rocks that form when existing rocks are subjected to enough heat and pressure to cause partial melting of the minerals present.

meter the standard unit of length in the metric system; 1 meter = 39.37 inches.

millibar a unit of air pressure in the atmosphere, commonly used on weather maps.

mineral a naturally occurring substance that is always made of the same elements in a fixed proportion.

moraine a large deposit formed when a glacier melts and leaves behind the material it is carrying.

observation a description of what is perceived by the senses. It can often be made more accurate by using instruments.

occluded front a weather front that forms when a cold front moves in behind a warm front, lifting the warm front off the ground.

orbit the path followed by one object as it revolves around another; for example, the orbit of the Earth around the Sun.

orogeny a period of extensive mountain building.

permeability the ability of water to penetrate through a material. The permeability of a soil depends upon the size of the soil grains and the amount of space between the grains.

physical weathering the breakdown of rocks due to physical changes, such as changes in temperature or the freezing and thawing of water, that fills cracks in the rock. *See also* **chemical weathering**.

plateau a mountain formed when horizontal rock layers are uplifted and rivers carve valleys across the layers.

polar air mass an air mass that forms over the polar regions and therefore contains relatively cold air.

pollution a condition of the air, water, or land in which there is a surplus of materials present that may be harmful to living things.

porosity the percentage of open space in a soil sample.

potential evapotranspiration the maximum amount of moisture that can be returned to the atmosphere at a given location by evaporation and transpiration. *See also* **actual evapotranspiration**.

precipitation all forms of moisture that reach the Earth's surface from the atmosphere, including rain, snow, hail, and sleet.

prevailing westerlies the wind belt stretching across the United States in which the general direction of wind movement is from southwest to northeast.

principle a generally accepted scientific law for which there is extensive supporting evidence. A theory becomes a law or principle when much time has passed and there is no evidence to disprove it.

***P*-wave (primary wave)** the compression type of wave emitted by an earthquake that travels through both liquids and solids. *See also* **S-wave**.

radiation the method of energy transfer by which energy from the Sun reaches the Earth.

radioactive dating the use of the radioactive isotope to determine the age of a fossil or rock layer. This technique is possible because each radioactive isotope decays at a unique and fixed rate.

radioactive decay the process by which a radioactive element breaks down to emit particles and radiation that form a new element.

radioactive isotope an unstable form of an element that breaks down by radioactive decay. For example, carbon-14 is a radioactive isotope of carbon that breaks down by emitting electrons to form nitrogen-14.

rain gauge an instrument that collects atmospheric precipitation so that the amount of rainfall can be measured.

relative humidity the percentage of moisture present in the air as compared to the maximum amount of moisture the air can hold at the prevailing temperature.

residual soil soil that is formed by the weathering of local rock material.

Richter scale a scale with a range of 1 to 10 that indicates the magnitude of an earthquake. For each successive number the magnitude increases by a factor of 10.

rock naturally occurring materials that are composed of one or more minerals.

rock cycle the process by which each of the three rock types (igneous, metamorphic, and sedimentary) can be converted into each other type.

runoff excess precipitation reaching the Earth's surface that flows into rivers and streams because the surface soil is saturated.

salinity a measure of the amount of salt dissolved in water.

saturated air air that contains all the moisture it can hold at the prevailing temperature.

saturation temperature *See* **dewpoint temperature**.

sea breeze a wind blowing from over a lake or ocean to over land. It occurs when the water temperature is cooler than the land temperature. *See also* **land breeze**.

sea floor spreading the concept that portions of the ocean floor are moving away from a central ridge because new material moving upward at the ridge is pushing the old material outward.

sedimentary referring to rocks that form by the compaction and cementation of sediments.

seismograph an instrument used to measure the disturbances caused by an earthquake.

sill a rock layer that forms when molten rock material forces its way between existing rock layers and then cools and solidifies.

sling psychrometer an instrument used to determine relative humidity.

solar eclipse the phenomenon that occurs when the Moon passes between the Earth and the Sun, causing light from the Sun to be blocked from reaching the surface of the Earth. *See also* **lunar eclipse**.

solar system the Sun and the various objects that orbit it, including the planets and their moons, comets, and asteroids.

solstice one of the two times a year, about June 22 and December 22, when the Sun is directly overhead at 23½° north and south latitude. The solstices represent the longest and shortest days of the year.

specific heat the relative amount of energy needed to raise the temperature of one gram of a material by one degree Celsius.

spectroscope an instrument used to break down the visible portion of the electromagnetic spectrum into individual wavelengths or colors.

stationary front a front that forms when the opposing warm and cold air masses are of equal energy.

station model a pattern of symbols on a weather map that is used to describe local weather conditions such as temperature, pressure, humidity, wind speed and direction, and cloud cover.

stratosphere the layer of the atmosphere directly above the troposphere.

streak a more accurate method of identifying the color of a mineral by rubbing it against a plate, causing powder to form.

stream load the total amount of material carried by a stream, including material that is carried in solution or suspension or is pushed along the bottom.

subsoil the layer of soil just below the topsoil. It does not contain the organic matter found in the topsoil and has not been as extensively weathered.

sunspots dark areas at the surface of the Sun that occur because the temperature of the surface is lower at these locations than in the surrounding area.

S-wave (shear wave) the transverse type of wave emitted by an earthquake that travels through solids but is absorbed by liquids. *See also* **P-wave**.

syncline a series of folded rock layers that dips downward near the center. *See also* **anticline**.

terminal moraine the material deposited by the meltwater of a glacier at the point of its farthest advance.

texture the grain size of a rock. A course texture represents large grains; a fine texture, small grains.

theory an explanation for scientific observations. A theory is formed from a hypothesis when there is substantial evidence to support it.

till the material found in a glacial moraine and usually representing a wide range of particle sizes.

topographic (contour) map a map of an area with contour lines to show the elevations at all locations, as well as other geologic features such as rivers, streams, and mountain peaks.

topsoil the uppermost layer of soil. It contains the most highly weathered rock fragments, as well as organic remains.

transpiration the process by which plants release moisture to the atmosphere.

transported soils soils formed from weathered rock material that has been carried from other locations and deposited.

trench a large valley on the ocean floor.

tropical air mass an air mass that has formed over the tropics and therefore contains relatively warm air.

Tropic of Cancer the line of latitude at $23^1/_2°$ North, which represents the farthest north that the noon Sun can be overhead.

Tropic of Capricorn the line of latitude at $23^1/_2°$ South, which represents the farthest south that the noon Sun can be overhead.

troposphere the layer of the atmosphere closest to the Earth's surface. All weather phenomena occur within this zone.

unconformity a gap in a sequence of rock layers, resulting either from the removal of layers by erosion or failure of layers to form for long periods.

uniformitarianism the principle that the geologic processes acting today are the same as those that occurred in the past.

valley glacier a glacier that forms in a valley between mountain peaks.

volcano a mountain formed when lava erupts from beneath the surface and then cools and solidifies.

volume a measure of the amount of space that matter occupies.

warm front the boundary between two air masses of different temperatures; the point where the warmer air mass moves over the colder air mass and pushes it backward. *See also* **cold front**.

water cycle the cyclic process during which water leaves the Earth's surface by evaporation and transpiration to enter the atmosphere. It returns to the surface as precipitation.

water table the upper boundary of saturated rock or soil beneath the Earth's surface.

weathering the processes in nature by which rock materials are broken down into smaller pieces to form sand, soil, and so on. *See also* **chemical weathering; physical weathering**.

wind belts zones around the Earth in which the winds blow in the same general direction.

Regents Examinations, Answers, and Self-Analysis Charts

Examination January 1997

Earth Science

PART I

Answer all 55 questions in this part. [55]

Directions (1–55): For *each* statement or question, select the word or expression that, of those given, best completes the statement or answers the question. Record your answers in the spaces provided.

1. The primary purpose of a classification system is to enable people to

  1 make measurements that are very accurate
  2 eliminate inaccurate inferences
  3 organize observations in a meaningful way
  4 extend their powers of observation

1 _____

2 A quantity of water is frozen solid and then heated from 0°C to 10°C. Which statement best describes the properties of the water during this time?

  1 Mass and volume change.
  2 Volume and density change.
  3 Mass changes but volume remains constant.
  4 Volume changes but density remains constant.

2 _____

3 A scientist who is studying a stream would have the most difficulty determining the stream's

1 age in years
2 velocity
3 temperature
4 transported sediment size

3 ____

4 What is the diameter of the Earth? [Refer to the *Earth Science Reference Tables*.]

(1) 6,370 km
(2) 63,700 km
(3) 12,740 km
(4) 127,400 km

4 ____

5 Which statement best describes the stratosphere? [Refer to the *Earth Science Reference Tables*.]

1 It is warmer at the top than at the bottom.
2 It is located 75 kilometers above sea level.
3 It has greater pressure at the top than at the bottom.
4 It absorbs large amounts of water vapor from the troposphere.

5 ____

6 The data table below shows the stream discharge in April for a creek in the southern United States for a period of 8 days.

Day	Stream Discharge (ft³/sec)
1	20.0
2	6.0
3	269.0
4	280.0
5	48.0
6	21.0
7	14.0
8	5.0

Which graph most accurately shows stream discharge for the 8-day period?

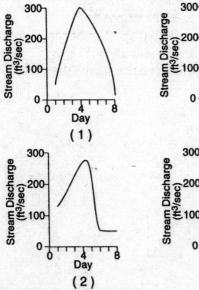

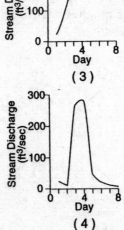

(1)

(3)

(2)

(4)

6 _____

7 The world map below shows latitude and longitude. Letters *A*, *B*, *C*, and *D* represent locations on the map.

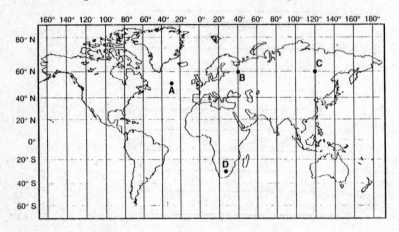

For which location are the correct latitude and longitude given?

(1) *A*: 45° S 30° W   (3) *C*: 60° N 120° E

(2) *B*: 40° N 60° W   (4) *D*: 30° S 30° W   7 _____

8 Which graph best represents the altitude of Polaris observed at northern latitude positions on the Earth's surface?

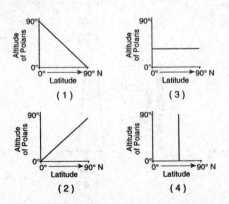

8 _____

9 The period of time a planet takes to make one revolution around the Sun is most dependent on the planet's average

1 rotation rate
2 mass
3 insolation from the Sun
4 distance from the Sun   9 _____

10 Which location on the Earth would the Sun's vertical rays strike on December 21?

1 Tropic of Cancer ($23\frac{1}{2}°$ N)
2 Equator ($0°$)
3 Tropic of Capricorn ($23\frac{1}{2}°$ S)
4 South Pole ($90°$ S)   10 _____

11 The best evidence that the Earth rotates on its axis comes from observations of

1 seasonal changes of constellations
2 the apparent motion of a Foucault pendulum
3 the changing altitude of the Sun at noon
4 the changing altitude of Polaris   11 _____

12 In our solar system, the orbits of the planets are best described as

1 circular, with the planet at the center
2 circular, with the Sun at the center
3 elliptical, with the planet at one of the foci
4 elliptical, with the Sun at one of the foci   12 _____

13 By which process is heat energy transferred when molecules within a substance collide?

1 conduction   3 radiation
2 convection   4 sublimation   13 _____

14 Water loses energy when it changes phase from

1 gas to liquid   3 solid to gas
2 solid to liquid   4 liquid to gas   14 _____

15 When do maximum surface temperatures usually occur in the Northern Hemisphere?

1 early June to mid-June
2 mid-July to early August
3 late August to mid-September
4 mid-September to early October   15 _____

16 Which graph best illustrates the relationship between the intensity of insolation and the angle of insolation?

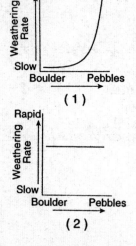

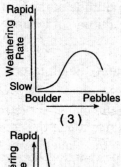

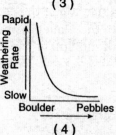

16 _____

17 The diagram below represents energy being absorbed and reradiated by the Earth.

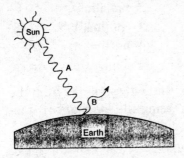

Which type of energy is represented by the radiation at *B*?

1 insolation   3 ultraviolet rays
2 visible light   4 infrared energy   17 _____

18 In the cartoon below, the large arrows represent surface winds.

"Residents of this tiny community are being urged to evacuate."

What feature is found at the location to which the meteorologist is pointing?

1 an anticyclone
2 an area of divergence
3 a low-pressure center
4 a high-pressure center   18 _____

19 According to the *Earth Science Reference Tables*, the prevailing winds at 45° S latitude are from the

  1 southwest   3 southeast

  2 northwest   4 northeast   19 _____

20 Liquid water will continue to evaporate from the Earth's surface, increasing the amount of atmospheric water vapor, until

  1 transpiration occurs

  2 the relative humidity falls below 50%

  3 the atmosphere becomes saturated

  4 the temperature of the atmosphere becomes greater than the dewpoint temperature   20 _____

21 Which weather station model indicates the highest relative humidity?

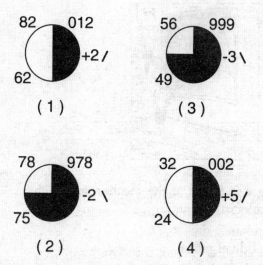

                                    21 _____

22 By which process are clouds, dew, and fog formed?

1 condensation   3 precipitation

2 evaporation   4 melting

22 _____

23 Which graph best represents the chemical weathering rate of a limestone boulder as the boulder is broken into pebble-sized particles?

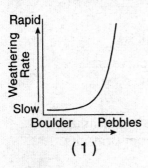

(1)

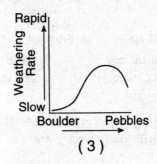

(3)

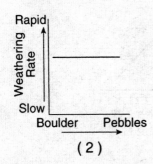

(2)

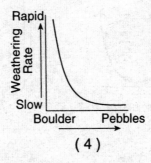

(4)

23 _____

24 Which water budget condition exists when precipitation is less than potential evapotranspiration and storage is depleted?

1 moisture surplus   3 moisture usage

2 moisture recharge   4 moisture deficit

24 _____

25 What is the source of most of the water vapor that enters the atmosphere?

1 lakes   3 soil
2 plants   4 oceans   25 ____

26 Runoff is usually greater than infiltration when the

1 soil is porous   3 rainfall is low
2 slope is steep   4 temperature is high   26 ____

27 The chief agent of erosion on Earth is

1 human beings   3 wind
2 running water   4 glaciers   27 ____

28 Which graph best represents the relationship between the particle size and the capillarity of a sample of soil?

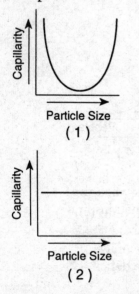

(1)

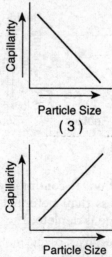

(3)

(2)

(4)   28 ____

29 The diagrams below represent two containers, each filled with a sample of nonporous particles of uniform size.

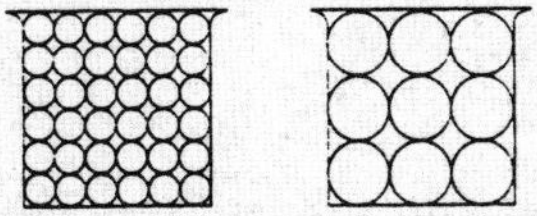

Compared to the sample of larger particles, the sample of smaller particles has

1 lower permeability   3 less porosity
2 higher permeability   4 more porosity   29 _____

30 Most metamorphic rocks are formed when

1 sediments are cemented and compacted
2 magma cools slowly, deep underground
3 flows of lava cool rapidly
4 rocks are subjected to heat and pressure   30 _____

31 The two pebbles shown below are dropped into a tank of water 1 meter deep.

Hematite   **Quartz**

Density = 6.5 g/cm^3 Density = 2.6 g/cm^3

Why does the hematite pebble settle faster than the quartz pebble?

1 Smaller objects settle faster than larger objects.
2 Flat objects settle faster than round objects.
3 Spherical objects have less gravitational attraction than flat objects.
4 Objects with higher density settle faster than objects with lower density.   31 _____

32 In the diagram below of a straight-flowing stream, the lengths of the arrows represent differences in relative stream velocity on the stream's surface.

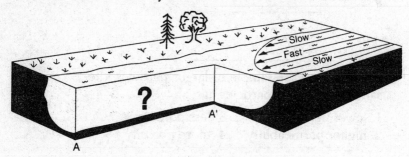

Which diagram best represents the relative stream velocity from the surface to the bottom of the stream for the cross section from A to A'?

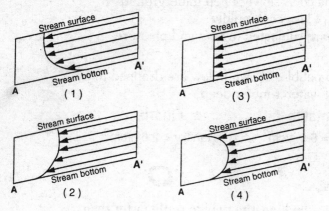

32 _____

33 Igneous, sedimentary, and metamorphic rocks are usually composed of

1 intergrown crystals   3 minerals
2 fossils   4 sediments   33 _____

34 Which feature is characteristic of sedimentary rocks?

   1 layering   3 distorted structure

   2 foliation   4 glassy texture   34 _____

Base your answers to questions 35 and 36 on the table below, which shows some characteristics of a rock-forming mineral.

Mineral	Cleavage	Hardness	Density (g/cm³)	Other Properties
Pyroxene (a complex family of minerals; augite is most common)	Two flat planes at nearly right angles	5–6	3.2–3.9	Found in igneous and metamorphic rocks; augite is dark green to black; other varieties are white to green

35 Which diagram best represents a sample of pyroxene?

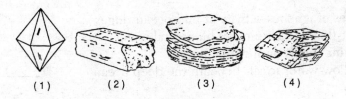

   (1)   (2)   (3)   (4)   35 _____

36 According to the *Earth Science Reference Tables*, an igneous rock containing large, visible crystals of pyroxene is best described as

   1 felsic and formed deep within the Earth's crust

   2 felsic and formed near the Earth's surface

   3 mafic and formed deep within the Earth's crust

   4 mafic and formed near the Earth's surface   36 _____

37 Most of the oceanic crust is composed of rock material similar to

1 basalt   3 sandstone
2 granite   4 limestone   37 _____

38 Which statement best explains why the *P*-wave of an earthquake arrives at a seismic station before the *S*-wave?

1 The *S*-wave originates from the earthquake focus.
2 The *S*-wave decreases in velocity as it passes through a liquid.
3 The *P*-wave originates from the earthquake epicenter.
4 The *P*-wave has a greater velocity than the *S*-wave.   38 _____

39 Evidence of crustal subsidence (sinking) is provided by

1 zones of igneous activity at mid-ocean ridges
2 heat-flow measurements on coastal plains
3 marine fossils on mountaintops
4 shallow-water fossils beneath the deep ocean   39 _____

40 An earthquake *P*-wave arrived at a seismograph station at 01 hour 21 minutes 40 seconds. The distance from the station to the epicenter is 3,000 kilometers. The earthquake's origin time was

(1) 01 h 11 min 40 sec
(2) 01 h 16 min 00 sec
(3) 01 h 20 min 20 sec
(4) 01 h 27 min 20 sec   40 _____

41 The map below shows the western part of the United States.

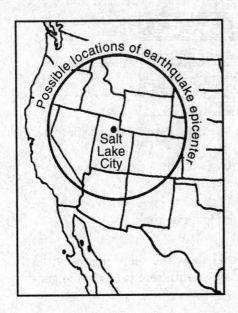

Which observation made at Salt Lake City would allow seismologists to determine that an earthquake had occurred somewhere along the circle shown on the map?

1 the relative strength of the *P*-waves and *S*-waves

2 the time interval between the arrival of the *P*-waves and *S*-waves

3 the difference in the direction of vibration of the *P*-waves and *S*-waves

4 the density of the subsurface bedrock through which the *P*-waves and *S*-waves travel

41 ____

42 Which feature in the geologic cross section below was formed by erosion?

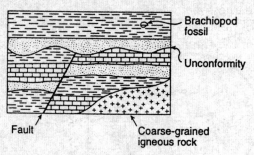

1 unconformity
2 fault
3 brachiopod fossil
4 coarse-grained igneous rock

42 _____

43 Diagrams I and II show the same region of the Earth's surface at two different times in the geologic past.

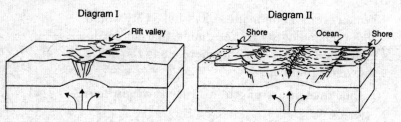

Which statement best explains the basic cause of the changes that occurred in this region?

1 Meteor impact on the crust caused widening of the valley.
2 Mantle convection currents caused crustal movement.
3 Climate changes caused flooding.
4 Temperature changes caused melting of polar ice caps.

43 _____

44 When did dinosaurs become extinct?

  1 before the earliest birds
  2 before the earliest mammals
  3 at the end of the Cretaceous Period
  4 at the end of the Cambrian Period   44 _____

45 Which locations are listed in order of the age of
   their surface bedrock, from oldest to youngest?

  1 Syracuse, Watertown, Elmira, Old Forge
  2 Elmira, Syracuse, Old Forge, Watertown
  3 Old Forge, Watertown, Syracuse, Elmira
  4 Syracuse, Elmira, Watertown, Old Forge   45 _____

Base your answers to questions 46 and 47 on the
table below, which shows the geologic ages of some
index fossils.

GEOLOGIC TIME	INDEX FOSSILS			
MISSISSIPPIAN	Spirifer	Muensteroceras	Crinoid Stem	Pentremites
DEVONIAN	Mucrospirifer		Phacops	
SILURIAN	Eospirifer			
ORDOVICIAN	Michelinoceras		Flexicalymene	

46 According to the *Earth Science Reference Tables*, which index fossil might be found in rock layers that are approximately 387 million years old?

Muensteroceras
(1)

Phacops
(2)

Eospirifer
(3)

Flexicalymene
(4)

46 _____

47 Which type of past environment is indicated by these index fossils?

1 equatorial rain forest   3 sandy desert
2 arctic tundra   4 ocean

47 _____

48 Two rock units contain the same radioactive substance. Rock *A* is buried deep underground; rock *B* is at the Earth's surface. Which statement best describes the half-life of the radioactive substance?

1 The radioactive substance in rock *A* has a longer half-life.
2 The radioactive substance in rock *B* has a longer half-life.
3 The radioactive substance has the same half-life in rock *A* and in rock *B*.
4 The radioactive substance's half-life has increased with time in rocks *A* and *B*.

48 _____

49 A marine fossil was found to contain one-half of its original quantity of carbon-14. According to the *Earth Science Reference Tables*, approximately how old is this fossil?

(1) 5,700 years   (3) 17,100 years
(2) 11,400 years   (4) 22,800 years   49 _____

50 In New York State, the Susquehanna River and the Delaware River both flow over a region classified as a

1 lowland   3 coastal plain
2 plateau   4 mountain range   50 _____

51 Which substance found in a soil sample collected in an arid region would most likely be absent in a soil sample collected in a humid region?

1 rock salt   3 obsidian
2 quartz   4 pyroxene   51 _____

52 Features such as mountains, plains, and plateaus divide continents into

1 crustal activity zones
2 natural resource zones
3 landscape regions
4 erosional activity areas   52 _____

53 According to the *Earth Science Reference Tables*, which New York State landscape surface is composed of gneisses, quartzites, marbles, and anorthositic bedrock?

1 Allegheny Plateau
2 Erie-Ontario Lowlands
3 the Catskills
4 Adirondack Mountains   53 _____

54 Which New York State landscape region could have surface bedrock containing dinosaur fossils?

1 Adirondack Highlands
2 Erie-Ontario Lowlands
3 St. Lawrence Lowlands
4 Newark Lowlands   54 _____

55 Which geologic cross section best represents the bedrock at Binghamton?

Key: Rock unit symbols are those found on the Generalized Bedrock Geology of New York State map. The symbol ∿∿∿ represents an unconformity.

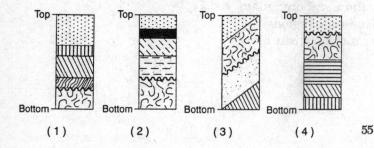

(1)   (2)   (3)   (4)   55 _____

PART II

This part consists of ten groups, ea
tions. Choose seven of these ten gr
answer all five questions in each gro
answers to these questions in the space

GROUP 1

If you choose this group, be sure to answer questions **56–60**.

Base your answers to questions 56 through 60 on the *Earth Science Reference Tables*, the table and information below, and your knowledge of Earth science.

The Bay of Fundy, located on the east coast of Canada, has the highest ocean tides in the world. The St. John River enters the Bay of Fundy at the city of St. John, where the river actually reverses direction twice a day at high tides. Data for the famous Reversing Falls of the St. John River are given below for high and low tides on June 26 through 28, 1994.

Tidal Record for Reversing Falls, St. John River				
Date	Time of First High Tide	Time of First Low Tide	Time of Second High Tide	Time of Second Low Tide
June 26	2:25 a.m.	8:45 a.m.	2:55 p.m.	9:05 p.m.
June 27	3:15 a.m.	9:35 a.m.	3:45 p.m.	9:55 p.m.
June 28	4:05 a.m.	10:25 a.m.	4:35 p.m.	10:45 p.m.

best represents the tides recorded on

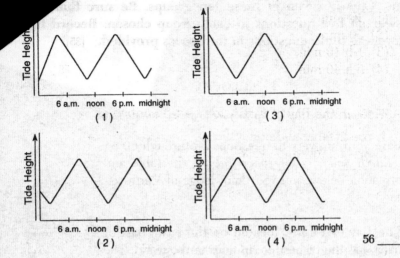

(1)

(3)

(2)

(4)

56 _____

57 Which model of the Sun, Earth (E), and Moon (M) best represents a position that would cause the highest ocean tides in the Bay of Fundy? [Sizes and distances are not drawn to scale.]

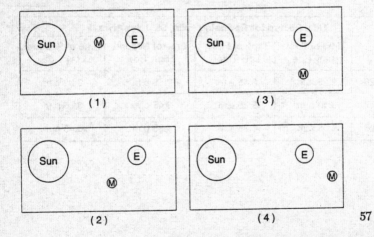

(1)

(3)

(2)

(4)

57 _____

58 Compared to the first high tide on June 26, how much later in the day did the first high tide occur on June 27?

(1) 10 min
(2) 50 min
(3) 1 h 10 min
(4) 5 h 40 min

58 _____

59 Tides in the Bay of Fundy are best described as

1 predictable and noncyclic
2 predictable and cyclic
3 unpredictable and noncyclic
4 unpredictable and cyclic

59 _____

60 The Moon has a greater effect on the Earth's ocean tides than the Sun has because the

1 Sun has a higher density than the Moon
2 Sun has a higher temperature than the Moon
3 Moon has a greater mass than the Sun
4 Moon is closer to the Earth than the Sun is

60 _____

GROUP 2

If you choose this group, be sure to answer questions **61–65**.

Base your answers to questions 61 through 65 on the topographic map below and on your knowledge of Earth science. Heavy dashed lines represent four hiking paths, *A*, *B*, *C*, and *D*. Point *P* is a location on the map.

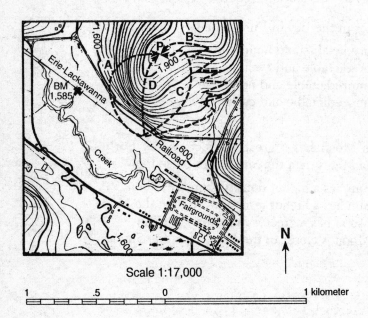

Scale 1:17,000

1   .5   0   1 kilometer

61 On this map, 1 centimeter represents how many centimeters on the surface of the Earth?

(1) 1,600   (3) 14,700

(2) 1,900   (4) 17,000   61 _____

62 What is the contour interval for this map?

(1) 10 ft   (3) 25 ft

(2) 20 ft   (4) 100 ft   62 _____

63 On what type of landscape are the fairgrounds located?

1 a cliff   3 a hilltop

2 a coastal plain   4 a floodplain   63 _____

64 What is the approximate length of the portion of the Erie-Lackawanna railroad tracks shown on the map?

(1) 2.5 km   (3) 3.0 km

(2) 2.0 km   (4) 3.5 km   64 _____

65 Which path climbs the steepest part of the hill from the railroad tracks to point *P*?

(1) *A*   (2) *B*   (3) *C*   (4) *D*   65 _____

GROUP 3

*If you choose this group, be sure to answer questions **66–70**.*

Base your answers to questions 66 through 70 on the *Earth Science Reference Tables*, the diagram below, and your knowledge of Earth science. The diagram represents a cross section of a portion of the Earth's crust. Points *A* through *D* represent locations in the bedrock. The rock layers have not been overturned.

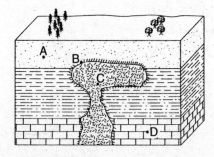

KEY

Limestone	Shale
Siltstone	Igneous intrusion (gabbro)
Sandstone	Contact metamorphism

66 Which rock is *least* likely to contain fossils?

    1 gabbro   3 limestone

    2 shale   4 siltstone   66 _____

67 Which rock formed most recently?

    1 limestone   3 shale

    2 siltstone   4 gabbro   67 _____

68 Which rock formed as a result of heat and pressure at point *B*?

    1 slate   3 marble

    2 quartzite   4 anthracite coal   68 _____

69 The limestone layer could have been formed primarily by

    1 foliation of mica during faulting

    2 chemical precipitation of calcite

    3 deposition of quartz fragments

    4 decomposition of plant remains   69 _____

70 Which two minerals will probably be most abundant in the igneous intrusion?

    1 quartz and calcite

    2 halite and gypsum

    3 potassium feldspar and biotite

    4 plagioclase feldspar and pyroxene   70 _____

GROUP 4

If you choose this group, be sure to answer questions **71–75**.

Base your answers to questions 71 through 75 on the *Earth Science Reference Tables*, the graph below, and your knowledge of Earth science. The graph shows the amount of heat energy (calories) needed to raise the temperature of 1-gram samples of four different materials.

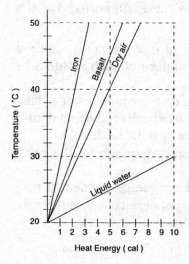

71 What is the total number of calories of heat needed to raise the temperature of the liquid water from 20°C to 30°C?

(1) 1 cal   (3) 20 cal
(2) 10 cal   (4) 30 cal   71 _____

72 Which of these materials has the highest specific heat?

1 liquid water   3 basalt
2 dry air   4 iron   72 _____

73 If all four materials were heated to 100°C and then allowed to cool, which material would show the most rapid drop in temperature?

1 basalt
2 iron
3 dry air
4 liquid water

73 _____

74 Which statement is best supported by the graph?

1 The same amount of heat energy is required to raise the temperature of each material by 10 Celsius degrees.
2 The temperature of a material with a high specific heat is raised faster than that of a material with a low specific heat.
3 Three of the four materials have the same specific heat.
4 The amount of heat energy needed to produce an equal temperature change varies with the materials heated.

74 _____

75 Which graph below correctly shows where the line representing the temperature of a fifth material, lead, would be located?

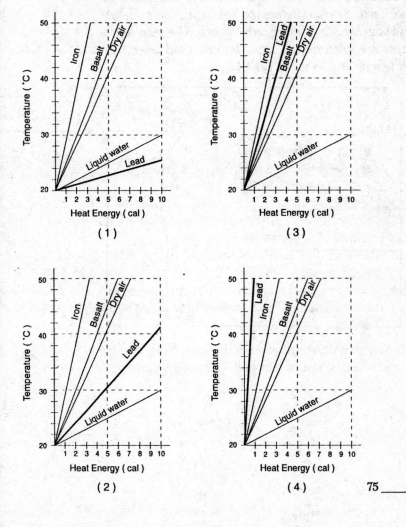

75 _____

GROUP 5

*If you choose this group, be sure to answer questions **76–80**.*

Base your answers to questions 76 through 80 on the *Earth Science Reference Tables*, the map below, and your knowledge of Earth science. The map represents a high-pressure air mass centered over New York State at 10 a.m. on June 16.

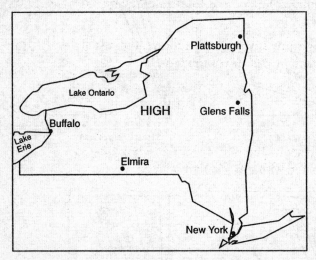

76 Which pattern represents the most likely surface wind direction of this high-pressure system?

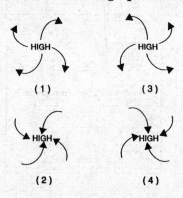

76 _____

77 At Syracuse, the weather is most likely

    1 cold, with snow   3 clear and cool

    2 hot and humid   4 cloudy, with rain   77 _____

78 Which type of air mass usually causes these high-pressure centers over New York State?

    1 continental tropical   3 maritime tropical

    2 continental polar   4 maritime polar   78 _____

79 As this high-pressure center follows the usual air-mass track, it will travel toward

    1 Buffalo   3 Lake Ontario

    2 Elmira   4 Glens Falls   79 _____

80 Which map shows the most likely pattern of isobars associated with this weather system?

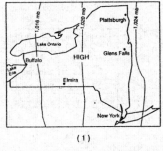

(2)

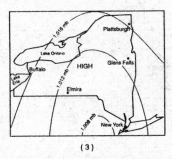

(3)

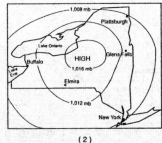

(4)

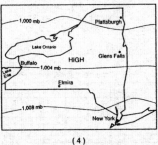

(1)

80 _____

GROUP 6

If you choose this group, be sure to answer questions 81–85.

Base your answers to questions 81 through 85 on the *Earth Science Reference Tables*, the map and data table below, and your knowledge of Earth science. The map represents an imaginary continent on the Earth. Letters *B* through *G* are locations on the map for which climate ratios are given. The climate ratio is determined by dividing the average yearly precipitation by the average yearly potential evapotranspiration (P/E_p). The data table shows the monthly precipitation and potential evapotranspiration values for location *A*.

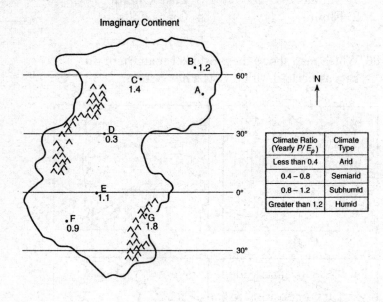

Imaginary Continent

Climate Ratio (Yearly P/E_p)	Climate Type
Less than 0.4	Arid
0.4 – 0.8	Semiarid
0.8 – 1.2	Subhumid
Greater than 1.2	Humid

Water Budget Data for Location A (mm)

	Jan.	Feb.	Mar.	Apr.	May	June	July	Aug.	Sept.	Oct.	Nov.	Dec.	Totals
Precipitation (*P*)	68	76	89	96	81	68	75	71	67	65	70	63	889
Potential Evapotranspiration (E_p)	5	10	35	60	85	155	170	159	82	60	34	10	865

81 At location A, the value of the yearly P/E_p ratio is 1089 mm/865 mm. What is the type of climate at location A?

    1 arid   3 subhumid

    2 semiarid   4 humid   81 _____

82 Which graph best shows the monthly precipitation and potential evapotranspiration values for location A?

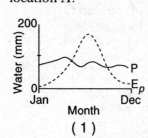

(1)

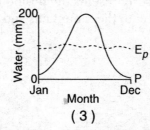

(3)

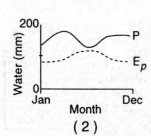

(2)

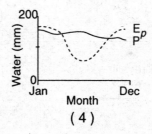

(4)   82 _____

83 The potential evapotranspiration (E_p) recorded at these locations depends primarily on the

    1 soil composition in the area

    2 height of the water table

    3 level of water in the streams

    4 amount of solar radiation absorbed   83 _____

84 Which climate condition is characteristic of both location *C* and location *D*?

　1 a large amount of yearly precipitation
　2 a large yearly temperature range
　3 the same yearly potential evapotranspiration
　4 the same number of months of moisture surplus

84 _____

85 Location *G* has a cold, humid climate. Which profile best represents the position of location *G* with respect to the mountains and the prevailing winds?

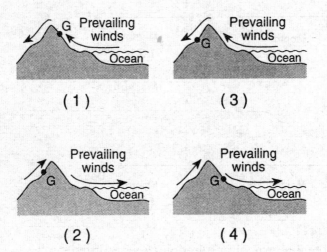

(1)　　　　　　　(3)

(2)　　　　　　　(4)

85 _____

GROUP 7

If you choose this group, be sure to answer questions **86–90**.

Base your answers to questions 86 through 90 on the *Earth Science Reference Tables*, the data table below, and your knowledge of Earth science. The table shows the percentages of different grain sizes deposited as sediment in the Mississippi River 100 miles to 1,000 miles downstream from Cairo, Illinois. The river's mouth empties into the Gulf of Mexico 1,000 miles from Cairo.

Grain Size Percentages

Grain Size	Miles Downstream from Cairo					
	100 mi	300 mi	500 mi	700 mi	900 mi	1,000 mi
Small Pebbles	29%	8%	14%	5%	none	none
Coarse Sand	30%	22%	9%	8%	1%	none
Medium Sand	32%	50%	46%	44%	26%	9%
Fine Sand	8%	19%	28%	41%	70%	69%
Silt	trace	trace	2%	1%	2%	10%
Clay	trace	trace	1%	trace	1%	10%

86 Which graph best shows the percentage of each grain size found at a location 300 miles downstream from Cairo?

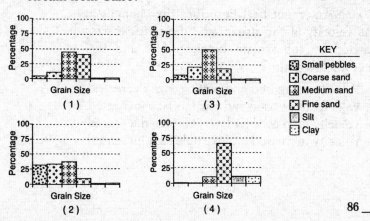

KEY
▨ Small pebbles
□ Coarse sand
▨ Medium sand
▧ Fine sand
▨ Silt
□ Clay

86 _____

87 At the river's mouth, 1,000 miles downstream from Cairo, small pebbles and coarse sand are *not* being deposited because these particles are

  1 dissolved in the river water
  2 deposited before reaching this location
  3 being carried in suspension
  4 being rolled along the stream bottom   87 _____

88 What was the *minimum* water velocity needed to transport the sediments deposited 900 miles downstream from Cairo?

  (1) 20 cm/sec   (3) 170 cm/sec
  (2) 50 cm/sec   (4) 200 cm/sec   88 _____

89 As a pebble travels downstream from Cairo toward the Gulf of Mexico, it most likely becomes more

  1 dense   3 flattened
  2 angular   4 rounded   89 _____

90 Which statement best describes the changes in the velocity of the stream and the particle sizes carried by the stream between 500 and 900 miles below Cairo?

  1 Velocity decreased and particle size decreased.
  2 Velocity decreased and particle size increased.
  3 Velocity increased and particle size decreased.
  4 Velocity increased and particle size increased.   90 _____

GROUP 8

If you choose this group, be sure to answer questions **91–95**.

Base your answers to questions 91 through 95 on the *Earth Science Reference Tables*, the world map below, and your knowledge of Earth science. The small dots on the map represent earthquake epicenters. The letters on the map represent locations.

Earthquake Epicenters

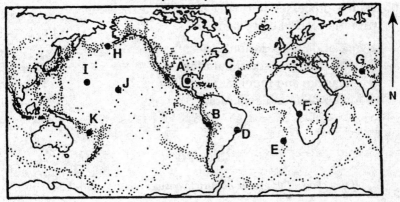

91 Where do most of these earthquakes occur?

   1 in the centers of the continents
   2 in specific belts within the crust
   3 randomly throughout the mantle
   4 along the core-mantle interface   91 _____

92 Locations *H* and *K* are found at tectonic plate boundaries referred to as

   1 divergent zones
   2 rift zones
   3 ridges
   4 trenches   92 _____

93 At which location do rocks, minerals, and fossils most closely match those at location *D*?

  (1) *H*   (2) *B*   (3) *E*   (4) *F*   93 _____

94 At which location did earthquakes occur as a result of the Nazca plate sliding under the South American plate?

  (1) *A*   (2) *B*   (3) *C*   (4) *D*   94 _____

95 Which processes normally occur in association with the plotted earthquakes?

  1 glaciation and erosion
  2 deposition and sedimentation
  3 volcanism and mountain building
  4 fossilization and evolution   95 _____

GROUP 9

If you choose this group, be sure to answer questions **96–100**.

Base your answers to questions 96 through 100 on the *Earth Science Reference Tables*, the diagram below, and your knowledge of Earth science. The diagram shows a cross section of a landscape and a nearby sea. Letters *A* and *B* indicate locations on the landscape surface. The geologic age of three of the rock types is shown.

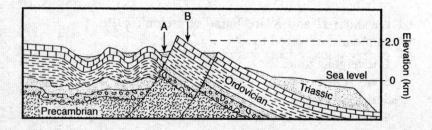

96 The landscape near locations *A* and *B* is considered mountainous because it contains

  1 horizontal rock layers of low relief
  2 horizontal rock layers of high relief
  3 deformed rock layers of low relief
  4 deformed rock layers of high relief   96 _____

97 Which particle size and type of fossil would be characteristic of the Ordovician rock layer?

  1 silt-sized particles and shark fossils
  2 silt-sized particles and coral fossils
  3 clay-sized particles and shark fossils
  4 clay-sized particles and coral fossils   97 _____

98 The structural feature shown in the bedrock between locations *A* and *B* is a

  1 volcano
  2 fault
  3 glacial moraine
  4 plateau   98 _____

99 Which activity caused the limestone bedrock layer to become folded?

  1 slow movement of volcanic rock as it was deposited
  2 crustal movement that occurred after deposition
  3 deposition of sediments with different densities
  4 deposition of loose fragments in angular beds   99 _____

100 Which change would occur at location *B* when uplifting forces dominate leveling forces?

  1 Streams would decrease in velocity.
  2 Erosion and deposition would decrease.
  3 Hillslopes would increase in steepness.
  4 Topographic features would become smoother.   100 _____

GROUP 10

*If you choose this group, be sure to answer questions **101–105**.*

Base your answers to questions 101 through 105 on the *Earth Science Reference Tables* and on your knowledge of Earth science.

101 Which graph best represents the percentage by volume of the elements making up the Earth's hydrosphere?

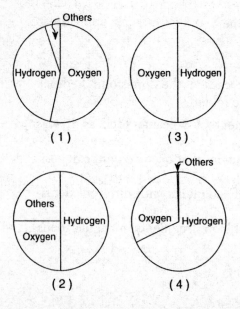

(1)   (3)

(2)   (4)   101 _____

102 At which temperature would an object radiate the most electromagnetic energy?

(1) 50°C   (3) 310 K

(2) 80°C   (4) 140°F   102 _____

103 Which information in the *Earth Science Reference Tables* is an inference rather than an observation?

1 Temperature decreases as elevation in the troposphere increases.

2 Saturn's period of rotation is 10 hours 14 minutes.

3 A *P*-wave travels 5,600 kilometers in 9 minutes.

4 The Earth's outer core is made of iron.   103 _____

104 An observer incorrectly measured the mass of a rock as 428.7 grams. The actual mass was 450.0 grams. What was the observer's approximate percentage of error?

(1) 5.0%   (3) 4.3%

(2) 2.1%   (4) 4.7%   104 _____

105 When the height of the mercury in a barometer is 29.92 inches, the barometric pressure is

(1) 1,000.0 mb   (3) 1,013.2 mb

(2) 1,005.5 mb   (4) 1,020.4 mb   105 _____

Answers
January 1997
Earth Science

Answer Key

PART I

1.	3	12.	4	23.	1	34.	1	45.	3
2.	2	13.	1	24.	4	35.	2	46.	2
3.	1	14.	1	25.	4	36.	3	47.	4
4.	3	15.	2	26.	2	37.	1	48.	3
5.	1	16.	2	27.	2	38.	4	49.	1
6.	4	17.	4	28.	3	39.	4	50.	2
7.	3	18.	3	29.	1	40.	2	51.	1
8.	2	19.	2	30.	4	41.	2	52.	3
9.	4	20.	3	31.	4	42.	1	53.	4
10.	3	21.	2	32.	1	43.	2	54.	4
11.	2	22.	1	33.	3	44.	3	55.	1

PART II

56.	1	66.	1	76.	1	86.	3	96.	4
57.	1	67.	4	77.	3	87.	2	97.	4
58.	2	68.	2	78.	2	88.	2	98.	2
59.	2	69.	2	79.	4	89.	4	99.	2
60.	4	70.	4	80.	2	90.	1	100.	3
61.	4	71.	2	81.	4	91.	2	101.	4
62.	2	72.	1	82.	1	92.	4	102.	2
63.	4	73.	2	83.	4	93.	4	103.	4
64.	2	74.	4	84.	2	94.	2	104.	4
65.	4	75.	4	85.	1	95.	3	105.	3

Answers Explained

PART I

1. **3** A classification system is useful, for example, for comparing rock samples. Rocks are classified, according to how they formed, as igneous, sedimentary, or metamorphic. There are further subdivisions within each of these categories. Classification systems allow us to organize observations in a meaningful way.

2. **2** Between 0°C and 4°C, water contracts when it is heated. Above 4°C, water expands with heat. The net effect between 0°C and 10°C is that water expands. As it expands, the volume increases and the density decreases. The decrease in density occurs because the mass stays the same as the volume increases.

3. **1** There is no method for determining the absolute age of a stream. As a stream gets older, increased erosion will tend to cause the stream to become wider and more curved. This is evidence, however, only of the relative age of a stream. It does not enable a scientist to determine how long the stream has actually been flowing.

4. **3** Find the Astronomy Measurements chart in the Physical Constants section of the *Earth Science Reference Tables*. Note that the radius of the Earth is 6370 km. The diameter is twice the radius or 12,740 km.

5. **1** Find the graphs entitled Selected Properties of Earth's Atmosphere in the *Earth Science Reference Tables*. According to the graph of temperature zones, at the bottom of the stratosphere the temperature is about –55°C. At the top of the stratosphere the temperature is about 0°C. Therefore the stratosphere is warmer at the top than at the bottom.

WRONG CHOICES EXPLAINED:
(2) According to the scale, the elevation at the base of the stratosphere is about 11 km and the elevation at the top is about 50 km.
(3) According to the graph that shows atmospheric pressure, the pressure *decreases* as the elevation increases.
(4) The graph of water vapor shows that the concentration of water vapor reaches zero near the boundary between the troposphere and the stratosphere.

Day	Stream Discharge (ft³/sec)
1	20.0
2	6.0
3	269.0
4	280.0
5	48.0
6	21.0
7	14.0
8	5.0

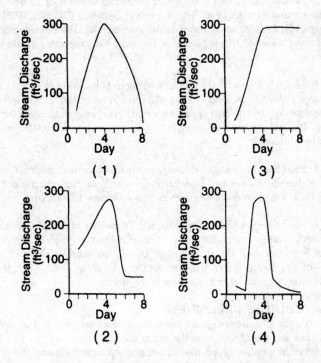

6. **4** The table shows that there is a drop in discharge between days 1 and 2 and a big increase on days 3 and 4. A big drop in discharge on day 5 is followed by a gradual decrease over the remaining days. This pattern is illustrated in graph (4).

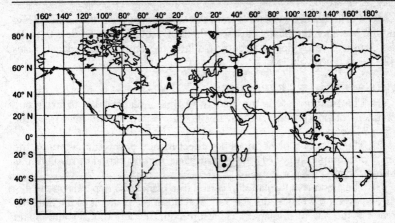

7. **3** Location *C* has a latitude of 60°N and a longitude of 120°E on the world map. It is the only location for which the correct coordinates are given.

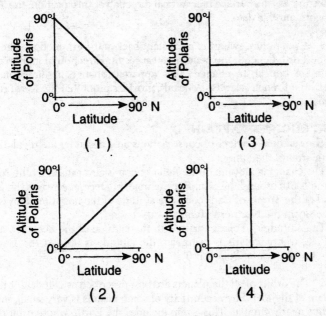

8. **2** Since Polaris is located over the North Pole, its altitude to an observer at the North Pole is 90°. At the Equator the altitude of Polaris is 0°. Graph (2) shows the altitude of Polaris increasing from 0° to 90° with a corresponding increase in latitude.

9. **4** The farther a planet is from the Sun, the longer is its period of revolution. Mercury, the planet closest to the Sun, requires only 88 days to make one complete revolution. In contrast, Pluto, the most distant planet, takes 247.7 years per revolution.

WRONG CHOICES EXPLAINED:

(1) The rotation rate is the amount of time a planet takes to make one complete rotation around its axis. For the Earth the rotation rate is a little under 24 hours.

(2) The mass of a planet does not affect its rate of revolution around the Sun. For example, Jupiter is the largest planet but its period of revolution is neither the greatest nor the least.

(3) The amount of insolation received by a planet *also* depends on the distance between the planet and the Sun. The closer the planet is to the Sun, the greater is the rate of insolation. Therefore both the period of revolution and the rate of insolation depend on the distance between a planet and the Sun.

10. **3** On December 21 the Sun is overhead at $23\frac{1}{2}°$ S. The rays of the Sun strike the Earth's surface in a vertical direction at this location, the Tropic of Capricorn, on this date.

11. **2** A pendulum swings in a constant back-and-forth motion. The rotation of the Earth beneath the pendulum causes the direction of motion of the pendulum to *appear* to change. This apparent change in direction, first noticed by the French scientist Foucault, provides proof that the Earth rotates on its axis.

WRONG CHOICES EXPLAINED:

(1) The seasonal changes of constellations are caused by the revolution of the Earth around the Sun.

(3) The changing altitude of the Sun at noon is also caused by the revolution of the Earth around the Sun. For example, in December when the Sun is overhead at the Tropic of Capricorn, the altitude of the noontime Sun is lowest in the sky in the Northern Hemisphere.

(4) The altitude of Polaris varies with the latitude of the observer. At the North Pole, where Polaris is overhead, the altitude is 90°. At the Equator, however, the altitude of Polaris is 0°.

12. **4** The orbits of all the planets are best described as elliptical, with the Sun as one of the foci. The eccentricity of some planets is very small, making their orbits nearly circular. This group includes the Earth, whose orbit drawn to scale would look circular. Conversely the orbits of Mercury and Pluto are highly eccentric. Scale models of their orbits would appear oval.

13. **1** When an object is heated, the molecules in the object speed up. When these molecules strike other molecules, these other molecules also accelerate. This process continues throughout the object until the average velocity of all the molecules in the object is greater. This process of energy transfer is called conduction.

WRONG CHOICES EXPLAINED:

(2) Convection is the process of energy transfer in fluids. When a fluid is heated, it expands and rises. The rising fluid is replaced by cooler, denser fluid. As this cooler fluid is heated, it too rises, forming a circular pattern of heat transfer called convection.

(3) Radiation is the method by which energy is transferred by the Sun to the Earth. Also, when a light bulb is turned on, it releases energy by radiation that we see as light.

(4) Sublimation is the process by which a solid changes directly into a vapor. Dry ice (carbon dioxide) is an example of a common substance that undergoes this type of transformation at room temperature.

14. **1** When a gas changes to a liquid or a liquid changes to a solid, energy is released. Energy is absorbed when changes occur in the opposite direction.

15. **2** The maximum duration of insolation or the greatest number of daylight hours occurs around June 21 in the Northern Hemisphere. From mid-July to early August the number of daylight hours still exceeds the number of hours of darkness. As a result the Earth's surface continues to heat up and maximum surface temperatures occur.

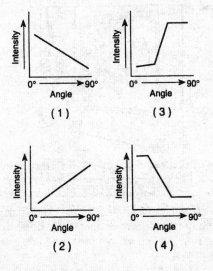

16. **2** As the angle of insolation increases, the intensity of insolation also increases. This pattern is illustrated in graph (2). For example, when the Sun is highest in the sky, the intensity of insolation is greatest and temperatures are warmer.

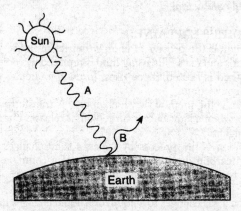

17. **4** The type of energy represented by the radiation at *B*, which is energy radiated by the Earth back to space, falls within the infrared portion of the electromagnetic spectrum. This radiation has longer wavelengths than the ultraviolet radiation and visible light emitted by the Sun. The wavelengths of energy emitted by the Earth are longer than the wavelengths emitted by the Sun because the surface temperature of the Earth is lower than the surface temperature of the Sun.

"Residents of this tiny community are being urged to evacuate."

18. **3** The arrows on the weather map in the cartoon are pointing inward. This is characteristic of a low-pressure center. Here the winds blow inward because winds blow from regions of high pressure to regions of low pressure.

WRONG CHOICES EXPLAINED:

(1), (4) An anticyclone is another name for a high-pressure center. Here the winds blow outward from the center, where the pressure is greatest.

(2) An area of divergence occurs where the air circulation is outward because the pressure is higher near the center.

19. **2** Find in the *Earth Science Reference Tables* the diagram entitled Planetary Wind and Moisture Belts in the Troposphere. Notice the belt labeled "N.W. WINDS" in the zone between 30°S and 60°S. The arrows show that the winds there are blowing from the northwest.

20. **3** Liquid water will continue to evaporate from the Earth's surface until the atmosphere becomes saturated. When the air is saturated, it contains the maximum amount of water vapor that it can hold at that temperature. In the atmosphere any excess moisture beyond the point of saturation tends to condense to form clouds.

WRONG CHOICES EXPLAINED:

(1) Transpiration is the process by which plants release moisture to the atmosphere.

(2) Liquid water will continue to evaporate until the relative humidity reaches 100%. At this point the air is saturated; it contains as much moisture as it can hold. Relative humidity is a comparison between the actual amount of moisture in the air and the maximum amount it can hold at a certain temperature. When the two are equal, the air is saturated and the relative humidity is 100%.

(4) When the temperature of the atmosphere is greater than the dewpoint temperature, the air is considered to be supersaturated. If condensation particles are present, the excess moisture will condense. Clouds or fog will form, depending on whether the excess moisture is near the ground or high in the sky.

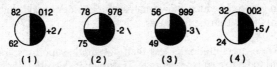

21. **2** Find the Station Model diagram under Weather Map Information in the *Earth Science Reference Tables*. Note the locations of the air temperature and the dewpoint. The air has the highest relative humidity when the two temperatures are most nearly the same. The air temperature at station model (2) is 78°C, and the dewpoint temperature is 75°C. Of the choices given, these two readings are closest to the same value.

22. **1** Clouds, dew, and fog form when the air cools and becomes saturated. The excess moisture changes from a vapor to a liquid. This is the process of condensation.

WRONG CHOICES EXPLAINED:

(2) If the air temperature rises, the liquid water in clouds, dew, and fog may evaporate and the clouds, fog, or dew will disappear.

(3) Precipitation is released from clouds when the size of the droplets of water in the clouds exceeds the level at which they can remain suspended.

(4) Melting occurs when a solid such as ice is heated beyond its melting temperature.

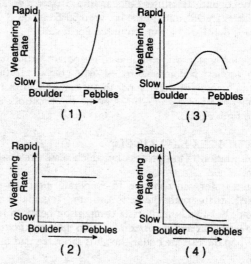

23. **1** The rate of weathering of the limestone depends on the amount of exposed surface area, which is where the weathering occurs. As weathering continues, smaller particles are produced. These smaller particles have greater amounts of surface area, so the rate of weathering increases. This pattern is illustrated in graph (1) which shows that, as the boulder breaks down into pebbles, the rate of weathering increases.

24. **4** When precipitation is less than potential evapotranspiration, moisture will tend to be drawn from storage. If storage is depleted, then a moisture deficit exists. There is no water in the ground to make up for the shortage.

25. **4** The oceans cover most of the Earth's surface, and therefore most of the water vapor that enters the atmosphere comes from this source. Much smaller amounts are contributed by soil, by plants, and by lakes and other water bodies.

26. **2** When the slope of a surface is steep, water will tend to run off before it has a chance to infiltrate the ground.

WRONG CHOICES EXPLAINED:

(1) If the soil is porous, water can more easily infiltrate the soil and runoff is less.

(3) Runoff will tend to be less if rainfall is low.

(4) The amount of runoff is not directly affected by temperature.

27. **2** Running water is responsible for more erosion on the Earth's surface than any other agent. Running water causes more erosion than the combined effects of glaciers, wind, and gravity.

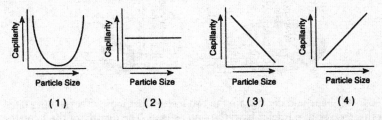

28. **3** Capillarity is the rate at which water rises in soil above the water table by adsorbing to the soil particles. The smaller the particle size, the greater is the height to which the water will rise. Graph (3) shows capillarity increasing as particle size decreases.

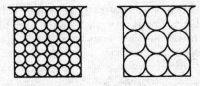

29. **1** Permeability is the rate at which water infiltrates a porous material. When the particle size is smaller, as in the container on the left, the permeability is lower.

WRONG CHOICES EXPLAINED:

(3), (4) Porosity, the amount of space between the particles, remains about the same for particles of uniform size even when their size changes. The smaller particles have smaller spaces between them but there are more spaces. The porosity of samples in the two containers is the same.

30. **4** Metamorphic rocks are formed when enough heat and pressure are applied to existing rocks to cause partial melting. This partial melting causes new minerals to form and may also change the arrangement of the minerals. Sandstone, a sedimentary rock containing minerals that are cemented together, undergoes metamorphism to form intergrown crystals of quartzite.

WRONG CHOICES EXPLAINED:

(1) Sedimentary rocks form when sediments are cemented and compacted.

(2) When magma cools slowly, deep underground, an igneous rock with large grains forms.

(3) Rapid cooling of lava produces an igneous rock with small grains.

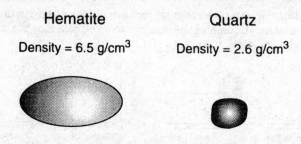

Hematite

Density = 6.5 g/cm^3

Quartz

Density = 2.6 g/cm^3

31. **4** Shape and density are the two factors that most affect settling rates. In this case both particles have the same shape. The density of the hematite pebble, however, is greater, so it will settle faster.

WRONG CHOICES EXPLAINED:

(1) The size of an object has less of an effect on settling rates than does density.

(2) The more nearly round an object is, the faster it will settle.

(3) Gravitational attraction is dependent on the mass of an object, not on its shape. Two objects with the same mass but different shapes have the same gravitational attraction.

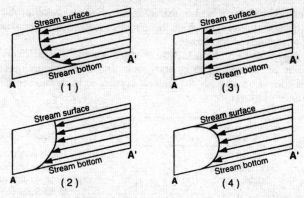

32. **1** The velocity of a stream is least at the sides and bottom of the stream. Diagram (1) shows that the stream velocity, for the cross section from A to A', is greatest near the surface and decreases near the bottom of the stream.

33. **3** Igneous, sedimentary, and metamorphic rocks are usually composed of minerals. The type of rock can be inferred from the types of minerals present and how they are arranged.

WRONG CHOICES EXPLAINED:

(1) Not all rocks contain intergrown crystals. For example, sandstone is a sedimentary rock in which the mineral grains are cemented together rather than intergrown.

(2) Fossils are never found in igneous rocks, which form from molten rock material. Fossils may be found in sedimentary or metamorphic rocks.

(4) Sediments may become cemented together to form sedimentary rock. Igneous and metamorphic rocks are not formed from sediments.

34. **1** Sedimentary rocks are often characterized by layering because the sediments from which they have formed were deposited in layers. When these sediments are converted to rock, the original layering pattern is often preserved.

WRONG CHOICES EXPLAINED:

(2) Foliation is characteristic of some metamorphic rocks. In these rocks some of the minerals are arranged in layers. This occurs because heat and pressure cause some minerals in the rock to melt and the molten mineral material to come together to form layers.

(3) Distorted structure is also characteristic of some metamorphic rocks. The high heat and pressure to which these rocks were subjected can cause the shape of the minerals to be changed.

(4) A glassy texture is characteristic of igneous rocks that cool very quickly. Obsidian is an example of an igneous rocks that is often found near volcanoes. It has cooled too quickly for mineral crystals to form and therefore has the appearance of glass.

Mineral	Cleavage	Hardness	Density (g/cm^3)	Other Properties
Pyroxene (a complex family of minerals; augite is most common)	Two flat planes at nearly right angles	5–6	3.2–3.9	Found in igneous and metamorphic rocks; augite is dark green to black; other varieties are white to green

(1)   (2)   (3)   (4)

35. **2** Diagram (2) shows a mineral sample with two flat planes at nearly right angles. According to the table, this is the cleavage pattern for pyroxene.

36. **3** Find the Scheme for Igneous Rock Identification in the *Earth Science Reference Tables*. Note the abundance of pyroxene in rocks that appear toward the right side of the table. According to the table, these rocks are classified as mafic. An igneous rock with large crystals forms deep within the Earth's crust, where cooling occurs slowly.

37. **1** The oceanic crust is composed mostly of rock material similar to basalt. The crust under the oceans is much thinner than the crust under the continents, which is composed mostly of granitic rock.

38. **4** *P*-waves travel faster than *S*-waves. Both waves leave the earthquake focus at the same time. Since the *P*-waves have greater velocity, however, they reach seismic stations first.

WRONG CHOICES EXPLAINED:

(1) Both P-waves and S-waves originate from the earthquake focus and at the same time.

(2) S-waves are absorbed by liquids. Only P-waves can travel through both solids and liquids.

(3) Both P-waves and S-waves originate from the earthquake focus, which is the point of origin of an earthquake. The epicenter of an earthquake is the point at the Earth's surface that is directly above the earthquake.

39. **4** If the remains of shallow-water animals become buried by overlying sediment, they may become converted into fossils. Sometimes these fossils are later discovered at great oceanic depths. When this occurs, scientists infer that the Earth's crust must have subsided since the fossils were first formed.

WRONG CHOICES EXPLAINED:

(1) Igneous activity at mid-ocean ridges causes material to be *uplifted* and pushed outward.

(2) Heat-flow measurements on coastal plains provide evidence of crustal movement, but not of crustal subsidence.

(3) Marine fossils on mountaintops are evidence of uplift rather than subsidence. The fossils were formed under water and raised to their current positions when uplifting occurred.

40. **2** On the Earthquake P-wave and S-wave Travel Time graph in the *Earth Science Reference Tables*, find a distance of 3,000 km along the horizontal axis. Trace upward to the P-wave curve and read across to the vertical axis. The travel time is 5 min 40 sec. Subtract 5 min 40 sec from 1 hr 21 min 40 sec, the time the earthquake P-wave reached the seismograph station. The result is 1 hr 16 min 00 sec.

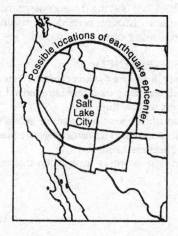

41. **2** The difference in arrival time between the *P*-waves and the *S*-waves determines the distance between the epicenter and the recording station. The greater the difference in time, the farther the recording station is from the epicenter. Also, because *P*-waves travel faster than *S*-waves, the longer the waves travel the farther apart they become. Based on the observations made at Salt Lake City, scientists can infer the distance between the earthquake epicenter and Salt Lake City. The earthquake must have occurred at a point along the curve on the map in the question. This circle represents all points at a given distance from Salt Lake City.

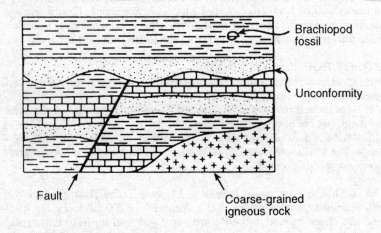

42. **1** An unconformity is a gap in the rock record. Note in the diagram the irregular shape of the rock layers along the unconformity. This irregular shape is evidence of past erosion. Erosion carried old material away before new material was deposited, leaving a gap in the record.

WRONG CHOICES EXPLAINED:

(2) A fault is a break in the rock layers caused by movement of the layers. On one side of the fault line the rock layers have moved upward in relation to the position of the layers on the other side.

(3) A brachiopod fossil forms when a plant or animal becomes buried and is rapidly covered with sediment. The remains of the plant or animal are preserved, and over time, become fossils.

(4) Coarse-grained igneous rocks form when molten rock material cools slowly.

Diagram I

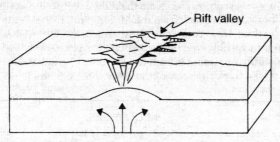

Diagram II

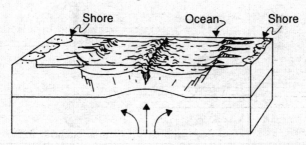

43. **2** Convection cells in the mantle caused rock material in the Earth's crust to be pushed upward and then outward. This caused the widening of the rift valley in diagram I that contains the ocean shown in diagram II.

WRONG CHOICES EXPLAINED:

(1) Meteor impact produces only a local effect and is not likely to cause such extensive widening. In addition, the shape of the crater formed by the meteor would be circular.

(3) Climate changes could cause flooding, which could cause the valley to become wider. However, climate changes would not cause the formation of the ridge that runs through the center of diagram II.

(4) A rise in temperature would produce melting of polar ice caps. This, in turn, could cause an increase in flooding, which could cause the valley walls to widen, but would not result in formation of the ridge shown in diagram II.

44. **3** Find the chart entitled Geologic History of New York State at a Glance in the *Earth Science Reference Tables*. Locate the column headed "Life on Earth." Note that dinosaurs became extinct at the end of the Cretaceous Period.

45. **3** Find the Generalized Bedrock Geology of New York State map in the *Earth Science Reference Tables*. Note that Old Forge in the northern part of the state is located on bedrock from the Middle Proterozoic Period. This is the oldest bedrock in New York State. Next comes the Ordovician bedrock underlying Watertown, followed by the Silurian bedrock underlying Syracuse. The bedrock underlying Elmira is from the Devonian Period and therefore is younger than the bedrock from any of the other three periods.

GEOLOGIC TIME	INDEX FOSSILS			
MISSISSIPPIAN	Spirifer	Muensteroceras	Crinoid Stem	Pentremites
DEVONIAN	Mucrospirifer		Phacops	
SILURIAN	Eospirifer			
ORDOVICIAN	Michelinoceras		Flexicalymene	

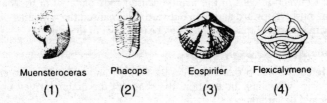

Muensteroceras   Phacops   Eospirifer   Flexicalymene

    (1)   (2)   (3)   (4)

46. **2** Find the chart entitled Geologic History of New York State at a Glance in the *Earth Science Reference Tables*. Note that 387 million years ago the Earth was in the Devonian Period. The table in the question shows that Phacops fossils might be found in rock layers from this period.

47. **4** All of the fossils shown in the table came from animals that lived in the ocean. In fact, geologic evidence suggests that all the early forms of animal life on the Earth were ocean dwellers. Remains of land animals have been found only for later periods. This fact may be inferred from the chart entitled Geologic History of New York State at a Glance in the *Earth Science Reference Tables*.

48. **3** The half-life of a radioactive substance is not affected by environmental factors, including such variables as the surrounding rock, the depth below the surface, the temperature, and the pressure conditions. In addition, the rate of decay, or the half-life, remains constant over time.

49. **1** Find the Radioactive Decay Data chart in the Physical Constants section of the *Earth Science Reference Tables*. Note that the half-life of carbon-14 is 5,700 years, meaning that after this period of time one-half the original sample will have decayed. Therefore, a marine fossil that contains one-half of its original quantity of carbon-14 is approximately 5,700 years old.

50. **2** Find the map entitled Generalized Bedrock Geology of New York State in the *Earth Science Reference Tables*. Note the location of the Susquehanna and Delaware rivers in the south central part of the state. Now find the corresponding location on the map entitled Generalized Landscape Regions of New York State in the *Earth Science Reference Tables*. Both rivers flow over in the region designated as the Allegheny Plateau.

51. **1** Rock salt is the only one of the substances listed that is very soluble in water. Over time, any rock salt in the soil of a humid region will be dissolved by the water present. In an arid region such as a desert, however, rock salt is often deposited as water evaporates.

52. **3** Continents are divided into landscape regions based on the surface features, such as mountains, plains, and plateaus, in an area. In this manner the Rocky Mountains are classified as a different landscape region from the Great Plains of the Midwest.

53. **4** Find the Generalized Bedrock Geology of New York State map in the *Earth Science Reference Tables*. Note that the bedrock of the Middle Proterozoic Period contains gneisses, quartzite, marbles, and anorthositic rocks. This bedrock is found in the northern part of the state. Now turn to the map entitled Generalized Landscape Regions of New York State in the *Earth Science Reference Tables*, and note that the corresponding location on this map is designated as the Adirondack Mountains.

54. **4** Find the chart entitled Geologic History of New York State at a Glance in the *Earth Science Reference Tables*. Note that the dinosaurs lived from the Triassic Period to the Cretaceous Period. The Generalized Bedrock Geology of New York State map in the *Earth Science Reference Tables* shows that bedrock of this age is found only in the southern part of the state. Now find the corresponding area on the map entitled Generalized Landscape Regions of New York State in the *Earth Science Reference Tables*. Note that this includes the landscape region designated as the Newark Lowlands.

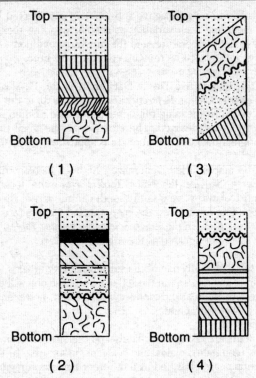

Top   Bottom

(1)

(3)

(2)

(4)

55. **1** Find the Generalized Bedrock Geology of New York State map in the *Earth Science Reference Tables*. Normally in a sequence of rock layers the youngest layer is on the top and the oldest is on the bottom. According to the symbols on the map, the top layer in each sequence shown in the cross sections is from the Devonian Period. In cross section (1) the next layer down is from the Silurian Period. This is followed by layers from the Ordovician and Proterozoic periods. According to the map, this cross section correctly represents a sequence from youngest layer on the top to oldest on the bottom.

WRONG CHOICES EXPLAINED:

(2) In this cross section there is a layer of Triassic rock below the surface Devonian bedrock. The Triassic Period occurred after the Devonian Period, so the Triassic rock should be on top.

(3), (4) In both of these cross sections a layer of Proterozoic rock lies above younger rock. The Proterozoic rock should be on the bottom because it is the oldest.

PART II

GROUP 1

Tidal Record for Reversing Falls, St. John River			
Date	Time of First High Tide	Time of First Low Tide	Time of Second High Tide	Time of Second Low Tide
June 26	2:25 a.m.	8:45 a.m.	2:55 p.m.	9:05 p.m.
June 27	3:15 a.m.	9:35 a.m.	3:45 p.m.	9:55 p.m.
June 28	4:05 a.m.	10:25 a.m.	4:35 p.m.	10:45 p.m.

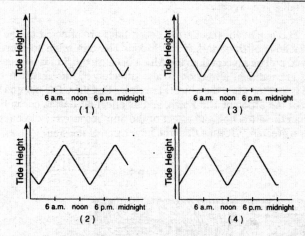

56. **1** On June 28 the first high tide occurred at 4:05 a.m., and the first low tide occurred at 10:25 a.m. These times, as well at the times of the second set of tides (4:35 p.m. and 10:45 p.m.), are indicated on graph (1).

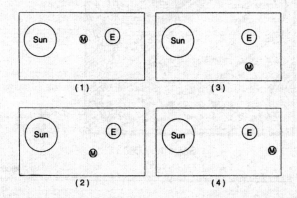

57. **1** The highest tides occur when the Sun, Earth, and Moon are aligned in a straight line, as shown in model (1). At this time the direction of gravitational attraction on the Earth is the same for the Sun and the Moon, resulting in the highest tides.

58. **2** The first high tide on June 26 occurred at 2:25 a.m. The first high tide on June 27 occurred at 3:15 a.m. The time difference is 50 min.

59. **2** The tides in the Bay of Fundy follow a pattern of repeating high and low tides, which can be considered cyclic. Also, the time span between tides is regular, so that they are predictable. It may therefore be inferred that the tides are predictable and cyclic.

60. **4** The heights of the Earth's ocean tides are affected by the gravitational attractions of the Sun and the Moon. The Sun tends to exert greater gravitational pull on objects than the Moon because the Sun is so much larger. However, gravitational attraction is also affected by the distance between objects; it is greater when the objects are closer together. The net gravitational attraction of the Moon on the Earth, and its effect on Earth's ocean tides, are greater than the attraction and effect of the Sun because the closeness of the Moon has a greater influence than the larger size of the Sun.

GROUP 2

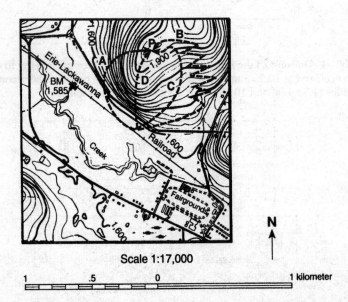

Scale 1:17,000

N

61. **4** The map shows a scale of 1:17,000; that is, 1 cm on the map represents 17,000 cm on the Earth. This scale applies to all measurements. For example, 1 in. on the map would represent 17,000 in. on the Earth.

62. **2** Note in the upper right portion of the map two dark curved lines between the lines marked 1,600 and 1,900. These represent 1,700 ft. and 1,800 ft., respectively, so the interval between dark lines is 100 ft. Between two dark lines there are four light lines. Each of these lines must therefore represent 20 ft., which is the contour interval for this map.

63. **4** The fact that no contour lines pass through the area of the fairgrounds means that this area is fairly level with a difference in elevation of less than 20 ft. The level surface and the closeness to the creek suggest that the fairgrounds are located on a floodplain.

64. **2** Mark off the length of the railroad tracks, starting near the upper left corner of the map and ending near the lower right. Apply this length against the scale at the bottom of the map. The distance is about 2.0 km. Note that the total length of the scale illustrated is 2.0 km since it increases by 1 km in each direction from a reading of zero in the middle.

65. **4** The path will be steepest when the contour lines are closest together, as occurs along path *D*. The closeness of the contour lines means that the elevation is increasing over the shortest distance—in other words, that the path is steepest.

GROUP 3

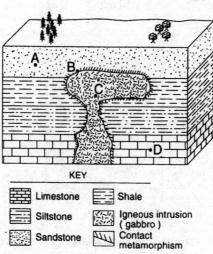

KEY

Limestone		Shale	
Siltstone		Igneous intrusion (gabbro)	
Sandstone		Contact metamorphism	

66. **1** Gabbro is an igneous rock. Igneous rocks form when molten rock material cools and solidifies. Since igneous rocks form directly from molten material, they do not contain fossils.

67. **4** Note in the diagram that there is contact metamorphism between the gabbro layer and all the other rock layers. This means that these other layers must have been present when the igneous intrusion occurred. It may therefore be inferred that the gabbro formed most recently.

68. **2** As the key to the diagram indicates, rock layer A contains sandstone. At point B the sandstone underwent contact metamorphism caused by the igneous intrusion; the heat and pressure converted the sandstone to the metamorphic rock quartzite. If you look in the Scheme for Metamorphic Rock Identification in the *Earth Science Reference Tables,* you will note that the metamorphism of sandstone produces quartzite.

69. **2** Find the Scheme for Sedimentary Rock Identification in the *Earth Science Reference Tables.* Under "Chemically and/or Organically Formed Sedimentary Rocks," note that, when chemical precipitation of calcite occurs, chemical limestone is formed.

70. **4** Find the Scheme for Igneous Rock Identification in the *Earth Science Reference Tables.* Locate gabbro, which formed as a result of the igneous intrusion, on the chart, and read downward to determine which minerals are present in this rock. According to the chart the two most abundant minerals in gabbro are plagioclase feldspar and pyroxene.

GROUP 4

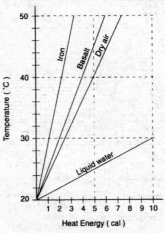

71. **2** Find the line on the graph that represents liquid water. Note that it rises from a temperature of 20° C to a temperature of 30° C. According to the values on the horizontal axis, this change requires a heat energy of 10 cal.

72. **1** Note that the line on the graph for liquid water has the smallest slope. This means that the rate of increase in temperature is lower for 1 g of water than for 1 g of iron, basalt, or dry air. The specific heat of a substance is the amount of heat required to raise the temperature of 1 g of that substance 1 degree. The fact that the rate of increase in temperature is lowest for water means that water has the highest specific heat of the materials illustrated.

73. **2** Of the four substances illustrated, the curve for iron has the steepest slope, indicating that iron heats up the fastest. The substance that heats up fastest will also cool the fastest.

74. **4** The slopes of the lines on the graph vary. This means that different amounts of heat energy are required to produce an equal temperature change for different substances.

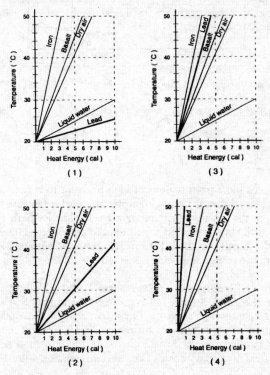

75. **4** Find the Physical Constants section in the *Earth Science Reference Tables*. Note in the Specific Heats of Common Materials table that lead has the lowest specific heat of the substances listed. Therefore, less heat is required to raise the temperature of an equal amount of lead. Since lead has the lowest specific heat, the slope of the graph for lead would be the steepest, as shown in graph (4).

GROUP 5

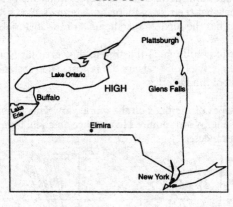

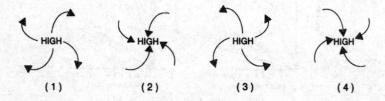

76. **1** Winds blow from regions of high pressure to regions of low pressure. Winds blow outward from a high-pressure center because the pressure is higher there than in the surrounding area. In the Northern Hemisphere, where New York State is located, winds tend to be deflected toward the right because of the Earth's rotation. Pattern (1) shows the surface winds blowing outward from the center and being deflected toward the right.

77. **3** Locate Syracuse on the Generalized Bedrock Geology of New York State map in the *Earth Science Reference Tables*, and note that it is near the high-pressure center indicated on the map given in the question. High-pressure centers are usually characterized by dry air, so there tends to be less cloud cover, causing the sky to be clear. Also, the temperatures near high-pressure centers are usually cooler than the temperatures where the pressure is low.

78. **2** The cool, dry air over Syracuse suggests that the air mass is continental polar. Polar air masses contain cooler air than tropical air masses. Also, because continental air masses form over land, they tend to be drier than maritime air masses, which form over water.

79. **4** The general direction of movement of air masses over New York State is from west to east, following the prevailing wind belt as shown in the Planetary Wind and Moisture Belts in the Troposphere diagram in the *Earth Science Reference Tables*. As a result the air mass will tend to move eastward toward Glens Falls.

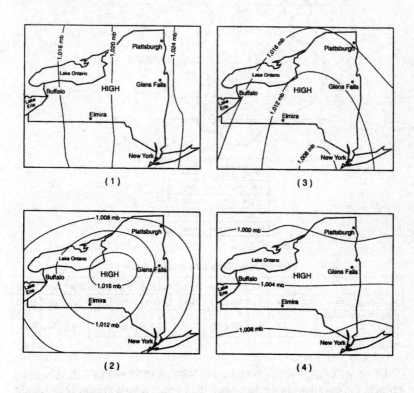

80. **2** The isobars around a high-pressure center tend to follow a circular pattern as the pressure decreases outward from the center. This circular pattern of isobars with decreasing pressure outward is illustrated in map (2).

GROUP 6

Imaginary Continent

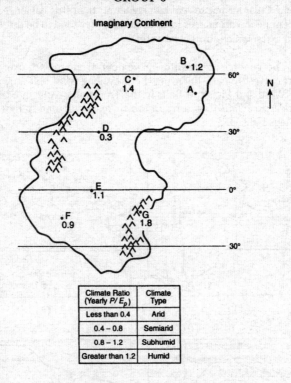

Climate Ratio (Yearly P/E_p)	Climate Type
Less than 0.4	Arid
0.4 – 0.8	Semiarid
0.8 – 1.2	Subhumid
Greater than 1.2	Humid

Water Budget Data for Location A (mm)

	Jan.	Feb.	Mar.	Apr.	May	June	July	Aug.	Sept.	Oct.	Nov.	Dec.	Totals
Precipitation (P)	68	76	89	96	81	68	75	71	67	65	70	63	889
Evapotranspiration (E_p)	5	10	35	60	85	155	170	159	82	60	34	10	865

81. **4** At location A the P/E_p ratio is 1089 mm/865 mm or 1.26. The chart given in the question shows that, when the climate ratio is greater than 1.2, the climate type is classified as humid. [Choice (3) has also been accepted as an answer because a subhumid climate extends to a ratio of 1.2.]

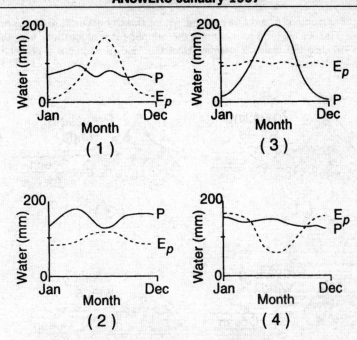

82. **1** According to the data in the water budget table, the potential evapotranspiration (E_p) values range for location A from 5 mm in January to a high of 170 mm in July and back down to a value of 10 mm in December. This pattern is illustrated in graph (1). Graph (1) also shows the fairly consistent pattern of precipitation that the water budget table indicates is characteristic of location A.

83. **4** Potential evapotranspiration (E_p) depends on the temperature and the amount of vegetation. When the temperatures are highest and the amount of vegetation is greatest, the potential evapotranspiration is greatest. The temperatures are highest during the summer months, when the path of the Sun in the sky is at its highest and the greatest amount of insolation is absorbed. Thus the E_p values recorded at the locations shown on the map depend primarily on the amount of solar radiation absorbed.

84. **2** Locations *C* and *D* are inland, where annual temperature ranges are greater. The oceans tend to moderate the temperatures of locations near the coasts because the water is warmer than the land in winter and colder in summer.

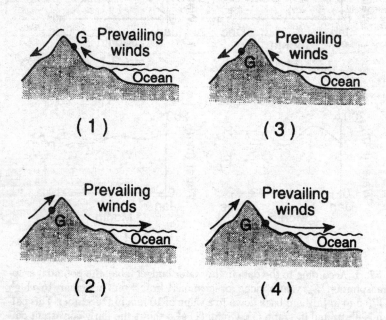

85. **1** Find the Planetary Wind and Moisture Belts in the Troposphere diagram in the *Earth Science Reference Tables*. The latitude of location *G*, as shown on the "Imaginary Continent" map, is about 12°S, which places it in the "S.E. WINDS" belt. Note in the planetary wind drawing that the direction of movement of the air in this belt is from southeast to northwest. Diagram (1) shows the winds blowing from east to west.

GROUP 7

Grain Size Percentages

Grain Size	Miles Downstream from Cairo					
	100 mi	300 mi	500 mi	700 mi	900 mi	1,000 mi
Small Pebbles	29%	8%	14%	5%	none	none
Coarse Sand	30%	22%	9%	8%	1%	none
Medium Sand	32%	50%	46%	44%	26%	9%
Fine Sand	8%	19%	28%	41%	70%	69%
Silt	trace	trace	2%	1%	2%	10%
Clay	trace	trace	1%	trace	1%	10%

KEY

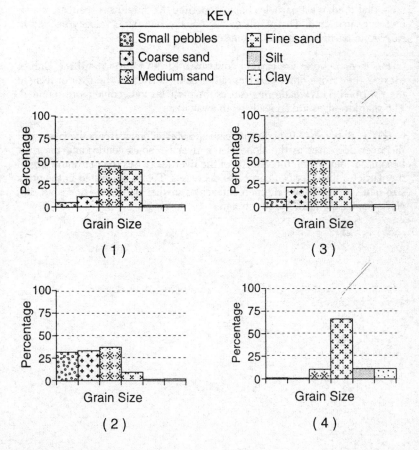

(1)

(2)

(3)

(4)

86. **3** On the chart in the question the data for a location 300 miles downstream are found in the second column. At this location the deposits contain 8% small pebbles, 22% coarse sand, 50% medium sand, and so on. This pattern of percentages is illustrated in graph (3).

87. **2** According to the data chart, no small pebbles or coarse sand are being deposited 1000 mi downstream. These represent the largest grain sizes. In a river the largest particles are deposited first as the river slows down. Therefore, the small pebbles and coarse sand have been completely deposited before reaching the river's mouth, 1000 mi downstream from Cairo.

88. **2** Find the graph entitled Relationship of Transported Particle Size to Water Velocity in the *Earth Science Reference Tables*. Note in the grain size table that the largest particles being deposited 900 mi downstream are classified as coarse sand. The graph shows that the minimum velocity needed to carry sand particles is about 50 cm/sec.

89. **4** As pebbles are carried downstream from Cairo toward the Gulf of Mexico, they strike other particles either being carried by the river or lying in the river bed. This weathering causes the particles to become more rounded. The sharper edges are typically worn away first.

90. **1** Between 500 mi downstream and 900 mi downstream the velocity of the river decreases as the slope of the river becomes gentler and the river becomes wider. When the velocity of the river decreases the maximum size of the particles the river can carry also decreases. This is illustrated in the grain size table, which shows the percentages of smaller particles—fine sand, silt, clay—increasing farther downstream.

GROUP 8

Earthquake Epicenters

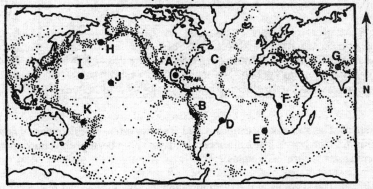

91. **2** Notice in the map that the heaviest earthquake activity occurs in distinct zones on the Earth. For example, the large clusters of dots indicate a zone of heavy earthquake activity along the continental borders surrounding the Pacific Ocean.

92. **4** Locations *H* and *K* are situated along zones of high earthquake activity. Find these two locations on the Tectonic Plates map in the *Earth Science Reference Tables*. Note that they are situated along areas designated as trenches.

93. **4** At a time in the geologic past the continents formed one large landmass. Then, locations *D* and *F*, now separated by a great distance because the continents have drifted apart, were near each other. Therefore, the rocks, minerals, and fossils of location *F* most closely match those of location *D*.

94. **2** Find the Tectonic Plates map in the *Earth Science Reference Tables*. On the map locate the Nazca plate and the South American plate. Location *B* lies along the boundary of these two plates. Earthquakes occurred here because of the Nazca plate sliding under the South American plate.

95. **3** Zones of earthquake activity often correlate with zones of volcanism and mountain building. In these areas, volcanic material is often being released and crustal material is being pushed upward to form mountains.

GROUP 9

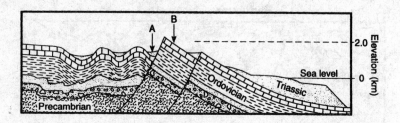

96. **4** The area near locations *A* and *B* shows extensive folding and faulting of the rock layers. The deformed rock layers have been pushed upward in some places to form mountains.

97. **4** The symbols in the diagram show that the Ordovician rock layer is composed of shale. Shale is a sedimentary rock composed of clay-sized particles, as indicated in the Scheme for Sedimentary Rock Identification in the *Earth Science Reference Tables*. For information about the fossil types present during the various periods of geologic history, turn to the chart entitled Geologic History of New York State at a Glance in the *Earth Science Reference Tables*. The column headed "Life on Earth" shows that corals first appeared during the Ordovician Period. Marine animals, such as sharks, date from the Cambrian period.

98. **2** Between locations *A* and *B* the rock layers have split apart and moved in opposite directions. This type of break is a structural feature referred to as a fault.

99. **2** The limestone rock layer can be identified by the symbols appearing in the Scheme for Sedimentary Rock Identification in the *Earth Science Reference Tables*. Sedimentary rock such as the limestone shown in the diagram is normally deposited in horizontal layers. Now, however, the layers appear to be curved or folded. It may therefore be inferred that crustal movement that occurred after deposition caused them to become distorted to their current shape.

100. **3** When uplifting forces dominate over leveling forces, there is a net increase in elevation. One likely effect at location *B* is that hillslopes would become taller and steeper.

GROUP 10

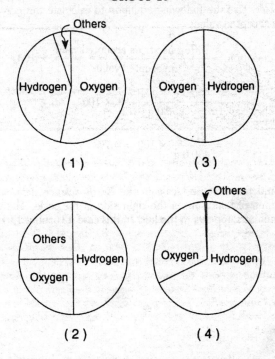

101. **4** Find the chart entitled Average Chemical Composition of Earth's Crust, Hydrosphere, and Troposphere in the *Earth Science Reference Tables*. Note that the hydrosphere contains 33% oxygen and 66% hydrogen. These percentages are best illustrated in graph (4).

102. **2** Objects radiate the most electromagnetic energy when the temperature is highest. To determine which of the temperatures in the choices is the highest, refer to the Temperature scale in the *Earth Science Reference Tables*. Find each of these temperatures on the scale and note that 80° C is the highest.

103. **4** Observations are made directly with the senses. Each of choices (1) through (3) can be determined either by direct observation or by observation with the aid of instruments. The composition of the Earth's outer core, however, cannot be observed. That it is made of iron is an inference based on various other measurements and observations, such as the magnetic properties of the Earth, its size, and its mass.

104. **4** Find the Equations and Proportions chart in the *Earth Science Reference Tables*. Use the following equation to calculate the percent deviation from the accepted value:

$$\text{Percent deviation} = \frac{\text{difference from accepted value}}{\text{accepted value}} \times 100$$

$$= \frac{450.0 \text{ g} - 428.7 \text{ g}}{450.0 \text{ g}} \times 100$$

$$= \frac{21.3 \text{ g}}{450.0 \text{ g}} \times 100 = 4.7\%$$

105. **3** Find the Pressure scale in the *Earth Science Reference Tables*. Locate a reading of 29.92 in. on the right side of the scale. The equivalent pressure in millibars appears to the left. In this case it is about 1,013.2 mb.

Topic	Question Numbers (total)	Wrong Answers (x)	Grade
Observation/Measurement of the Environment; The Changing Environment	1–3, 59, 61, 64, 102–105 (10)		$\dfrac{100(10-x)}{10} = \%$
Measuring the Earth	4, 5, 7, 8, 62, 65, 80, 101 (8)		$\dfrac{100(8-x)}{8} = \%$
Earth Motions	9–12, 56–58, 60 (8)		$\dfrac{100(8-x)}{8} = \%$
Energy in Earth's Processes	13, 14, 71–75, (7)		$\dfrac{100(7-x)}{7} = \%$
Insolation and the Earth's Surface	15–17 (3)		$\dfrac{100(3-x)}{3} = \%$
Energy Exchanges in the Atmosphere	18, 20–22, 25, 76–79 (9)		$\dfrac{100(9-x)}{9} = \%$
Moisture and Energy Budgets	6, 19, 24, 26, 28, 29, 81–85 (11)		$\dfrac{100(11-x)}{11} = \%$
Processes of Erosion	23, 27, 32, 89, 90 (5)		$\dfrac{100(5-x)}{5} = \%$
Processes of Deposition	31, 86–88 (4)		$\dfrac{100(4-x)}{4} = \%$
The Formation of Rocks	30, 33–37, 51, 68–70 (10)		$\dfrac{100(10-x)}{10} = \%$
The Dynamic Crust	38–41, 43, 91–95 (10)		$\dfrac{100(10-x)}{10} = \%$
Interpreting Geologic History	42, 44–49, 54, 55, 66, 67, 97–99 (14)		$\dfrac{100(14-x)}{14} = \%$
Landscape Development	50, 52, 53, 63, 96, 100 (6)		$\dfrac{100(6-x)}{6} = \%$

To further pinpoint your weak areas, use the Topic Outline in the front of the book.

Examination
June 1997
Earth Science

PART I

Answer all 55 questions in this part. [55]

Directions (1–55): For *each* statement or question, select the word or expression that, of those given, best completes the statement or answers the question. Record your answers in the spaces provided.

1 The circumference of the Earth is about 4.0×10^4 kilometers. This value is equal to

(1) 400 km   (3) 40,000 km

(2) 4,000 km   (4) 400,000 km   1 _____

2 The diagram below shows a process of weathering called frost wedging.

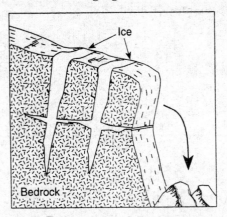

Frost wedging breaks rocks because as water freezes it increases in

1 mass   3 density
2 volume   4 specific heat   2 _____

3 A student calculates the period of Saturn's revolution to be 31.33 years. What is the student's approximate deviation from the accepted value?

(1) 1.9%   (3) 6.3%
(2) 5.9%   (4) 19%   3 _____

4 Which object best represents a true scale model of the shape of the Earth?

1 a Ping-Pong ball   3 an egg
2 a football   4 a pear   4 _____

5 Which graph best represents the most common relationship between the amount of air pollution and the distance from an industrial city?

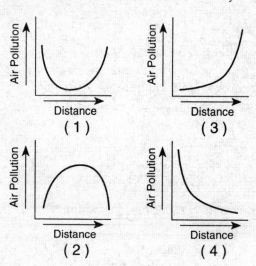

6 Isolines on the topographic map below show elevations above sea level, measured in meters.

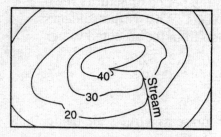

What could be the highest possible elevation represented on this map?

(1) 39 m   (3) 45 m
(2) 41 m   (4) 49 m

6 _____

7 Oxygen is the most abundant element by volume
  in the Earth's

  1 inner core   3 hydrosphere
  2 crust   4 troposphere   7 _____

    Base your answers to questions 8 and 9 on the
diagram below and on your knowledge of Earth
science. The diagram represents the path of a planet
orbiting a star. Points A, B, C, and D indicate four
orbital positions of the planet.

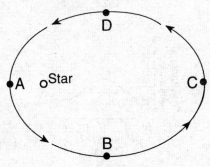

8 When viewed by an observer on the planet, the
  star has the largest apparent diameter at position

  (1) A   (3) C
  (2) B   (4) D   8 _____

9 Which graph best represents the gravitational attraction between the star and the planet?

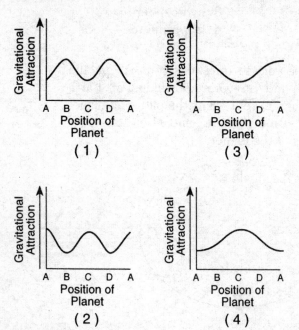

10 Which statement best explains why different phases of the Moon can be observed from the Earth?

1 The size of the Earth's shadow falling on the Moon changes.
2 The Moon moves into different parts of the Earth's shadow.
3 Differing amounts of the Moon's sunlit surface are seen because the Moon revolves around the Sun.
4 Differing amounts of the Moon's sunlit surface are seen because the Moon revolves around the Earth.

10 _____

11 The diagram below represents an activity in which an eye dropper was used to place a drop of water on a spinning globe. Instead of flowing due south toward the target point, the drop followed a curved path and missed the target.

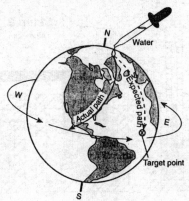

The actual path results from

1 the tilt of the globe's axis
2 the Coriolis effect
3 the globe's revolution
4 dynamic equilibrium

11 _____

Base your answers to questions 12 and 13 on the diagrams below, which show laboratory equipment setups A and B being used to study energy transfer in a classroom laboratory.

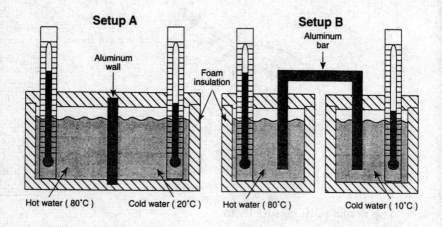

12 In both A and B, most of the heat transferred from the hot water to the cold water is transferred by

1 convection   3 radiation

2 conduction   4 gravity   12 _____

13 Which laboratory setup is more efficient at transferring heat energy from the hot water to the cold water?

(1) A, because less energy is lost to the surrounding environment

(2) A, because the hot water has a higher temperature

(3) B, because the aluminum bar is bigger than the aluminum wall

(4) B, because the cold water has a lower temperature   13 _____

14 Which diagram best shows how air inside a greenhouse warms as a result of insolation from the Sun?

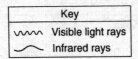

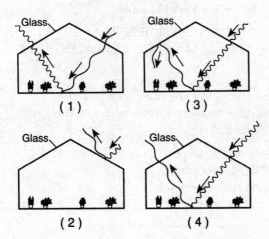

14 _____

15 Why do the locations of sunrise and sunset vary in a cyclical pattern throughout the year?

1 The Earth's orbit around the Sun is an ellipse.
2 The Sun's orbit around the Earth is an ellipse.
3 The Sun rotates on an inclined axis while revolving around the Earth.
4 The Earth rotates on an inclined axis while revolving around the Sun.

15 _____

16 On a clear April morning near Rochester, New York, which surface will absorb the most insolation per square meter?

1 a calm lake   3 a white-sand beach
2 a snowdrift   4 a freshly plowed farm field

16 _____

Base your answers to questions 17 and 18 on the weather instrument shown in the diagram below.

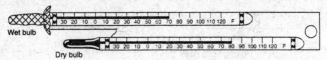

17 What are the equivalent Celsius temperature readings for the Fahrenheit readings shown?

  1 wet 21°C, dry 27°C   3 wet 70°C, dry 80°C
  2 wet 26°C, dry 37°C   4 wet 158°C, dry 176°C   17 _____

18 Which weather variables are most easily determined by using this weather instrument and the *Earth Science Reference Tables?*

  1 air temperature and windspeed
  2 visability and wind direction
  3 relative humidity and dewpoint
  4 air pressure and cloud type   18 _____

19 All of the containers shown below contain the same volume of water and are at room temperature. In a two-day period, from which container will the *least* amount of water evaporate?

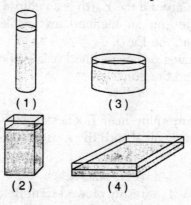

    (1)   (3)

    (2)   (4)   19 _____

20 Which statement best explains why precipitation occurs at the frontal surfaces between air masses?

  1 Warm, moist air sinks when it meets cold, dry air.

  2 Warm, moist air rises when it meets cold, dry air.

  3 Cold fronts move faster than warm fronts.

  4 Cold fronts move slower than warm fronts.   20 _____

21 By which process does water vapor leave the atmosphere and form dew?

  1 condensation

  2 transpiration

  3 convection

  4 precipitation   21 _____

22 The diagram below represents a section of a weather map showing high- and low-pressure systems. The lines represent isobars.

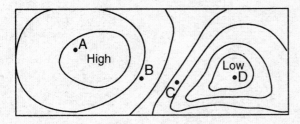

At which point is the windspeed greatest?

  (1) A

  (2) B

  (3) C

  (4) D   22 _____

23 A low-pressure system is shown on the weather map below.

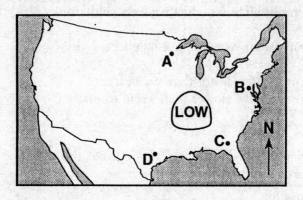

Toward which point will the low-pressure system move if it follows a typical storm track?

(1) A
(2) B
(3) C
(4) D

23 _____

24 The diagram below shows equal volumes of loosely packed sand, clay, and small pebbles placed in
identical funnels. The soils are dry, and the
beakers are empty.

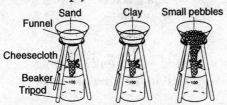

A 100-milliliter sample of water is poured into
each funnel at the same time and allowed to seep
through for 15 minutes. Which diagram best
shows the amount of water that passes through
each funnel into the beakers?

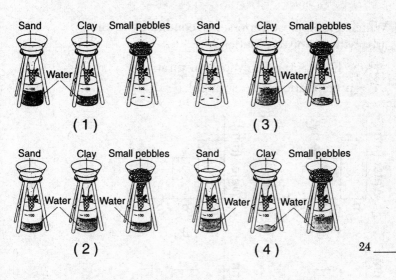

24 _____

25 The upward movement of water through tiny
spaces in soil or rock is called

  1 water retention   3 porosity
  2 capillary action   4 permeability

25 _____

26 A deposit of rock particles that are angular, scratched, and unsorted has most likely been transported and deposited by

1 ocean waves   3 a glacier
2 running water   4 wind   26 _____

27 When the soil is saturated in a gently sloping area, any additional rainfall in the area will most likely

1 become ground water
2 become surface runoff
3 cause a moisture deficit
4 cause a higher potential evapotranspiration   27 _____

28 Which graph best represents conditions in an area with a moisture deficit?

[Key: E_p = potential evapotranspiration; P = precipitation; St = storage]

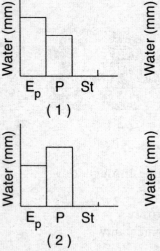

(1)

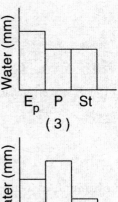

(3)

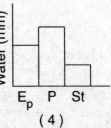

(2)   (4)   28 _____

29 The diagram below shows a residual soil profile formed in an area of granite bedrock. Four different soil horizons, *A*, *B*, *C*, and *D*, are shown.

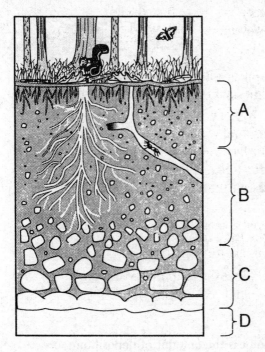

Which soil horizon contains the greatest amount of material formed by biological activity?

(1) *A*
(2) *B*
(3) *C*
(4) *D*

29 _____

Base your answers to questions 30 and 31 on the diagram below. The diagram shows points *A*, *B*, *C*, and *D* on a meandering stream.

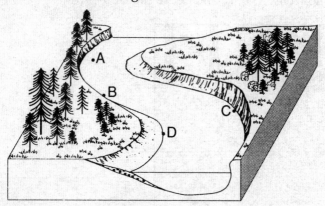

30 Which material is most likely to be transported in suspension during periods of slowest stream velocity?

1 gravel   3 silt

2 sand   4 clay   30 _____

31 At which point is the amount of deposition more than the amount of erosion?

(1) *A*   (3) *C*

(2) *B*   (4) *D*   31 _____

32 Which rock is usually composed of several different minerals?

1 gneiss   3 rock gypsum

2 quartzite   4 chemical limestone   32 _____

33 Which granite sample most likely formed from magma that cooled and solidified at the slowest rate?

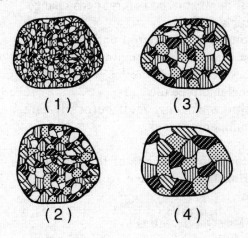

(1)   (3)

(2)   (4)

33 _____

34 The diagram below shows the results of one test for mineral identification.

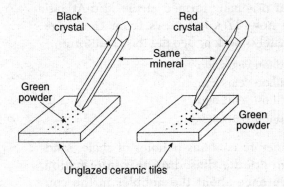

Which mineral property is being tested?

1 density   3 streak
2 fracture   4 luster

34 _____

35 Which statement about the formation of a rock is best supported by geologic evidence?

  1 Magma must be weathered before it can change to metamorphic rock.

  2 Sediment must be compacted and cemented before it can change to sedimentary rock.

  3 Sedimentary rock must melt before it can change to metamorphic rock.

  4 Metamorphic rock must melt before it can change to sedimentary rock.   35 _____

36 The scale below shows the age of the sea-floor crust in relation to its distance from the Mid-Atlantic Ridge.

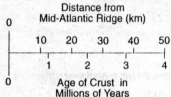

Crust that originally formed at the Mid-Atlantic Ridge is now 37 kilometers from the ridge. Approximately how long ago did this crust form?

  (1) 1.8 million years ago

  (2) 2.0 million years ago

  (3) 3.0 million years ago

  (4) 4.5 million years ago   36 _____

37 A conglomerate contains pebbles of shale, sandstone, and granite. Based on this information, which inference about the pebbles in the conglomerate is most accurate?

  1 They were eroded by slow-moving water.

  2 They came from other conglomerates.

  3 They are all the same age.

  4 They had various origins.   37 _____

38 The diagrams below show demonstrations that represent the behavior of two seismic waves, A and B.

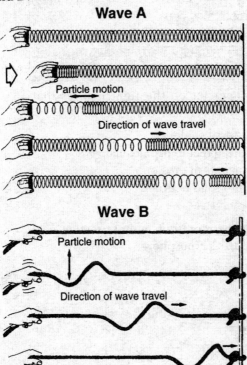

Wave A

Particle motion

Direction of wave travel

Wave B

Particle motion

Direction of wave travel

Which statement concerning the demonstrated waves is correct?

1 Wave A represents a compressional wave, and wave B represents a shear wave.

2 Wave A represents a shear wave, and wave B represents a compressional wave.

3 Wave A represents compressional waves in the crust, and wave B represents compressional waves in the mantle.

4 Wave A represents shear waves in the crust, and wave B represents shear waves in the mantle.

38 ____

39 The cartoon below presents a humorous view of Earth science.

PEANUTS   BY CHARLES M. SCHULZ

..AND A THOUSAND YEARS FROM NOW PEOPLE WILL LOOK AT WHAT WE HAVE BUILT HERE TODAY, AND BE TOTALLY AMAZED...

The cartoon character on the right realizes that the sand castle will eventually be

1 preserved as fossil evidence
2 deformed during metamorphic change
3 removed by agents of erosion
4 compacted into solid bedrock

39 _____

40 The diagram below represents a cross section of a portion of the Earth's lithosphere.

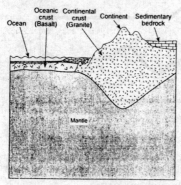

Which statement about the Earth's crust is best supported by the diagram?

1 The crust is thicker than the mantle.
2 The continental crust is thicker than the oceanic crust.
3 The crust is composed primarily of sedimentary rock.
4 The crust is composed of denser rock than the mantle.

40 _____

41 The epicenter of an earthquake is 6,000 kilometers from an observation point. What is the difference in travel time for the *P*-waves and *S*-waves?

(1) 7 min 35 sec   (3) 13 min 10 sec
(2) 9 min 20 sec   (4) 17 min 00 sec   41 _____

42 At a depth of 2,000 kilometers, the temperature of the stiffer mantle is inferred to be

(1) 6,500°C   (3) 3,500°C
(2) 4,200°C   (4) 1,500°C   42 _____

43 Hot springs on the ocean floor near the mid-ocean ridges provide evidence that

1 climate change has melted huge glaciers
2 marine fossils have been uplifted to high elevations
3 meteor craters are found beneath the oceans
4 convection currents exist in the asthenosphere   43 _____

44 The diagram below represents the skull of a saber-toothed tiger that died 30,000 years ago.

The age of the skull could be determined most accurately by using

1 uranium-238   3 potassium-40
2 rubidium-87   4 carbon-14   44 _____

45 A geologist collected the fossils shown below from locations in New York State.

Which sequence correctly shows the fossils from oldest to youngest?

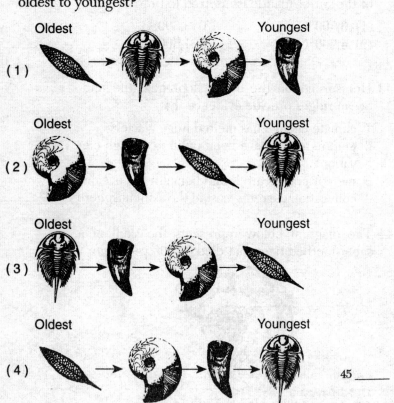

45 _____

46 Present-day corals live in warm, tropical ocean water. Which inference is best supported by the discovery of Ordovician-age corals in the surface bedrock of western New York State?

1 Western New York State was covered by a warm, shallow sea during Ordovician time.
2 Ordovician-age corals lived in the forests of western New York State.
3 Ordovician-age corals were transported to western New York State by cold, freshwater streams.
4 Western New York State was covered by a continental ice sheet that created coral fossils of Ordovician time.

46 _____

47 The diagram below represents the radioactive decay of uranium-238.

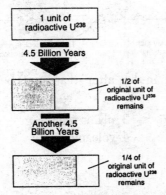

According to the *Earth Science Reference Tables*, shaded areas on the diagram represent the amount of

1 undecayed radioactive rubidium-87 (Rb^{87})
2 undecayed radioactive uranium-238 (U^{238})
3 stable lead-206 (Pb^{206})
4 stable carbon-14 (C^{14})

47 _____

48 In which way are index fossils and volcanic ash deposits similar?

　　1 Both can usually be dated with radiocarbon.
　　2 Both normally occur in nonsedimentary rocks.
　　3 Both strongly resist chemical weathering.
　　4 Both often serve as geologic time markers.　　48 _____

49 The block diagrams below show a landscape region before and after uplift and erosion.

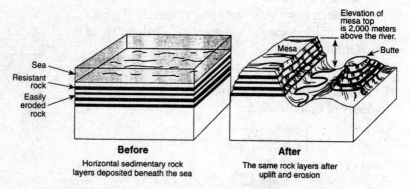

Before
Horizontal sedimentary rock
layers deposited beneath the sea

After
The same rock layers after
uplift and erosion

The landscape shown in the "after" diagram is best classified as a

　　1 folded mountain　　　3 plains region
　　2 plateau region　　　　4 volcanic dome　　　　49 _____

50 Which statement correctly describes an age relationship in the geologic cross section below?

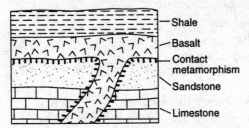

1 The sandstone is younger than the basalt.
2 The shale is younger than the basalt.
3 The limestone is younger than the shale.
4 The limestone is younger than the sandstone.   50 _____

51 The landscape around Old Forge, New York, consists of

1 flat, low-lying volcanic rock
2 hills of horizontal sedimentary rock layers
3 mountains of metamorphic rock
4 eroded, tilted sedimentary rock   51 _____

52 The diagrams below represent geologic cross sections from two widely separated regions.

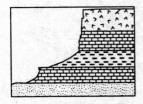 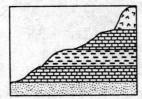

The layers of rock appear very similar, but the hillslopes and shapes are different. These differences are most likely the result of

1 volcanic eruptions   3 soil formation
2 earthquake activity   4 climate variations   52 _____

53 Which bedrock characteristics most influence landscape development?

1 composition and structure
2 structure and age
3 age and color
4 color and composition   53 _____

54 The map below shows the location of an ancient sea, which evaporated to form the Silurian-age deposits of rock salt and rock gypsum now found in some New York State bedrock.

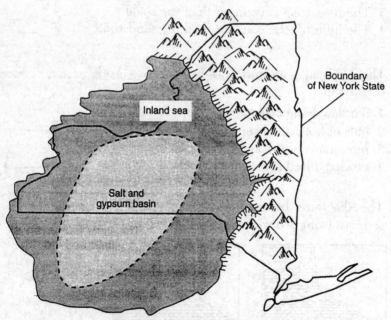

Within which two landscape regions are these large salt and rock gypsum deposits found?

1 Hudson Highlands and Taconic Mountains
2 Tug Hill Plateau and Adirondack Mountains
3 Erie-Ontario Lowlands and Allegheny Plateau
4 Catskills and Hudson-Mohawk Lowlands   54 _____

55 The diagram below represents a cross section of the bedrock and land surface in part of Tennessee. The dotted lines indicate missing rock layers.

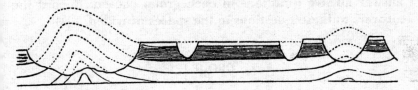

Which statement is best supported by the diagram?

1 Rocks are weathered and eroded evenly.
2 Folded rocks are more easily weathered and eroded.
3 Deposits of sediments provide evidence of erosion.
4 Climate differences affect the amount of erosion.

55 _____

PART II

This part consists of ten groups, each containing five questions. Choose seven of these ten groups. Be sure that you answer all five questions in each group chosen. Record the answers to these questions in the spaces provided. [35]

GROUP 1

If you choose this group, be sure to answer questions 56–60.

Base your answers to questions 56 and 60 on the table below and on your knowledge of Earth science. The table shows data for a student's collection of rock samples *A* through *I*, which are classified into groups *X*, *Y*, and *Z*. For each rock sample, the student recorded mass, volume, density, and a brief description. The density for rock *D* has been left blank.

Rock Collection

Group	Rock	Mass (g)	Volume (cm³)	Density (g/cm³)	Description
X	A	82.9	34.4	2.41	Grey, smooth, rounded
	B	114.2	42.6	2.68	Brown, smooth, rounded
	C	144.7	63.2	2.29	Black, smooth, rounded
Y	D	159.4	59.7		Black and grey crystals, angular
	E	87.7	33.1	2.65	Clear and pink crystals, angular
	F	59.6	21.0	2.84	White, grey, and black crystals; angular
Z	G	201.1	68.4	2.94	Grey, shiny, flat
	H	85.1	39.8	2.14	Brown, sandy feel, flat
	I	110.2	47.3	2.33	Dark grey, flaky, flat

56 The student's classification system is based on

  1 density   3 color
  2 shape   4 mass   56 _____

57 The approximate density of rock sample D is

  (1) 2.67 g/cm^3 (3) 3.32 g/cm^3
  (2) 2.75 g/cm^3 (4) 3.75 g/cm^3 57 _____

58 Which statement is an inference rather than an observation?

  1 Rock H is flat.
  2 Rock B has been rounded by stream action.
  3 Rock E has a volume of 33.1 cm^3.
  4 Rock G is the same color as rock I.   58 _____

59 To obtain the data recorded in the column labeled "Description," the student used

  1 a triple-beam balance
  2 an overflow can
  3 a calculator
  4 her senses   59 _____

Note that question 60 has only three choices.

60 The student broke rock G into two pieces. Compared to the density of the original rock, the density of one piece would most likely be

  1 less
  2 greater
  3 the same   60 _____

GROUP 2

If you choose this group, be sure to answer questions **61–65**.

Base your answers to questions 61 through 65 on the topographic map below and on your knowledge of Earth science. Points *A, B, C, D, E, F, X,* and *Y* are locations on the map. Elevation is measured in feet.

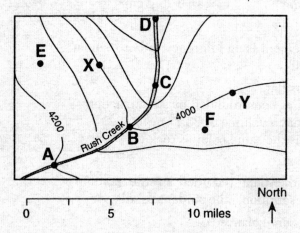

61 What is the contour interval used on this map?

   (1) 20 ft   (3) 100 ft

   (2) 50 ft   (4) 200 ft   61 _____

62 Which locations have the greatest difference in elevation?

   (1) *A* and *D*   (3) *C* and *F*

   (2) *B* and *X*   (4) *E* and *Y*   62 _____

63 Between points *C* and *D*, Rush Creek flows toward the

   1 north   3 east

   2 south   4 west   63 _____

64 The gradient between points *A* and *B* is closest to

(1) 20 ft/mi
(2) 40 ft/mi
(3) 80 ft/mi
(4) 200 ft/mi

64 _____

65 Which diagram best represents the profile along a straight line between points *X* and *Y*?

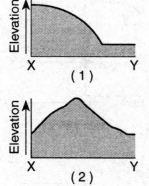

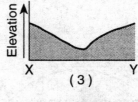

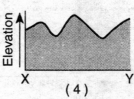

65 _____

GROUP 3

If you choose this group, be sure to answer questions 66–70.

Base your answers to questions 66 through 70 on the *Earth Science Reference Tables,* the map below, and your knowledge of Earth science. The map represents a view of the Earth looking down from above the North Pole (N.P.), showing the Earth's 24 standard time zones. The Sun's rays are striking the Earth from the right. Points *A, B, C,* and *D* are locations on the Earth's surface.

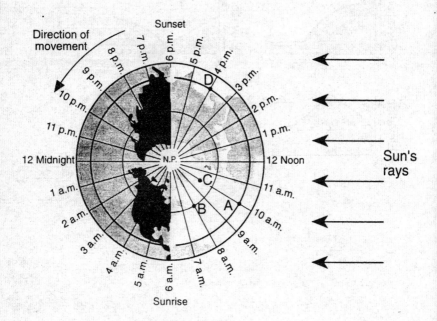

66 The timekeeping system shown in the diagram is
based on the

   1 Sun's revolution   3 Earth's revolution

   2 Sun's rotation   4 Earth's rotation   66 _____

67 Which date could this diagram represent?

   1 January 21   3 June 21

   2 March 21   4 August 21   67 _____

68 Which two points have the same longitude?

   (1) A and C (3) A and D

   (2) B and C (4) B and D 68 _____

69 Areas within a time zone generally keep the
same standard clock time. In degrees of longitude, approximately how wide is one standard
time zone?

   (1) $7\frac{1}{2}°$ (3) $23\frac{1}{2}°$

   (2) $15°$ (4) $30°$ 69 _____

70 At which position would the altitude of the
North Star (Polaris) be greatest?

   (1) A (3) C

   (2) B (4) D 70 _____

GROUP 4

If you choose this group, be sure to answer questions 71–75.

Base your answers to questions 71 through 75 on the *Earth Science Reference Tables*, the weather map below, and your knowledge of Earth science. The map shows a low-pressure system over the eastern part of the United States. Weather data is given for cities *A* through *I*. The temperature at city *E* has been left blank.

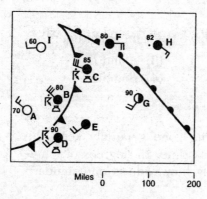

Miles 0   100   200

71 Which map correctly shows the locations of the cP and mT air-mass labels?

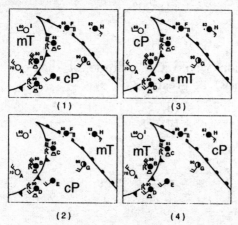

(1)   (3)

(2)   (4)

71 _____

72 The symbol ⌂ represents a cumulonimbus cloud. What is the most probable explanation for the absence of this cloud symbol at city A?

1 City A's atmosphere lacks the necessary moisture.
2 City A is located ahead of the cold front.
3 Cumulonimbus clouds form only at temperatures higher than 70°F.
4 Cumulonimbus clouds form only when a location has southwesterly winds.

72 _____

73 What is the most probable temperature for city E?

(1) 60°F   (3) 75°F
(2) 70°F   (4) 88°F

73 _____

74 Which city is *least* likely to have precipitation in the next few hours?

(1) A   (3) C
(2) F   (4) H

74 _____

75 Which map correctly shows arrows indicating the surface wind pattern?

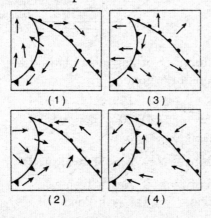

(1)   (3)

(2)   (4)

75 _____

GROUP 5

If you choose this group, be sure to answer questions 76–80.

Base your answers to questions 76 through 80 on the *Earth Science Reference Tables,* the diagram below, and your knowledge of Earth science. The diagram shows the latitude zones of the Earth.

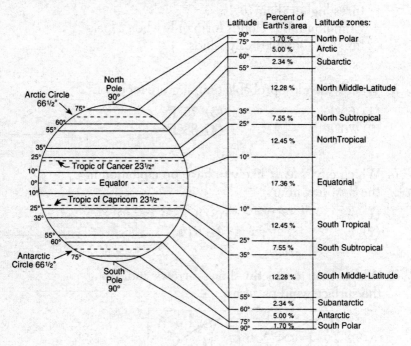

76 What is the total number of degrees of latitude covered by the Equatorial zone?

  (1) 0°   (3) 17°

  (2) 10°   (4) 20°   76 _____

77 In which latitude zone is New York State located?

  1 North Middle-Latitude   3 North Tropical

  2 North Subtropical   4 North Polar   77 _____

78 Which graph best represents the relationship between average yearly temperatures and latitude?

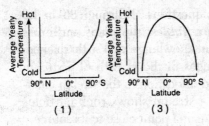

(1)   (3)

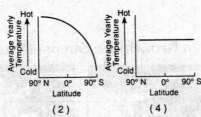

(2)   (4)

78 _____

79 Which zone receives the greatest intensity of sunlight on June 21?

1 North Tropical   3 South Tropical
2 Equatorial   4 North Polar

79 _____

80 The locations of the Tropic of Cancer and the Tropic of Capricorn are set at $23\frac{1}{2}°$ from the Equator because the

1 Earth is slightly bulged at the equatorial region
2 direct rays of the Sun move between these latitudes
3 Arctic Circle and the Antarctic Circle are $23\frac{1}{2}°$ from the poles
4 center of the Earth's gravitational field is located $23\frac{1}{2}°$ from the Equator

80 _____

GROUP 6

If you choose this group, be sure to answer questions **81–85**.

Base your answers to questions 81 through 86 on the *Earth Science Reference Tables*, the maps and cross section below, and your knowledge of Earth science. The maps show the stages in the growth of a stream delta. Point *X* represents a location in the stream channel. The side view of a stream shows rock particles transported in the stream at a point close to its source.

Maps: Stages in Growth of a Stream Delta

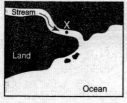

Early stage

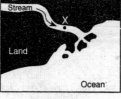

Middle stage

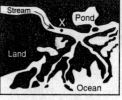

Late stage

Side View of a Stream

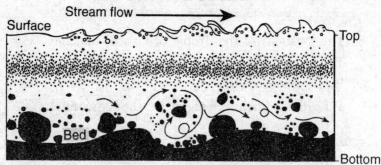

81 The rock materials transported in the stream are most likely transported by which methods?

    1 in solution, only
    2 in suspension, only
    3 in solution and in suspension, only
    4 in solution, in suspension, and by rolling   81 _____

82 The velocity of the stream at location X is controlled primarily by the

    1 amount of sediment carried at location X
    2 distance from location X to the stream source
    3 slope of the stream at location X
    4 temperature of the stream at location X 82 _____

83 Which graph best illustrates the effect that changes in stream discharge have on stream velocity at location X?

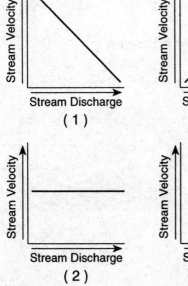
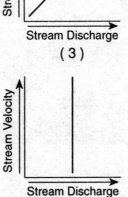

(1)   (3)

(2)   (4)   83 _____

84 A decrease in the velocity of the stream at location X will usually cause an increase in

   1 downcutting by the stream
   2 deposition within the stream channel
   3 the size of the particles carried by the stream
   4 the amount of material carried by the stream   84 _____

85 Which characteristics are most likely shown by the sediments in the delta?

   1 jagged fragments deposited in elongated hills
   2 unsorted mixed sizes deposited in scattered piles
   3 large cobbles deposited in parallel lines
   4 round grains deposited in layers   85 _____

GROUP 7

If you choose this group, be sure to answer questions 86–90.

Base your answers to questions 86 through 90 on the *Earth Science Reference Tables*, the information below, and your knowledge of Earth science. The map shows surface geology of a portion of the Schoharie Valley in New York State. Patterns and letters are used to indicate bedrock of different ages. The Schoharie Valley contains mostly horizontal rock structure in which overturning has not occurred. The table provides information about the rocks shown on the map.

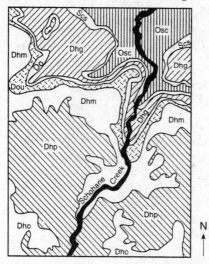

Age of the Rock	Symbol	Composition of the Rock
Middle Devonian	Dho	Shale, sandstone
	Dhp	Shale, siltstone, sandstone
	Dhm	Shale, sandstone, limestone
Early Devonian	Dou	Limestone, shale, siltstone
	Do	Sandstone, limestone
	Dhg	Limestone, dolostone
Late Silurian	Scs	Limestone, shale, dolostone
Middle–Late Ordovician	Osc	Sandstone, siltstone, shale

86 What is the age of the surface bedrock in this portion of the Schoharie Valley?

   (1) 320–374 million years
   (2) 374–478 million years
   (3) 478–505 million years
   (4) 505–540 million years   86 ____

87 The surface bedrock of this portion of the Schoharie Valley is composed mainly of

   1 metamorphic rock
   2 sedimentary rock
   3 extrusive igneous rock
   4 intrusive igneous rock   87 ____

88 Which fossil could be found in the surface bedrock of this portion of the Schoharie Valley?

   1 figlike leaf   3 brachiopod
   2 mastodont   4 coelophysis   88 ____

89 Which cross section represents a possible arrangement of rock units in a cliff along this portion of Schoharie Creek?

90 Based on the age of the bedrock, between which two rock units is an unconformity representing the longest period of time located?

   (1) Dhg and Scs   (3) Do and Dhg
   (2) Dhm and Dou   (4) Scs and Osc   90 ____

GROUP 8

If you choose this group, be sure to answer questions **91–95**.

Base your answers to questions 91 through 95 on the *Earth Science Reference Tables*, the graph below, and your knowledge of Earth science. The graph shows the results of a laboratory activity in which a 200-gram sample of ice at –50°C was heated in an open beaker at a uniform rate for 70 minutes and was stirred continually.

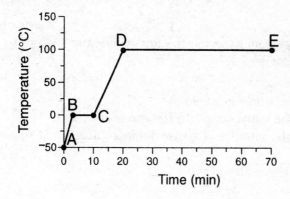

91 What was the temperature of the water 17 minutes after the heating began?

  (1) 0°C   (3) 75°C

  (2) 12°C   (4) 100°C   91 ____

92 Which change occurred between point *A* and point *B*?

  1 Ice melted.   3 Water froze.

  2 Ice warmed.   4 Water condensed.   92 ____

93 What was the total amount of energy absorbed by the sample during the time between points B and C on the graph?

(1) 200 calories
(3) 10,800 calories
(2) 800 calories
(4) 16,000 calories

93 _____

94 During which time interval was the greatest amount of energy added to the water?

(1) A to B
(3) C to D
(2) B to C
(4) D to E

94 _____

95 Which change could shorten the time needed to melt the ice completely?

1 using colder ice
2 stirring the sample more slowly
3 reducing the initial sample to 100 grams of ice
4 reducing the number of temperature readings taken

95 _____

GROUP 9

If you choose this group, be sure to answer questions 96–100.

Base your answers to questions 96 through 100 on the *Earth Science Reference Tables,* the map below, and your knowledge of Earth science. The map shows many of the major faults and fractures in the surface bedrock of New York State.

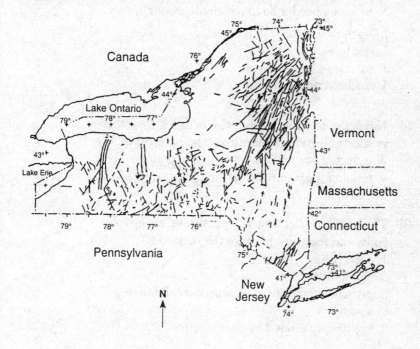

96 In which landscape region do the faults and fractures appear most concentrated?

   1 Adirondack Mountains
   2 the Catskills
   3 Atlantic Coastal Plain
   4 St. Lawrence Lowlands   96 _____

97 Most of the faults found along the New York-New Jersey border lie in which direction?

   1 north to south
   2 east to west
   3 northwest to southwest
   4 northeast to southwest   97 _____

98 The faults and fractures in the bedrock have the greatest effect on the locations and patterns of

   1 streams   3 precipitation
   2 residual soils   4 temperature zones   98 _____

99 The surface of Long Island shows no visible faults and fractures because the surface is

   1 a flat plain
   2 old bedrock
   3 primarily composed of unconsolidated sediments
   4 extensively eroded by ocean waves   99 _____

100 A large earthquake associated with one of these faults occurred at 45° N 75° W on September 5, 1994. Which location in New York State was closest to the epicenter of the earthquake?

   1 Buffalo   3 Albany
   2 Massena   4 New York City   100 _____

GROUP 10

*If you choose this group, be sure to answer questions **101–105**.*

101 The diagram below shows air rising from the Earth's surface to form a thunderstorm cloud.

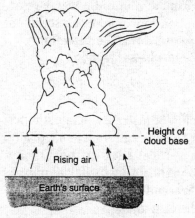

According to the Lapse Rate chart, what is the height of the base of the thunderstorm cloud when the air at the Earth's surface has a temperature of 30°C and a dewpoint of 22°C?

(1) 1.0 km   (3) 3.0 km
(2) 1.5 km   (4) 0.7 km   101 _____

102 Which list shows atmospheric layers in the correct order upward from the Earth's surface?

1 thermosphere, mesosphere, stratosphere, troposphere

2 troposphere, stratosphere, mesosphere, thermosphere

3 stratosphere, mesosphere, troposphere, thermosphere

4 thermosphere, troposphere, mesosphere, stratosphere   102 _____

103 Which planet takes longer for one spin on its axis than for one orbit around the Sun?

1 Mercury   3 Earth

2 Venus   4 Mars   103 _____

104 What is the approximate minimum water velocity needed to maintain movement of a sediment particle with a diameter of 5.0 centimeters?

(1) 75 cm/sec   (3) 150 cm/sec

(2) 100 cm/sec   (4) 200 cm/sec   104 _____

105 The diagram below illustrates Eratosthenes' method of finding the circumference of a planet. At noon, when a vertical stick at the Equator casts no shadow, a vertical stick 2,500 kilometers away casts a shadow and makes an angle of 40° with the rays of the Sun as shown.

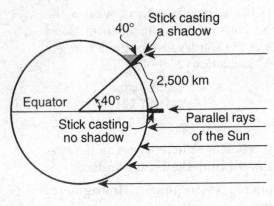

(Not drawn to scale)

What is the circumference of this planet?

(1) 2,500 km   (3) 22,500 km

(2) 20,000 km   (4) 45,000 km   105 _____

Answers
June 1997
Earth Science

Answer Key

PART I

1.	3	12.	2	23.	2	34.	3	45.	3
2.	2	13.	1	24.	4	35.	2	46.	1
3.	3	14.	3	25.	2	36.	3	47.	3
4.	1	15.	4	26.	3	37.	4	48.	4
5.	4	16.	4	27.	2	38.	1	49.	2
6.	4	17.	1	28.	1	39.	3	50.	2
7.	2	18.	3	29.	1	40.	2	51.	3
8.	1	19.	1	30.	4	41.	1	52.	4
9.	3	20.	2	31.	4	42.	2	53.	1
10.	4	21.	1	32.	1	43.	4	54.	3
11.	2	22.	3	33.	4	44.	4	55.	2

PART II

56.	2	66.	4	76.	4	86.	2	96.	1
57.	1	67.	2	77.	1	87.	2	97.	4
58.	2	68.	1	78.	3	88.	3	98.	1
59.	4	69.	2	79.	1	89.	1	99.	3
60.	3	70.	3	80.	2	90.	4	100.	2
61.	3	71.	3	81.	4	91.	3	101.	1
62.	1	72.	1	82.	3	92.	2	102.	2
63.	1	73.	4	83.	3	93.	4	103.	2
64.	2	74.	1	84.	2	94.	4	104.	3
65.	3	75.	2	85.	4	95.	3	105.	3

Answers Explained

PART I

1. **3** Since 10^4 is equal to 10,000, 4.0×10^4 kilometers is equal to 40,000 kilometers.

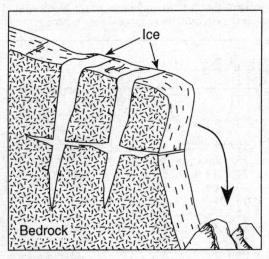

2. **2** Most natural materials contract when they change from liquids to solids, but water is an exception. As water freezes, it expands, so its volume increases as it changes from a liquid to a solid. As a result of this increase in volume, frost wedging breaks rocks.

WRONG CHOICES EXPLAINED:

(1) The mass of an object is the amount of matter it contains. When water freezes, the mass remains the same since the amount of matter does not change.

(3) The density of an object is its mass divided by its volume. Since the volume increases as water freezes, the density will decrease.

(4) The specific heat of a substance is the amount of heat required to raise the temperature of 1 gram of the substance 1 Celsius degree.

3. **3** Find the chart entitled Solar System Data in the *Earth Science Reference Tables*, and note that the period of revolution of Saturn is 29.46 years. Now, locate the equation for percent deviation from the accepted value in the Equations and Proportions section of the reference tables.

$$\text{Deviation (\%)} = \frac{\text{difference from accepted value}}{\text{accepted value}} \times 100$$

$$= \frac{31.33 \text{ years} - 29.46 \text{ years}}{29.46 \text{ years}} \times 100 = \frac{1.78}{29.46} \times 100$$

$$= 6.3\%$$

4. **1** When viewed from space, the Earth appears spherical, like a Ping-Pong ball. The reason is that the circumference of the Earth around the poles is nearly the same as the circumference around the equator.

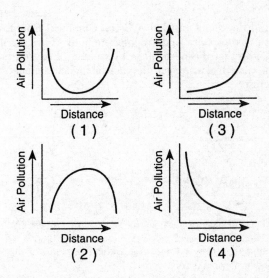

5. **4** Air pollution levels tend to be higher in an industrial city than in the surrounding area because of the presence of factories, which release air pollutants, and the higher concentration of automobiles, which emit air pollutants in their exhaust. This pattern is illustrated in graph (4), which shows that air pollution decreases as distance from an industrial city increases.

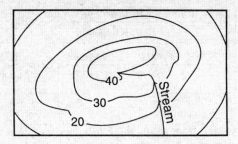

6. **4** The highest elevation on the map can be 49 meters because, if the elevation reached or exceeded 50 meters, a 50-meter contour line would be shown.

7. **2** Find the chart entitled Average Chemical Composition of the Earth's Crust, Hydrosphere, and Troposphere in the *Earth Science Reference Tables*, and note that the percentage by volume of oxygen in the crust is 94.04%. This value makes oxygen the most abundant element by volume in the Earth's crust.

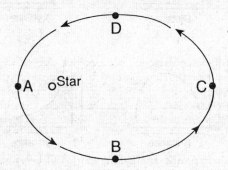

8. **1** The closer an object is to an observer, the larger it will appear. In the diagram the planet is closest to the star when it is in position *A*. At this time its apparent diameter will be greatest.

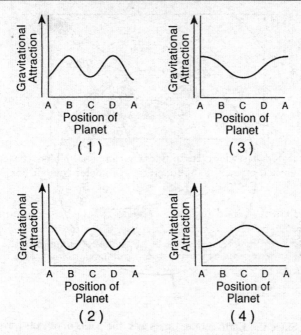

9. **3** The gravitational attraction between two objects is inversely related to the distance between them. According to the diagram, the gravitational attraction will be greatest when the planet is in position *A*, where the star and the planet are closest together. Also, the gravitational attraction will be least when the two objects are farthest apart, at position *C*. This pattern is illustrated in graph (3).

10. **4** As the Moon revolves around the Earth, different amounts of its surface appear illuminated by the Sun. As the Moon approaches the side of the Earth opposite the Sun, its surface appears completely illuminated. Then, as the Moon's path crosses between the Earth and the Sun, its surface appears darkened (i.e., the portion of the Moon's surface that is completely illuminated is facing the Sun, not the Earth). The fact that different amounts of the Moon's sunlit surfaces are seen on Earth as the Moon revolves around the Earth best explains the different phases of the Moon that are observed on Earth.

WRONG CHOICES EXPLAINED:
(1), (2) The Earth casts a shadow on the Moon only during a lunar eclipse, when the Moon passes between the Earth and the Sun.

(3) The Moon revolves around the Earth, which in turn revolves around the Sun.

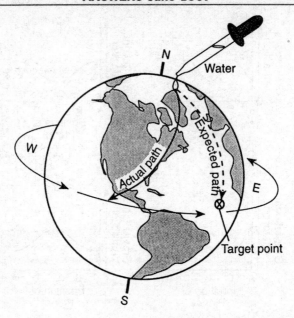

11. **2** Since the Earth rotates on its axis, the paths of objects traveling in a straight line in any direction except due east or west appear to curve. The rotation of the Earth causes the objects to be deflected toward the right in the Northern Hemisphere, and toward the left in the Southern Hemisphere. This phenomenon is called the Coriolis effect.

WRONG CHOICES EXPLAINED:

(1) The tilt of the Earth's axis in its orbit around the Sun causes the seasons. As a result of this tilt the angle of the Sun's rays changes during the year. If the axis were not tilted, all locations would have the same temperatures throughout the year. The tilt of the axis does not affect the paths of objects traveling at the surface of the Earth.

(3) The revolution of the Earth around the Sun, together with the tilt in its axis, causes the seasons. It does not affect the path of objects traveling at the surface of the Earth.

(4) A system is in dynamic equilibrium when opposite changes are occurring at the same rate. A beaker of water in a sealed container eventually reaches dynamic equilibrium when the rate of evaporation of water from the beaker equals the rate at which condensation of moisture in the surrounding air is returning water to the beaker. Dynamic equilibrium is not related to the movement of objects on the Earth's surface.

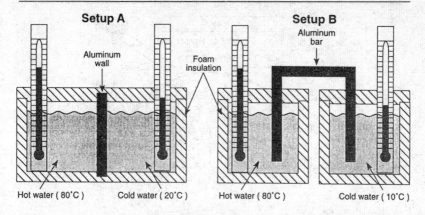

Setup A

Aluminum wall

Foam insulation

Hot water (80°C)   Cold water (20°C)

Setup B

Aluminum bar

Hot water (80°C)   Cold water (10°C)

12. **2** Conduction is the method of heat transfer in solids. When the outer atoms of aluminum in the metal bar are heated by the hot water, they speed up. As they strike nearby atoms, these atoms are also speeded up. This process continues across the bar into the cold water, where the heat energy that originated in the hot water is transferred to the cold water.

WRONG CHOICES EXPLAINED:

(1) Convection is the method of heat transfer in fluids. When air and water are heated, they expand and rise. The rising fluid is replaced by cooler, denser fluid, setting up a convection cell.

(3) Radiation is the method of energy transfer through empty space. Energy from the Sun reaches the Earth's surface by the process of radiation.

(4) Gravitational attraction, or gravity, exists between any two objects and causes them to tend to be pulled toward each other. The amount of attraction depends on the masses of the objects. Gravity is such a weak force that at least one of the objects must be extremely large for the attraction to be noticeable. Gravity is not related to heat transfer.

13. **1** In setup *A* the aluminum bar is completely immersed in the water, so less heat is lost than in setup *B*. In setup *B*, where the aluminum bar is in contact with the surrounding air, some of the heat being transferred from the hot water will be lost to the surrounding environment.

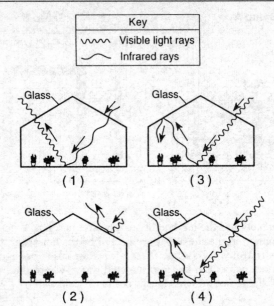

14. **3** Diagram (3) shows visible light passing through the glass walls of the greenhouse and striking the soil surface. Some of this energy is absorbed by the soil and then radiated back. The wavelength of this radiated energy will be in the longer infrared band of the electromagnetic spectrum because the temperature of the soil is lower than the temperature of the Sun, where the visible light originated. The infrared radiation cannot pass through the glass walls and is reflected back into the greenhouse. This "trapping" of energy causes the temperature within the greenhouse to rise—hence the term *greenhouse effect.*

15. **4** The Earth is tilted on its axis as it revolves around the Sun. As a result of this inclination, the angle at which the Sun's rays strike locations at the Earth's surface varies during the year. In the Northern Hemisphere in December, for example, the angle is small, causing the Sun to rise later and set earlier in the day. During the months of June, July, and August, when the angle is greatest, the Sun rises earlier and sets later at these same locations.

WRONG CHOICES EXPLAINED:
(1) The fact that the orbit of the Earth around the Sun is slightly elliptical does not affect the times of sunrise and sunset.
(2), (3) The Earth orbits around the Sun, not vice versa.

16. **4** Dark-colored surfaces absorb more energy than light colored surfaces. The freshly plowed farm field will absorb more energy because soil is darker in color than snow or white sand. Also, the water surface of a calm lake will tend to reflect more energy than a solid, dark surface such as soil.

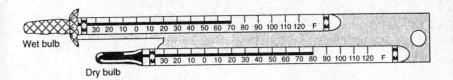

17. **1** Find the Temperature scale in the *Earth Science Reference Tables*. In the diagram the wet-bulb temperature is 70°F, and the dry-bulb temperature is 80°C. Locate these readings on the Fahrenheit temperature scale and read across to find the corresponding Celsius temperatures. The equivalent temperatures are 21°C for the wet bulb and 27°C for the dry bulb.

18. **3** The difference between the wet-bulb and dry-bulb temperature is an indication of the relative amount of moisture in the air. For example, readings that are close together indicate a high moisture content (i.e., a high relative humidity). Such readings also indicate that the dewpoint will be close to the air or dry-bulb temperature. To determine the dewpoint, both the weather instrument shown in the diagram and the Dewpoint Temperatures scale in the *Earth Science Reference Tables* would be used.

WRONG CHOICES EXPLAINED:
(1) Dry-bulb temperature *is* air temperature. Windspeed is measured with an instrument called an anemometer.

(2) Visibility is a measure of how clear the air appears and is indicated by how far an observer can see. Wind direction expresses the direction from which the wind is blowing, such as north or northeast.

(4) Air pressure is related to air density and indicates the force that the air is exerting on its surroundings. Higher pressures are often accompanied by lower temperatures. Cloud types are based on the size and shapes of cloud formations and on the amount of moisture present.

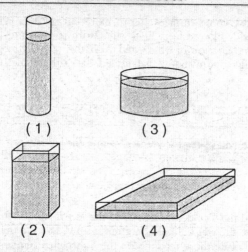

19. **1** Water evaporates at the surface of a body of water exposed to the air. Therefore, the rate of evaporation of water varies directly with the amount of exposed surface area. Since container (1) has the smallest exposed surface area, the rate of evaporation will be *least* for this container.

20. **2** A front is the boundary between two air masses. If a cold air mass comes into contact with a warm, moist air mass, the warm air mass rises and is cooled. If it cools enough, the excess moisture will begin to condense and precipitation at the frontal surfaces between the air masses may result.

WRONG CHOICES EXPLAINED:

(1) When warm, moist air meets cold, dry air, the warm air begins to rise because it is less dense than the cold air.

(3), (4) The speed at which fronts move depends on the relative energies of the two air masses. As a result, either type of front may move faster.

21. **1** When air containing moisture is cooled to the dewpoint temperature, the excess moisture begins to condense. When this condensation occurs high in the sky, the condensation produces clouds. When it occurs near the surface, it produces dew or frost.

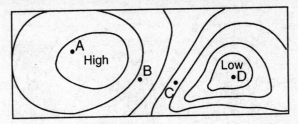

22. **3** The wind speed is greatest where the isobars are closest together, indicating that the air pressure is changing at the greatest rate. Windspeed is directly related to pressure differences. The rapid change in pressure at point *C* causes the greatest windspeed at this location.

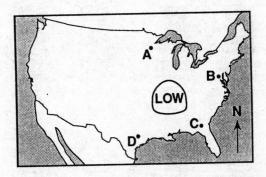

23. **2** Across the United States the general direction of air mass movement is from southwest to northeast, following the prevailing wind pattern. This pattern is illustrated in the map entitled Planetary Wind and Moisture Belts in the Troposphere in the *Earth Science Reference Tables*. Following this pattern, the low-pressure system will travel from its current location to point *B* if it follows a typical storm track.

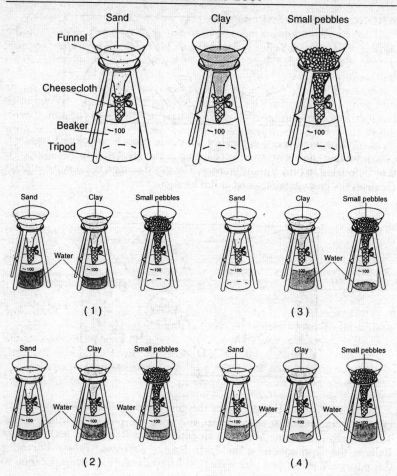

24. **4** The rate at which water infiltrates particles is directly related to particle size. The rate of infiltration becomes greater as the particle size increases. Diagram (4) shows that the least amount of water has infiltrated the clay, which has the smallest particle sizes. The greatest amount of water has filtered through the funnel with the small pebbles, which are the largest particles.

25. **2** Capillary action is the upward movement of water through tiny spaces in soil or rock. It occurs because the soil or rock exerts an attractive force on the water. The finer the soil, the higher the water rises because more surface area is exposed to the water to produce the attraction.

WRONG CHOICES EXPLAINED:

(1) Water retention is a measure of the amount of water remaining on the surface of soil particles after water has passed through. The smaller the particle size, the greater is the amount of water retention because more surface area is exposed.

(3) Porosity is the percentage of open space between particles in a soil or rock sample. If the particles are uniform in size and shape, the porosity remains the same as the particle size increases. When the particles are larger, the spaces between them become larger but also there are fewer spaces. As a result the porosity remains the same.

(4) Permeability is the rate at which water passes through a soil sample. When the particles are larger, the water passes through more quickly, so we say that the permeability is greater.

26. **3** Glaciers are capable of carrying larger particles than running water or wind can transport. When glacial ice melts, the deposit usually contains a wide range of particle sizes. As the glacier moves, the particles along the base and sides of the glacier become scratched and angular as they come into contact with local bedrock.

WRONG CHOICES EXPLAINED:

(1) Ocean waves can break down the bedrock along coastal areas. The sand found on many beaches is a result of this activity.

(2) The deposits produced by running water tend to be well sorted. As the water in a river or stream slows down, the largest particles are deposited first. Farther downstream, the deposited particles become smaller and smaller.

(4) Winds can carry only the smallest particles, so wind deposits tend to be well sorted with a very small range of particle size.

27. **2** There is a limit to the amount of water that soil can hold. When soil becomes saturated, any additional precipitation in the area will most likely become surface runoff into rivers and streams.

WRONG CHOICES EXPLAINED:

(1) Ground water is runoff that occurs when water filtering through soil reaches an impermeable rock layer.

(3) A moisture deficit occurs after all the water has evaporated from the soil. Since evaporation will normally continue, we say that a moisture deficit has occurred.

(4) Potential evapotranspiration is a measure of the maximum amount of moisture that can be expected to return to the atmosphere through evaporation and plant transpiration. It is directly related to temperature and the amount of plant cover.

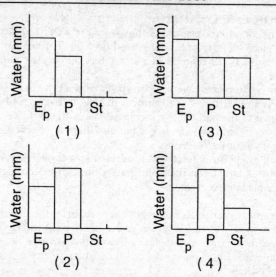

28. **1** Graph (1) shows an area where the potential evapotranspiration is greater than the amount of precipitation. Under these conditions, the shortage in water tends to be made up by evaporation of water held in the ground (i.e., water storage.) Since this graph shows that water storage is zero, the area is experiencing a moisture deficit.

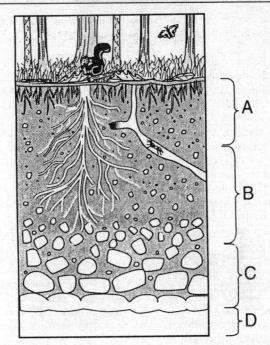

29. **1** Horizon *A* contains the greatest amount of material formed by biological activity because it is closest to the surface, below which most of the plant roots are located, and above which animal activity occurs.

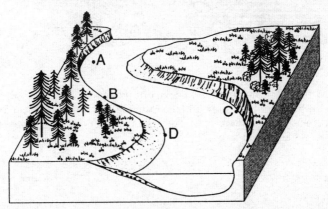

30. **4** The slower the stream velocity, the smaller are the particle sizes that the stream can carry. Of the particles listed, clay is the smallest and

therefore is the most likely to be transported in suspension during periods of slowest stream velocity.

31. **4** The water velocity speeds up around the outside of a curve in a stream and slows down along the inside of a curve. In the diagram, point *D* is located along the inside of a curve in the stream. As the water slows down there, the amount of deposition will exceed the amount of erosion.

32. **1** The composition of the various rock types can be found in the Schemes for Igneous, Sedimentary, and Metamorphic Rock Identification in the *Earth Science Reference Tables*. The charts show that gneiss is composed of several minerals, including pyroxene, garnet, and amphibole. The charts also show that quartzite is composed of quartz, gypsum is composed of the mineral gypsum, and limestone is composed of calcite. Of the four rock types, only gneiss is composed of more than one mineral.

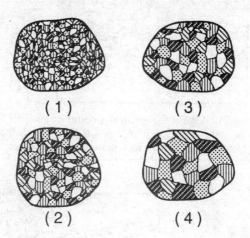

(1)   (3)

(2)   (4)

33. **4** The slower the rate of cooling of molten rock material, the larger will be the grains of the minerals in the resulting rock. Sample (4) contains the largest grains and therefore most likely formed from magma that cooled and solidified at the slowest rate.

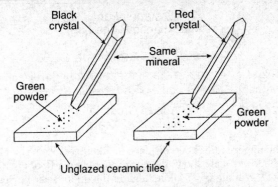

Black crystal — Same mineral — Red crystal

Green powder — Green powder

Unglazed ceramic tiles

34. **3** Different varieties of the same mineral may appear to be different colors (e.g., rose quartz, milky quartz). However, all samples of the same mineral will be the same color when reduced to a powder by being rubbed on a streak plate. Therefore, the mineral property being tested is streak.

WRONG CHOICES EXPLAINED:

(1) The density of a mineral is its mass divided by its volume. The mass of a mineral sample can be measured using a balance, and the volume can be measured by the amount of water it displaces.

(2) Fracture is the property of a mineral that describes how the mineral breaks. For example, the mineral halite breaks into cubes.

(4) The luster of a mineral describes its shine. For example, quartz has a glassy luster, while hematite is dull.

35. **2** Sedimentary rocks can be formed when sediments become buried beneath the surface. Material dissolved in water can act as a cement. This cementation, together with the compaction due to the pressure of the overlying material, can cause the sediments to be converted to rock. These inferences can be formed by observing the characteristics of some sedimentary rocks.

WRONG CHOICES EXPLAINED:

(1) Magma is converted into igneous rock when it cools and hardens.

(3) Sedimentary rock can be converted into metamorphic rock when there is enough heat and pressure to cause partial melting. If the rock completely melts and then cools and hardens, an igneous rock is produced, not a sedimentary rock.

(4) If metamorphic rock melts and then cools and hardens, it will form an igneous rock.

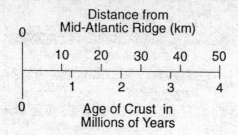

36. **3** The upper scale shows distance in kilometers from the Mid-Atlantic Ridge, and the lower scale shows the corresponding sea-floor crust age in millions of years. A reading of 37 kilometers on the upper scale corresponds to a crust age of about 3.0 million years.

37. **4** Conglomerate contains a wide range of particle sizes and compositions. The variation in composition of the pebbles in the question suggests that the individual particles came from different rock types and may have undergone varying amounts of weathering and erosion before being converted into rock.

WRONG CHOICES EXPLAINED:

(1) Slow-moving water can carry only the smallest particles, such as clay in suspension. It cannot carry particles as large as pebbles.

(2) No evidence is presented to suggest that the particles came from other conglomerates.

(3) No evidence that permits drawing inferences about the ages of the various types of pebbles is presented.

Wave A

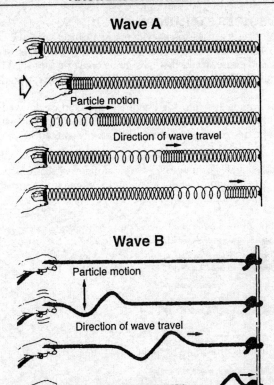

Wave B

38. **1** In compressional waves, the particles move back and forth, as illustrated in diagram *A*. In shear waves, the particles move from side to side, as illustrated in diagram *B*.

PEANUTS   BY CHARLES M. SCHULZ

39. **3** In the cartoon the character at the right sees the approaching rain cloud. When the rain comes, the sand castle will be removed by erosion as the sand is carried away.

WRONG CHOICES EXPLAINED:

(1) To be preserved as fossil evidence, an object must be covered with deep layers of sediment. This process would clearly destroy the sand castle.

(2) The sand castle would first have to be converted into sedimentary rock before it could undergo metamorphism. This conversion would destroy the sand castle.

(4) To be compacted into solid bedrock, the sand castle would have to be covered with thick layers of overlying sediment. Most likely all evidence of the castle would be destroyed when the first layers of sediment were deposited.

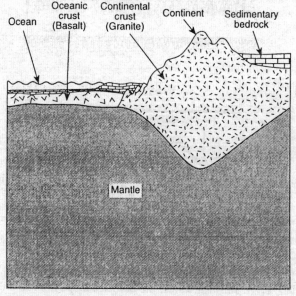

40. **2** In the diagram the crust under the continents extends farther down into the mantle than does the crust under the oceans. Therefore, the statement about the Earth's crust that is best supported by the diagram is "The continental crust is thicker than the oceanic crust."

WRONG CHOICES EXPLAINED:

(1) In the diagram the mantle is obviously thicker than the crust. Compared to the zones within the Earth, the crust represents a thin veneer at the Earth's surface.

(3) The diagram indicates that the crust is composed primarily of igneous rock (granite and basalt) and that only a small portion near the surface is composed of sedimentary rock.

(4) No evidence that would permit inferences about the relative densities of the rock in the crust and the rock in the mantle is presented in the diagram.

41. **1** Find the Earthquake *P*-wave and *S*-wave Travel Time graph in the *Earth Science Reference Tables*, and locate an epicenter distance of 6,000 kilometers along the horizontal axis. Note that for this distance the *S*-wave arrival time is about 17 minutes and the *P*-wave arrival time is about 9 minutes 25 seconds. The difference between these two times is about 7 minutes 35 seconds.

42. **2** Find the diagram entitled Inferred Properties of Earth's Interior in the *Earth Science Reference Tables*, and locate a depth of 2,000 kilometers along the horizontal axis of the temperature graph. According to the graph, the actual temperature reading is about 4,200°C.

43. **4** Convection currents in the asthenosphere cause warmer rock material to be pushed upward. This warmer material heats the water near the ocean floor, causing hot springs.

WRONG CHOICES EXPLAINED:

(1) The effect of widespread glacial melting would be to raise the ocean level. Glacial melting would not cause hot springs to form.

(2) Convection currents can cause portions of the ocean floor to be pushed upward, explaining the presence at high elevations of marine fossils that originally formed on the ocean floor.

(3) Meteor craters would not be expected to be found beneath the oceans. Most of the impact of meteors striking the oceans would be absorbed by the water. If a meteor did form a crater on the ocean floor, the shifting of sediment along the floor would level the area over time and remove evidence of the meteor's impact.

44. **4** Find the chart entitled Radioactive Decay Data in the Physical Constants section of the *Earth Science Reference Tables*. Note that the half-life (i.e., the amount of time required for half a sample to decay) of carbon-14 is 5,700 years, so that this isotope is useful for dating relatively recent remains.

The half-lives of the other choices, also given in the chart, are so long that too little of the substance would have decayed in 30,000 years for the isotope to be useful in radioactive dating.

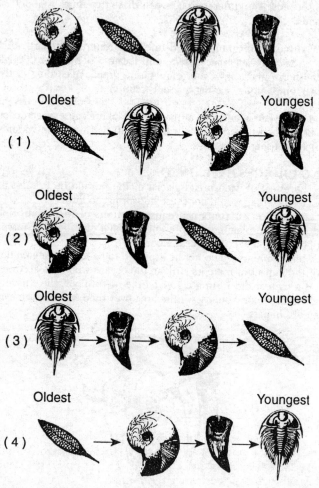

45. **3** Find the chart entitled Geologic History of New York State at a Glance in the *Earth Science Reference Tables*. Note in the column labeled "Important Fossils of New York" the locations of the four fossils depicted in the question. The oldest fossils appear at the bottom of the column; the fossils become progressively younger in going toward the top. According to the chart, choice (3) correctly depicts the fossils in order from oldest to youngest.

46. **1** Geologists infer that the environment in which fossil animals lived is similar to the environment in which similar species live today. Since modern-day corals live in warm, tropical waters, the presence of fossil corals in the surface bedrock of western New York State suggests that this area was covered by a warm, shallow sea during Ordovician time.

WRONG CHOICES EXPLAINED:

(2) Since modern-day corals live in a marine environment, it may be inferred that ancient corals also lived in this type of environment, rather than in forests.

(3) In most cases fossils are found in the preserved environment in which they lived. Occasionally fossils exposed to weathering and erosion may be transported to a different environment, but this is not a common occurrence.

(4) The continental ice sheets carried existing rock material and fossils to other locations. The ice sheets did not create the coral fossils.

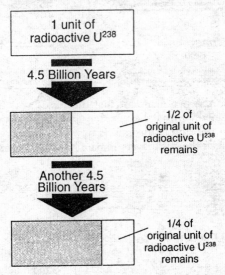

47. **3** Find the chart entitled Radioactive Decay Data in the Physical Constants section of the *Earth Science Reference Tables*. Note that when uranium-238 decays it produces lead-206 (Pb^{206}). This is the stable isotope represented by the shaded areas in the diagram.

48. **4** A fossil is classified as an index fossil if the plant or animal lived for a relatively short period of geologic time and if the fossil is found over widespread areas. These characteristics make index fossils useful for determining the geologic age of the surrounding rock. Volcanic ash is also deposited in a short time and over widespread areas, so that it too is a good geologic time marker.

WRONG CHOICES EXPLAINED:

(1) Radiocarbon dating is useful for determining the ages of organic remains of plants and animals. Volcanic ash contains little if any carbon since the high temperature of volcanic material destroys any surrounding life forms.

(2) Fossils are most commonly found in sedimentary rock. They are found in metamorphic rock only when the rock has formed from sedimentary rock.

(3) Index fossils are particularly susceptible to chemical weathering, and volcanic ash may also undergo this process.

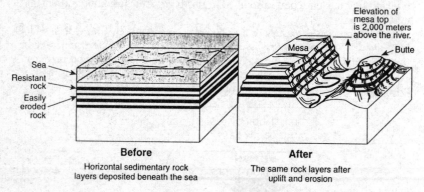

Before
Horizontal sedimentary rock layers deposited beneath the sea

After
The same rock layers after uplift and erosion

49. **2** Note in the two diagrams that the rock layers in the landscape region have remained horizontal even after uplift and erosion have occurred. The landscape in the "after" diagram is best classified as a plateau region because in a plateau the rock layers remain horizontal even after uplift has occurred.

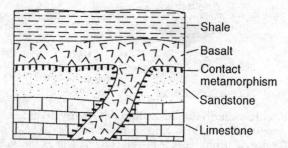

50. **2** In a series of rock layers that have not been overturned, the younger layers are on top. Note in the diagram that the shale lies above the basalt and therefore is the younger layer. In addition, the zone between the shale and the basalt has not undergone contact metamorphism. If, however, the molten material that formed the basalt had passed between the shale and the sandstone, then the shale, as well as the sandstone, would have undergone contact metamorphism and the basalt would be the younger layer.

WRONG CHOICES EXPLAINED:

(1) The sandstone is older than the basalt because it has undergone contact metamorphism. The sandstone must already have been in place when the molten material came in contact with it.

(3), (4) Since the limestone is located below both the shale and the sandstone, it may be inferred that it is older than either of these layers.

51. **3** Find the map entitled Generalized Bedrock Geology of New York State in the *Earth Science Reference Tables,* and locate Old Forge in the north central part of the state. According to the codes at the bottom of the map, this area contains mountains of metamorphic rock.

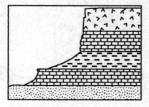 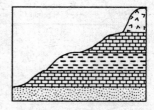

52. **4** Note that the slope of the rock layers in the diagram at the left is much steeper than the slope in the diagram at the right. Steep slopes are characteristic of arid regions, while gentler slopes are commonly found in more humid climates where there is more erosion by running water. Therefore, the differences in hillslopes and shapes are most likely the result of climatic variations.

53. **1** Landscape development is highly influenced by bedrock composition; some rock types are more resistant to weathering than others. Where the rock type is more resistant to weathering, the slope tends to be steeper. The structure of the bedrock also influences landscape development. Steeper slopes tend to weather more rapidly than more level landscapes.

WRONG CHOICES EXPLAINED:

.(2), (3), (4) Neither the age of the bedrock nor the color of the rock formations is directly related to landscape development.

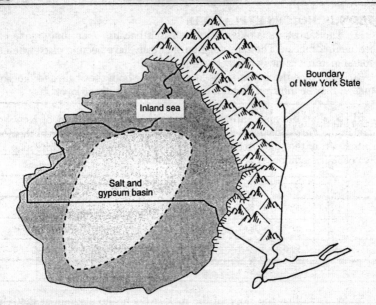

54. **3** Find the location of the "Salt and gypsum basin" shown on the map in the question on the map entitled Generalized Landscape Regions of New York State in the *Earth Science Reference Tables*. On this map the area falls within the Erie-Ontario Lowlands and the Allegheny Plateau.

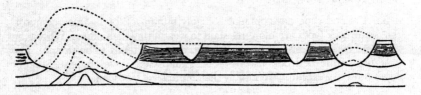

55. **2** The folded rock layers shown in the diagram have weathered and eroded faster than the surrounding areas because of their steep slopes. Over time, weathering and erosion tend to reduce hilly and mountainous areas to level plains.

WRONG CHOICES EXPLAINED:

(1) Weathering and erosion have not occurred evenly. If they had, the area would have retained the original elevation at all locations.

(3) There is no evidence in the diagram of deposition of sediments.

(4) Climate differences *do* affect the amount of erosion. In this case, however, the climate has been the same since the diagram represents one area.

PART II

GROUP 1

Rock Collection

Group	Rock	Mass (g)	Volume (cm³)	Density (g/cm³)	Description
X	A	82.9	34.4	2.41	Grey, smooth, rounded
	B	114.2	42.6	2.68	Brown, smooth, rounded
	C	144.7	63.2	2.29	Black, smooth, rounded
Y	D	159.4	59.7		Black and grey crystals, angular
	E	87.7	33.1	2.65	Clear and pink crystals, angular
	F	59.6	21.0	2.84	White, grey, and black crystals; angular
Z	G	201.1	68.4	2.94	Grey, shiny, flat
	H	85.1	39.8	2.14	Brown, sandy feel, flat
	I	110.2	47.3	2.33	Dark grey, flaky, flat

56. **2** All the rock samples in group X are rounded. Those in group Y are angular, and those in group Z are flat. The student has used shape to classify the rocks in her collection.

57. **1** To calculate the density of a rock sample, find the equation for density in the Equations and Proportions section in the *Earth Science Reference Tables*:

$$\text{Density} = \frac{\text{mass}}{\text{volume}} = \frac{159.4 \text{ g}}{59.7 \text{ cm}^3} = 2.67 \text{ g/cm}^3$$

58. **2** Changes in rock shape caused by stream action must be inferred based on observations because they happen too slowly to be observed directly. The shape, color, and volume of rock samples are examples of observations that can be made directly with the senses or with the aid of instruments.

59. **4** The shape, color, and luster (shine) of rocks can all be determined by observations made directly with the sense of sight. Some other properties, such as feel, can also be determined directly with the senses. Although still other observations require the use of instruments, such as a balance for mass or a graduated cylinder for volume, the student used her senses to obtain all the data in the column labeled "Description."

60. **3** The density of an object is the ratio of its mass and its volume. If a rock sample is split in half, both the mass and the volume are reduced by half. As a result the ratio between the two, that is, the density of each piece, remains the same.

GROUP 2

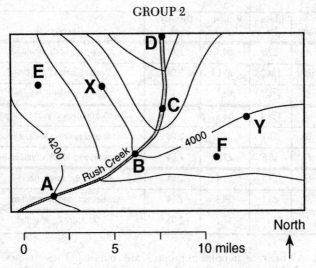

61. **3** Note that on the map there is one contour line between the 4000-foot contour line and the 4200-foot contour line. Therefore, each line represents a change in elevation of 100 feet, which is the contour interval for this map.

62. **1** The elevation at point *A* is 4200 feet, which is nearly the highest elevation on the map. The elevation decreases from the bottom to the top of the map as indicated by the contour lines. Point *D* is located past the 3800-foot contour line, in the lowest area on the map. Therefore, locations *A* and *D* have the greatest difference in elevation.

63. **1** According to the contour lines on the map, the elevation is decreasing between points *C* and *D*. Streams flow from areas of higher elevation to areas of lower elevation. Therefore, Rush Creek flows from point *C* to point *D*, or toward the north.

64. **2** Between points *A* and *B* the elevation drops by 200 feet. The scale at the bottom of the map indicates that the distance between these two points is about 5 miles. Therefore, the gradient between points *A* and *B* is closest to 40 feet/mile (200 feet/5 miles).

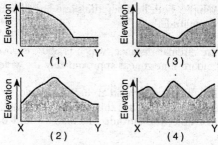

65. **3** Traveling from *X* to *Y* in a straight line, you cross the same lower contour line twice. Point *Y* lies along the same contour line as point *X*, and therefore has the same elevation, 4000 feet. Graph (3) shows the elevation first decreasing, then increasing, and finally returning to the same level.

GROUP 3

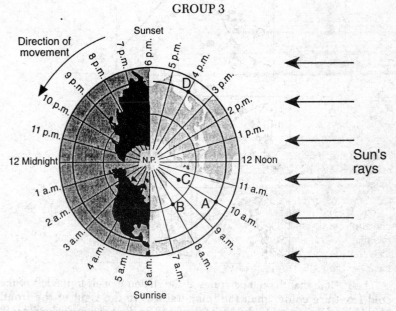

66. **4** The Earth makes one complete rotation on its axis every 24 hours. The timekeeping system shown in the diagram is based on this 24-hour pattern of the Earth's rotation.

67. **2** In the diagram the Earth's axis is tilted vertically with respect to its orbit around the Sun. This occurs on two days a year: about March 21 and September 21.

68. 1 Points *A* and *C* both lie along the same line of longitude and therefore have the same longitude.

69. 2 The entire circular distance, which is 360°, is covered in 24 hours. Therefore, each standard time zone is approximately 15° wide (360°/24 hours).

70. 3 Polaris lies directly overhead at the North Pole, so its altitude at this location is 90°. As one travels south, the altitude decreases. Since point *C* is closest to the North Pole, the altitude of Polaris would be greatest at this point.

GROUP 4

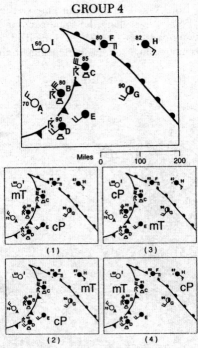

71. 3 Note that the temperatures at the station models to the left of the cold front are colder than the temperatures to the right of the front. Therefore, the air mass to the left of the front can be a continental polar (cP) air mass, and the air mass to the right can be a maritime tropical (mT) air mass. The cP and mT air-mass labels are shown correctly on map (3).

72. 1 Locate the station model symbols in the Weather Map Information section in the *Earth Science Reference Tables*, and note that the circle at the center of the station model indicates cloud cover. On the weather map in the question this circle for city *A* is clear, meaning that there is little, if any, cloud cover.

73. **4** Note that the temperatures at cities *C*, *D*, and *G* range from 85° to 90°F. Since city *E* is located within the same air mass, it can be expected to have a temperature within this range, that is, 88°F.

74. **1** The cities located near the advancing cold front are most likely to experience precipitation in the next few hours because, as warm air is cooled, condensation can occur. City *A*, which is located in the cold air mass and behind the front, is *least* likely to experience precipitation in the next few hours because air is not being cooled.

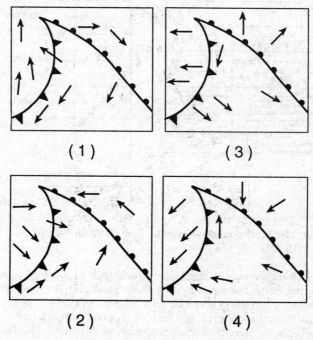

(1)   (3)

(2)   (4)

75. **2** The arrows on map (2) show the winds blowing across the front from the cold air to the warm air. This pattern occurs because warmer air is less dense and tends to rise. It is replaced by cooler and denser air, causing the winds to blow from the cold air to the warm air.

GROUP 5

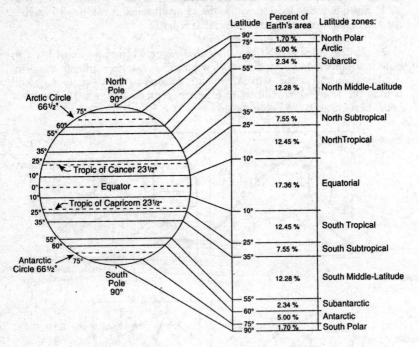

76. **4** The Equatorial zone extends from 10° S latitude to 10° N latitude for a span of 20°.

77. **1** Find the map entitled Generalized Bedrock Geology of New York State in the *Earth Science Reference Tables*, and note that New York State extends from a latitude of about 40° to a latitude of about 45°. In the diagram in the question, the latitude range from 35° N to 55° N is designated as North Middle-Latitude. New York State is located in this North Middle-Latitude zone.

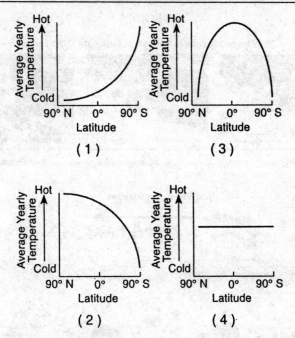

78. **3** In general, the coldest average yearly temperatures occur near the poles, and the warmest temperatures near the Equator. Graph (3) shows that the coldest temperatures occur near 90° N and 90° S (i.e., the poles) and the warmest temperatures near 0° latitude (i.e., the Equator). Therefore, this graph best represents the relationship between average yearly temperatures and latitude.

79. **1** The Sun is directly overhead at the Tropic of Cancer, 23½° N, about June 21, and this location would therefore receive the greatest intensity of sunlight on that date. The diagram in the question shows that the Tropic of Cancer lies within the North Tropical zone, which extends from 10° N to 25° N.

80. **2** The Sun is directly overhead at the Tropic of Cancer (latitude 23½° N) about June 21 and at the Tropic of Capricorn (latitude 23½° S) about December 21. Between these two dates the direct rays of the Sun move between these two latitudes.

GROUP 6

Maps: Stages in Growth of a Stream Delta

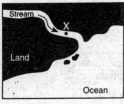

Early stage

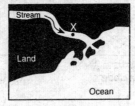

Middle stage

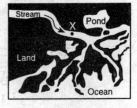

Late stage

Side View of a Stream

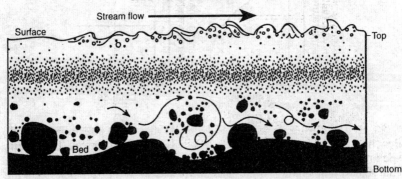

81. **4** In a stream, soluble material is carried in solution. The smallest particles are carried in suspension; the upper size limit of these particles depends directly on the velocity of the stream. Particles that are too large to be carried in suspension may roll along the streambed, pushed by the water flowing in the stream. Thus, rock materials are most likely transported in the stream by three methods—in solution, in suspension, and by rolling.

82. **3** The greater the slope of the stream, the greater will be the velocity of the water flowing in the stream. Therefore, the velocity of the stream at location X is controlled primarily by the slope of the stream at this location.

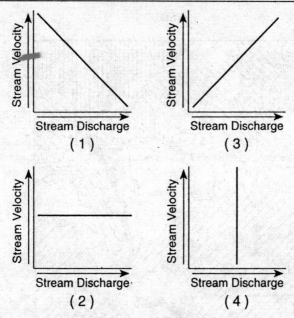

83. **3** The greater the volume of water flowing in the stream per unit of time (i.e., the discharge), the greater will be the velocity of water. This direct relationship is illustrated in graph (3), which shows that, as stream discharge increases at location X, stream velocity also increases.

84. **2** The largest size of particles that a stream can carry is directly related to the velocity of the stream. Therefore, if the velocity of the stream decreases at location X, the largest particles being carried will usually be deposited within the stream channel.

85. **4** When a stream reaches the ocean, the velocity of the water drops rapidly causing all but the smallest particles being carried to be deposited. This process of deposition produces the stream delta. The largest particles are deposited first followed by increasingly smaller particles as stream velocity continues to decrease. As time passes, the sediments will have the characteristics of round grains deposited in layers.

GROUP 7

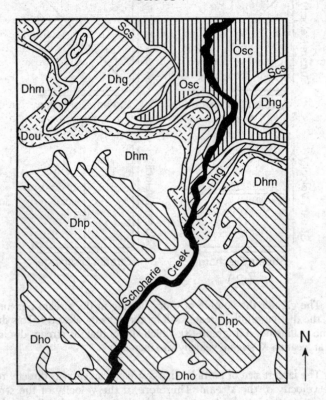

Age of the Rock	Symbol	Composition of the Rock
Middle Devonian	Dho	Shale, sandstone
	Dhp	Shale, siltstone, sandstone
	Dhm	Shale, sandstone, limestone
Early Devonian	Dou	Limestone, shale, siltstone
	Do	Sandstone, limestone
	Dhg	Limestone, dolostone
Late Silurian	Scs	Limestone, shale, dolostone
Middle–Late Ordovician	Osc	Sandstone, siltstone, shale

86. **2** The age of the bedrock in this portion of the Schoharie Valley extends from the Middle-Late Ordovician Period to the Middle Devonian Period. Find the chart entitled Geologic History of New York State at a Glance in the *Earth Science Reference Tables*. Note that the Middle Devonian Period ended about 374 million years ago and the Middle Ordovician Period began about 478 million years ago. This span—374–478 million years—represents the age of the surface bedrock in this area.

87. **2** All the rock types indicated in the chart—shale, sandstone, siltstone, limestone, and dolostone are types of sedimentary rock. These rock types appear in the scheme for sedimentary rock identification in the *Earth Science Reference Tables*.

88. **3** Find the chart entitled Geologic History of New York State at a Glance in the *Earth Science Reference Tables*. In the column headed "Important Fossils of New York," note that brachiopods were common during the Devonian Period, which represents the age of some of the bedrock in this area. The other fossil types listed represent animals that existed during much later periods.

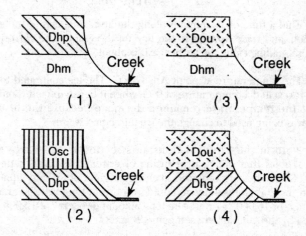

89. **1** Cross section (1) shows a Middle Devonian rock layer (Dhp) overlying an older rock layer (Dhm) from the same geologic period. In the other cross sections, older layers are on top of younger layers or layers are missing.

90. **4** Find the chart entitled Geologic History of New York State at a Glance in the *Earth Science Reference Tables*, and note the sequence of periods and epochs. The sequence between Late Silurian and Middle Devonian is continuous. Several epochs are missing, however, between Middle-Late

Ordovician and Late Silurian, which, as the chart in the question shows, are the ages of the Osc and Scs rock units, respectively. It may be inferred, therefore, that there is an unconformity or gap in the record between these two rock units.

GROUP 8

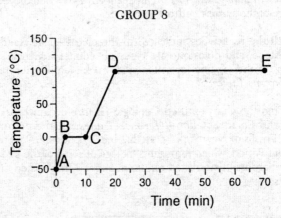

91. **3** Find a time of 17 minutes along the horizontal axis of the graph in the question, and trace upward to the line on the graph. The corresponding temperature reading along the vertical axis is about 75°C.

92. **2** The temperature at point A is –50°C. The ice continued to absorb heat until it reached a temperature of 0°C, when it began to melt (point B). At this point the temperature remained the same because the heat being absorbed was being used to change the ice into liquid water.

93. **4** Note in the Physical Constants section in the *Earth Science Reference Tables* that the heat of fusion of water is 80 calories per gram. Therefore, 80 calories are required to change 1 gram of ice into liquid water. To convert the 200 grams of ice to liquid water would require 200 grams × 80 calories/gram or 16,000 calories. The total amount of energy of 16,000 calories is absorbed by the sample between points B and C.

94. **4** The question indicates that heat was being added at a steady rate. Since the time period between points D and E is the largest period on the graph, this is the time interval during which the greatest amount of energy was being added to the water.

95. **3** Since heat was being added at a steady rate, reducing the sample to 100 grams of ice would reduce the amount of time required for the ice to melt. Lowering the initial temperature would increase the amount of time

until the ice began to melt, but the time span for the melting would remain the same. Stirring the sample more slowly might increase the melting time by reducing the rate at which the heat was being distributed. The number of temperature readings being taken has no effect on the rate of melting.

GROUP 9

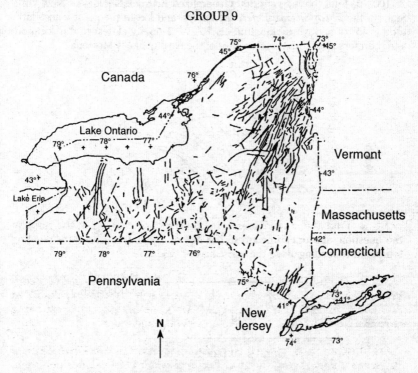

96. **1** Note that the highest concentration of faults and fractures occurs in the northeastern part of the state. Find the map entitled Generalized Landscape Regions of New York State in the *Earth Science Reference Tables*. The corresponding region on this map is the Adirondack Mountains.

97. **4** Note the angle of the fault lines in the southeastern part of the state along the New Jersey border. The lines are running from northeast to southwest.

98. **1** Faults and fractures in the bedrock alter the shape of the local landscape, thus affecting the locations and patterns of streams. There is little, if any, effect on residual soils and no effect on climate factors such as precipitation and temperature zones.

99. **3** The surface of Long Island is composed primarily of material deposited either directly by continental glaciers or by the runoff of water from the glaciers as they melted. This surface shows no visible faults and fractures because it is primarily composed of unconsolidated sediments.

100. **2** Find the map entitled Generalized Bedrock Geology of New York State in the *Earth Science Reference Tables*, and locate the point whose latitude is 45° N and whose longitude is 75° W. The city closest to this location and therefore closest to the epicenter of the earthquake is Massena.

GROUP 10

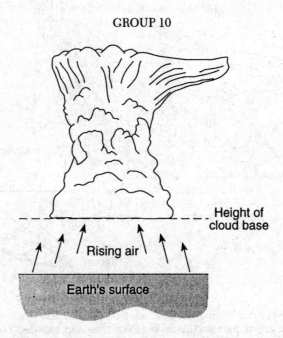

101. **1** Find the Lapse Rate chart in the *Earth Science Reference Tables*. Locate a temperature of 30°C along the horizontal axis, and trace upward along the solid line. Now, find a dewpoint temperature of 22°C along the same axis, and trace upward along the dashed line. Note where the two lines intersect, and read the altitude along the vertical axis. The altitude at this point is 1.0 kilometer, which represents the height of the base of the thunderstorm cloud.

102. **2** Find the graphs entitled Selected Properties of Earth's Atmosphere in the *Earth Science Reference Tables*. The graph at the right shows that the correct sequence of layers upward from the Earth's surface is troposphere, stratosphere, mesosphere, and thermosphere.

103. **2** Find the chart entitled Solar System Data in the *Earth Science Reference Tables*. Note that for Venus the period of revolution around the Sun is 224.7 days and the period of rotation on its axis is 243 days. For all the other planets the period of revolution is greater than the period of rotation.

104. **3** Find the graph entitled Relationship of Transported Particle Size to Water Velocity in the *Earth Science Reference Tables*, and locate a particle diameter of 5.0 centimeters along the vertical axis. Trace across to the curve on the graph, and then down to the horizontal axis. The minimum water velocity needed to carry a 5.0-centimeter-diameter particle is about 150 centimeter per second.

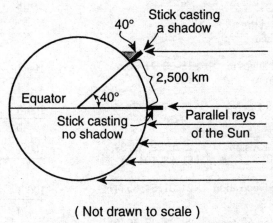

(Not drawn to scale)

105. **3** Find the equation for determining the circumference of the Earth using Eratosthenes' method in the Equations and Proportions section in the *Earth Science Reference Tables*.

$$\frac{\text{Shadow angle}}{360°} = \frac{\text{distance on surface}}{\text{circumference of planet}}$$

$$\frac{40°}{360°} = \frac{2,500 \text{ km}}{\text{circumference}}$$

$$\text{Circumference of planet} = \frac{2,500 \times 360}{40} = 22,500 \text{ km}$$

Topic	Question Numbers (total)	Wrong Answers (x)	Grade
Observation/Measure-ment of the Environment; The Changing Environment	3, 17, 57–60, 91 (7)		$\dfrac{100(7-x)}{7} = \%$
Measuring the Earth	1, 4, 6, 7, 61–65, 68–70, 76, 77, 100, 102, 105 (17)		$\dfrac{100(17-x)}{17} = \%$
Earth Motions	8–11, 15, 66, 67, 80, 103 (9)		$\dfrac{100(9-x)}{9} = \%$
Energy in Earth's Processes	12–14, 16, 92–95 (8)		$\dfrac{100(8-x)}{8} = \%$
Insolation and the Earth's Surface			
Energy Exchanges in the Atmosphere	18–23, 71–75, 101 (12)		$\dfrac{100(12-x)}{12} = \%$
Moisture and Energy Budgets	25, 27, 28, 78 79 (5)		$\dfrac{100(5-x)}{5} = \%$
Processes of Erosion	2, 26, 29, 30, 39, 81–83, 104 (9)		$\dfrac{100(9-x)}{9} = \%$
Processes of Deposition	24, 31, 84, 85 (4)		$\dfrac{100(4-x)}{4} = \%$
The Formation of Rocks	32–35, 37, 56 (6)		$\dfrac{100(6-x)}{6} = \%$
The Dynamic Crust	36, 38, 40–43 (6)		$\dfrac{100(6-x)}{6} = \%$
Interpreting Geologic History	44–48, 50, 51, 86–90 (12)		$\dfrac{100(12-x)}{12} = \%$
Landscape Development	5, 49, 52–55, 96–99 (10)		$\dfrac{100(10-x)}{10} = \%$

To further pinpoint your weak areas, use the Topic Outline in the front of the book.

Examination
January 1998
Earth Science

PART I

Answer all 55 questions in this part. [55]

Directions (1–55): For *each* statement or question, select the word or expression that, of those given, best completes the statement or answers the question. Record your answers in the spaces provided.

1 The diagram below is a cross section of an ice-covered lake in New York State during the month of January. Points *A*, *B*, *C*, and *D* are locations at various levels in the lake. The temperature of the water at location *D* is 4°C.

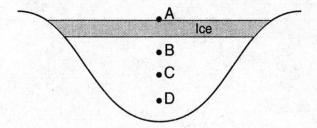

Which graph best represents the relationship between location and density of the ice or water?

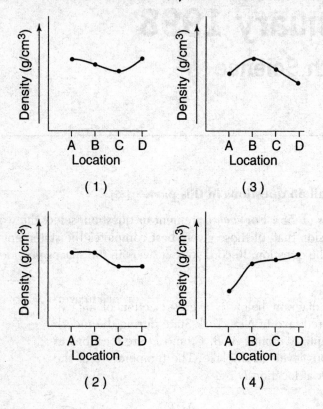

(1)

(3)

(2)

(4)

1____

Base your answers to questions 2 and 3 on the graph below, which shows the average daily precipitation for Paris, France, during an 8-year period.

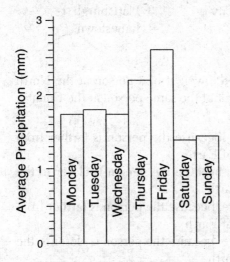

2 Which days showed the greatest difference in average precipitation during this 8-year period?

  1 Mondays and Tuesdays
  2 Wednesdays and Thursdays
  3 Thursdays and Fridays
  4 Fridays and Saturdays   2____

3 The average weekly precipitation total for Paris, France, during the 8-year period was approximately

  (1) 13 mm/week   (3) 30 mm/week
  (2) 2 mm/week   (4) 91 mm/week   3____

4 For an observer in New York State, the altitude of Polaris is 43° above the northern horizon. This observer's latitude is closest to the latitude of

1 New York City   3 Plattsburgh
2 Utica   4 Jamestown   4 _____

5 Compared to the weight of a person at the North Pole, the weight of the same person at the Equator would be

1 slightly less, because the person is farther from the center of Earth
2 slightly less, because the person is closer to the center of Earth
3 slightly more, because the person is farther from the center of Earth
4 slightly more, because the person is closer to the center of Earth   5 _____

6 The diagram below represents the approximate distances covered by one degree of longitude on Earth's surface at various latitudes.

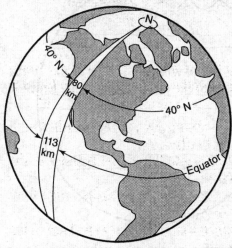

(Not drawn to scale)

What is the distance represented by one degree of longitude at Massena, New York?

(1) 78 km   (3) 90 km
(2) 85 km   (4) 113 km   6____

7 Most of the water vapor in the atmosphere is found in the

1 mesosphere   3 troposphere
2 thermosphere   4 stratosphere   7____

8 Which planet has the most eccentric orbit?

1 Mercury   3 Neptune
2 Venus   4 Pluto   8____

9 The planetary winds on Earth are indicated by the curving arrows in the diagram below.

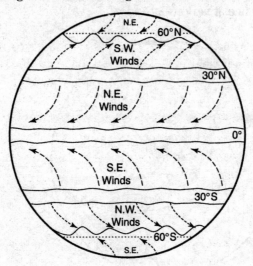

The curved paths of the planetary winds are a result of

1 changes in humidity
2 changes in temperature
3 Earth's rotation on its axis
4 Earth's gravitational force

9 _____

10 The Sun's apparent daily path through the daytime sky is best described by an observer in New York State as

1 a circle around the North Star
2 an arc that extends from east to west
3 a straight line that passes directly overhead
4 a random motion that varies with the seasons

10 _____

11 The Moon's cycle of phases can be observed from Earth because the Moon

　1 is smaller than Earth
　2 is tilted on its axis
　3 rotates on its axis
　4 revolves around Earth 　　　　　　　　11 _____

12 All forms of electromagnetic energy have

　1 transverse wave properties
　2 the same temperature
　3 the same wavelength
　4 their own half-life 　　　　　　　　　12 _____

13 The diagram below shows a solid iron bar that is being heated in a flame.

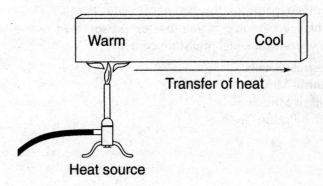

The primary method of heat transfer in the solid iron bar is

　1 convection　　　　3 absorption
　2 conduction　　　　4 advection 　　　　13 _____

14 An object that is a good absorber of electromagnetic energy is most likely a good

    1 convector   3 radiator

    2 reflector   4 refractor   14 _____

15 The hottest climates on Earth are located near the Equator because this region

    1 is usually closest to the Sun

    2 reflects the greatest amount of insolation

    3 receives the most hours of daylight

    4 receives the most nearly perpendicular insolation   15 _____

16 In which geographic region are air masses most often warm with a high moisture content?

    1 Central Canada

    2 Central Mexico

    3 Gulf of Mexico

    4 North Pacific Ocean   16 _____

17 The diagrams below represent Earth's tilt on its axis on four different dates. The shaded portion represents the nighttime side of Earth. Which diagram best represents the day on which the longest duration of insolation occurs in New York State?

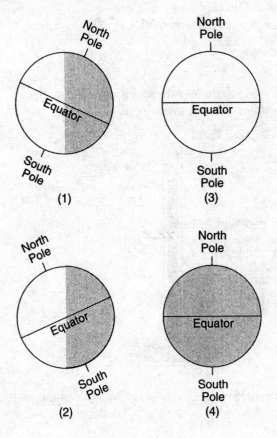

(1)

(3)

(2)

(4)

17 _____

18 If Earth's tilt were to increase from $23\frac{1}{2}°$ to $33\frac{1}{2}°$, the result would be

  1 shorter days and longer nights at the Equator
  2 less difference between winter and summer temperatures in New York State
  3 colder winters and warmer summers in the Northern Hemisphere
  4 an increase in the amount of solar radiation received by Earth

18 _____

Base your answers to questions 19 and 20 on the diagram below of a weather instrument.

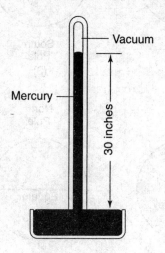

19 Which weather variable is this instrument designed to measure?

  1 visibility
  2 relative humidity
  3 dewpoint temperature
  4 air pressure

19 _____

20 In New York State, which weather conditions are most likely to exist when the height of the mercury in the tube is much greater than 30 inches?

1 cold, dry air with clear skies
2 warm, moist air with overcast skies
3 strong southerly winds with hail warnings
4 a violent storm associated with the autumn season

20 _____

21 Surface ocean currents resulting from the prevailing winds over the oceans illustrate a transfer of energy from

1 lithosphere to atmosphere
2 hydrosphere to lithosphere
3 atmosphere to hydrosphere
4 stratosphere to troposphere

21 _____

22 A low-pressure system near Utica, New York, causes heavy precipitation. If this system followed the usual track, which city most likely had the same weather conditions a few hours earlier?

1 Syracuse   3 Albany
2 Kingston   4 Plattsburgh

22 _____

Base your answers to questions 23 and 24 on the diagram below, which shows the frontal boundary between mT and cP air masses.

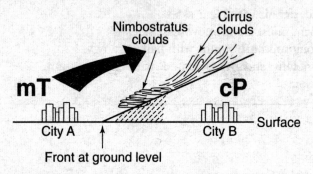

23 If the front at ground level is moving toward city *B*, which type of weather front is shown?

1 cold front   3 occluded front
2 warm front   4 stationary front   23 _____

24 Why do clouds and precipitation usually occur along the frontal surface?

1 The warm air rises, expands, and cools.
2 The warm air sinks, expands, and warms.
3 The cool air rises, compresses, and cools.
4 The cool air sinks, compresses, and warms.   24 _____

25 What is the name of the warm ocean current that flows along the east coast of the United States?

  1 California Current
  2 Florida Current
  3 Labrador Current
  4 North Pacific Current   25 _____

26 Why would a stream in New York State have a lower stream discharge in late summer than in spring?

  1 Potential evapotranspiration is less in late summer than in spring.
  2 Plants carry on more transpiration in spring than in late summer.
  3 The local water budget shows a surplus in late summer.
  4 The local water budget shows a deficit in late summer.   26 _____

27 The lines on the map below represent the average yearly amount of precipitation in inches.

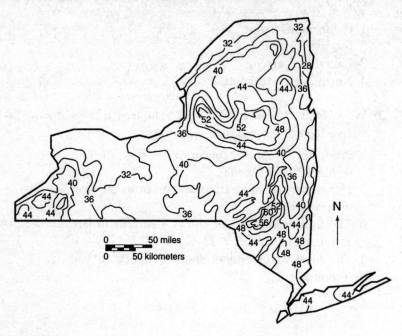

Which statement best explains the difference in yearly precipitation between Watertown and Plattsburgh?

1 Watertown receives more precipitation because it is farther from the Atlantic Ocean.

2 Plattsburgh receives more precipitation because it is closer to the Atlantic Ocean.

3 Watertown receives more precipitation·because of the effects of change in elevation and prevailing winds.

4 Plattsburgh receives more precipitation because of the effects of change in elevation and prevailing winds.

27 _____

28 Soil with the greatest porosity has particles that are

   1 poorly sorted and densely packed
   2 poorly sorted and loosely packed
   3 well sorted and densely packed
   4 well sorted and loosely packed   28____

29 Which factor has the most influence on the development of soil?
   1 climate
   2 longitude
   3 amount of rounded sediment
   4 slope of the landscape   29____

30 Which diagram best represents a cross section of the sediment deposited directly by a glacier in New York State?

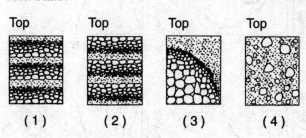

   (1)   (2)   (3)   (4)   30____

31 The diagram below shows a meandering stream. Measurements of stream velocity were taken along straight line *AB*.

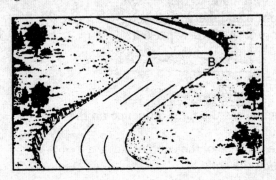

Which graph best shows the relative stream velocities across the stream from *A* to *B*?

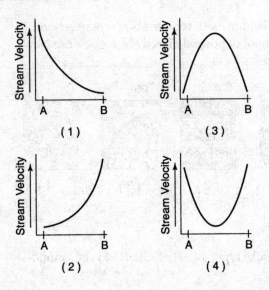

31 _____

32 The diagram below shows cobbles used in the construction of the walls of a cobblestone building.

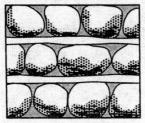

(Not actual size)

The shape and size of the cobbles suggest that they were collected from

1 the channel of a fast-flowing stream
2 volcanic ash deposits
3 a desert sand dune
4 the base of a cliff from which they had weathered

32 _____

33 Which mineral property is illustrated by the peeling of muscovite mica into thin, flat sheets?

1 luster   3 hardness
2 streak   4 cleavage

33 _____

34 Which sedimentary rock may form as a result of biologic processes?

1 shale   3 fossil limestone
2 siltstone   4 breccia

34 _____

35 Which rock type is most likely to be monomineralic?

1 rock salt   3 basalt
2 rhyolite   4 conglomerate

35 _____

36 The diagram below shows actual sizes and shapes of particles removed from a clastic sedimentary rock.

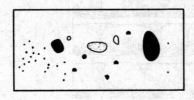

The sediments are from

1 chemical limestone   3 granite
2 conglomerate   4 sandstone   36____

37 Which rock forms by the recrystallization of unmelted rock material under conditions of high temperature and pressure?

1 granite   3 rock gypsum
2 gneiss   4 bituminous coal   37____

38 Seafloor spreading is occurring at the boundary between the

1 African plate and Antarctic plate
2 Nazca plate and South American plate
3 China plate and Philippine plate
4 Australian plate and Eurasian plate   38____

39 In which area of Earth's interior is the pressure most likely to be 2.5 million atmospheres?

1 asthenosphere   3 inner core
2 stiffer mantle   4 outer core   39____

Base your answers to questions 40 and 41 on the map below. The map shows point X, which is the location of an earthquake epicenter, and point A, which is the location of a seismic station.

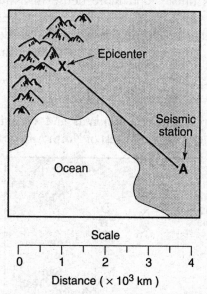

Scale

0   1   2   3   4

Distance (× 10³ km)

40 Approximately how long did the earthquake's P-wave take to arrive at the seismic station?

(1) 3 min 40 sec   (3) 6 min 20 sec
(2) 5 min 10 sec   (4) 11 min 5 sec   40 ____

41 Which statement best describes the arrival of the initial S-wave at the seismic station?

1 It arrived later than the P-wave because S-waves travel more slowly.

2 It arrived earlier than the P-wave because S-waves travel faster.

3 It arrived at the same time as the P-wave because S-waves and P-waves have the same velocity on Earth's surface.

4 It never reached location A because S-waves can travel only through a liquid medium.   41 ____

42 Which statement best describes Earth's crust and mantle?

  1 The crust is thicker and less dense than the mantle.
  2 The crust is thicker and more dense than the mantle.
  3 The crust is thinner and less dense than the mantle.
  4 The crust is thinner and more dense than the mantle.

42 _____

43 The map below shows continental and oceanic crustal plates along the west coast of North America.

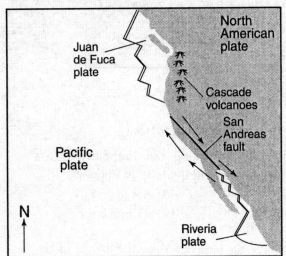

Which conclusion is best supported by the map?

  1 The boundary of the Pacific plate has very few faults.
  2 The Pacific plate has stopped moving.
  3 The west coast of North America is composed of the oldest rocks on the continent.
  4 The west coast of North America is a zone of frequent crustal movement.

43 _____

44 Which feature of a sandstone rock layer usually is the youngest?

  1 sand grains that make up the rock
  2 cement that binds the sand grains together
  3 fossils found in the rock
  4 faults that have broken the rock   44 _____

45 Which bedrock would be most likely to contain fossils?

  1 Precambrian granite
  2 Cambrian shale
  3 Pleistocene basalt
  4 Middle-Proterozoic quartzite   45 _____

46 Which life-forms existed on Earth during the same time period?

  1 trilobites and mastodonts
  2 ammonoids and Naples trees
  3 armored fish and flowering plants
  4 dinosaurs and early humans   46 _____

47 The cartoon below is a humorous interpretation of the results of an invention.

50,000 B.C.: Gak invents the first and last silent mammoth whistle.

The dominant life-forms existing on Earth at the time represented by the cartoon are classified as

1 mammals
2 reptiles
3 amphibians
4 invertebrates

47 _____

48 What is the age of most of the surface bedrock found in New York State at a latitude of 45°?

1 Precambrian Middle Proterozoic
2 Triassic and Jurassic
3 Silurian and Devonian
4 Cambrian and Ordovician   48 _____

49 Radioactive carbon-14 dating has determined that a fossil is 5.7×10^3 years old. What is the total amount of the original C^{14} still present in the fossil?

(1) 0%   (3) 50%
(2) 25%   (4) 75%   49 _____

50 Using radioactive dating methods and mathematical inferences, scientists have estimated the date of Earth's formation to be approximately

(1) 1.1×10^6 years ago
(2) 2.4×10^6 years ago
(3) 3.3×10^9 years ago
(4) 4.6×10^9 years ago   50 _____

51 A landscape region that has broad, U-shaped valleys with polished and grooved bedrock was most likely formed by

1 glaciers   3 wave action
2 wind   4 running water   51 _____

52 Continents are divided into landscape regions on the basis of

    1 bedrock fossils and depositional patterns
    2 rainfall and temperature changes
    3 surface features and bedrock structure
    4 boundaries of the drainage basins of major rivers   52 _____

53 In which New York State landscape region is Albany located?

    1 Catskills
    2 Taconic Mountains
    3 Hudson-Mohawk Lowlands
    4 Champlain Lowlands   53 _____

54 Bedrock in the area of Binghamton, New York, consists of

    1 plutonic igneous rock
    2 sedimentary rock layers
    3 faulted and tilted volcanic rock
    4 folded metamorphic rock   54 _____

55 Which New York State landscape region is composed primarily of metamorphic bedrock at its surface?

    1 Manhattan Prong
    2 Allegheny Plateau
    3 St. Lawrence Lowlands
    4 Erie-Ontario Lowlands   55 _____

PART II

This part consists of ten groups, each containing five questions. Choose seven of these ten groups. Be sure that you answer all five questions in each group chosen. Record the answers to these questions in the spaces provided. [35]

GROUP 1

If you choose this group, be sure to answer questions 56–60.

Base your answers to questions 56 through 60 on the *Earth Science Reference Tables*, the diagrams below, and your knowledge of Earth science. The diagrams represent four different mineral samples with different shapes and masses. Diagrams are not drawn to scale.

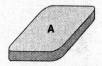

Mass = 12.0 g   Mass = 16.0 g   Mass = 20.0 g   Mass = 24.0 g
Volume = 4.0 cm³ Volume = 4.0 cm³ Volume = 4.0 cm³ Volume = 4.0 cm³

56 Which instrument was most likely used to find the volume of each sample?

  1 graduated cylinder   3 thermometer
  2 balance   4 psychrometer   56 _____

57 A second sample of mineral *A* has a mass of 48 grams. What is the volume of this sample?

  (1) 24.0 cm³   (3) 12.0 cm³
  (2) 16.0 cm³   (4) 4.0 cm³   57 _____

58 A student finds the mass of sample B to be 17.5 grams. What is the student's approximate percent deviation (percentage of error)?

(1) 1.5%　　　　　(3) 8.8%

(2) 6.7%　　　　　(4) 9.4%　　　　　58 ____

59 Which sample would most likely be the slowest to settle in a quiet body of water?

(1) A　　　　　(3) C

(2) B　　　　　(4) D　　　　　59 ____

60 Which graph best represents the density of each sample?

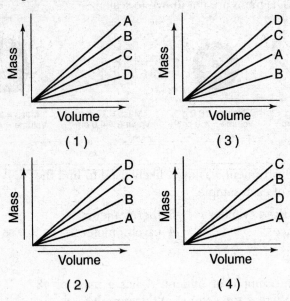

60 ____

GROUP 2

If you choose this group, be sure to answer questions **61–65**.

Base your answers to questions 61 through 65 on the topographic map below and on your knowledge of Earth science. Letters *A* through *F* represent locations on the map.

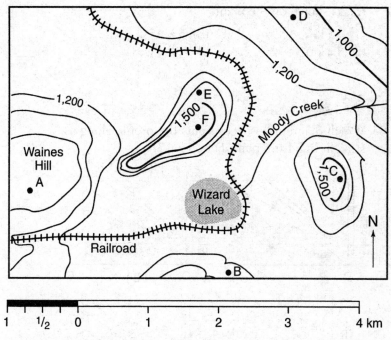

61 What is the contour interval of this map?

   (1) 10 m   (3) 100 m

   (2) 50 m   (4) 150 m   61 _____

62 Toward which direction does Moody Creek flow?

   1 southwest   3 northeast

   2 northwest   4 southeast   62 _____

63 Which location has the lowest elevation?

    (1) *A*   (3) *C*

    (2) *E*   (4) *D*   63 _____

64 What is the approximate length of the railroad tracks shown on the map?

    (1) 15 km   (3) 8 km

    (2) 12 km   (4) 4 km   64 _____

65 Which diagram best represents the profile along a straight line from point *D* to point *C*?

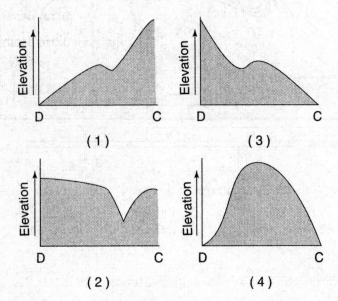

65 _____

GROUP 3

If you choose this group, be sure to answer questions **66–70**.

Base your answers to questions 66 through 70 on the graph below and on your knowledge of Earth science. The graph shows temperature data taken at four different depths in the soil at one location on Long Island, New York, for one year.

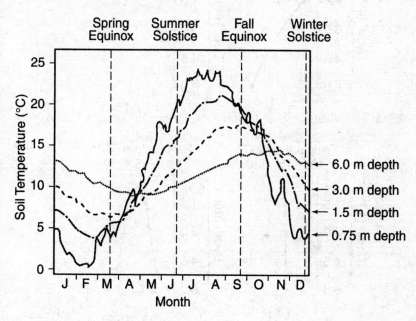

66 Which graph best represents the relationship between soil temperature and depth in the soil on June 21?

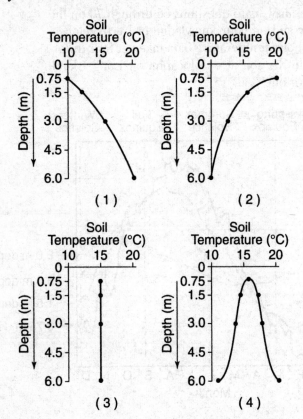

67 When did maximum soil temperature occur at depths in the soil of less than 5 meters?

1 at the spring equinox
2 between the spring equinox and the summer solstice
3 at the summer solstice
4 between the summer solstice and the fall equinox

66 _____

67 _____

68 On what date was the temperature at the 3-meter depth greater than the temperature at any of the other three depths?

1 July 11   3 November 1
2 August 31   4 December 21   68 _____

69 The general yearly pattern shown in the diagram of temperature changes at each depth in the soil is best described as

1 a one-way change
2 a random change
3 an unpredictable change
4 a cyclic change   69 _____

Note that question 70 has only three choices.

70 The graph shows that as depth increases, the annual temperature range

1 decreases
2 increases
3 remains the same   70 _____

GROUP 4

If you choose this group, be sure to answer questions **71–75.**

Base your answers to questions 71 through 75 on the *Earth Science Reference Tables*, the diagram below, and your knowledge of Earth science. The diagram represents a laboratory stream table.

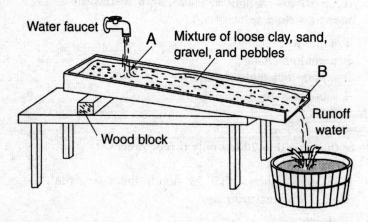

71 Which equation should be used to determine the gradient of the stream?

1 $\text{gradient} = \dfrac{\text{height of faucet above the floor (cm)}}{\text{water discharge from the tube (m}^3\text{/sec)}}$

2 $\text{gradient} = \dfrac{\text{falling distance of runoff water (cm)}}{\text{volume of water collected in pan (mL)}}$

3 $\text{gradient} = \dfrac{\text{volume of water in the stream between points } A \text{ and } B \text{ (mL)}}{\text{height of the wood block (cm)}}$

4 $\text{gradient} = \dfrac{\text{difference in elevation between points } A \text{ and } B \text{ (cm)}}{\text{distance between points } A \text{ and } B \text{ (m)}}$

72 Which particles are transported most easily by the water in this stream?

1 clay   3 silt
2 sand   4 pebbles   72 _____

Note that question 73 has only three choices.

73 When stream volume increases after the faucet is opened, stream velocity will

1 decrease
2 increase
3 remain the same   73 _____

74 Water flowing in the stream can move sediments along the stream channel because of an exchange of energy from the

1 channel to the water
2 water to the sediment
3 sediment to the channel
4 channel to the sediment   74 _____

75 How do streams transport sediments?

1 in suspension, only
2 by rolling, only
3 in suspension and by rolling, only
4 in solution, in suspension, and by rolling 75 _____

GROUP 5

If you choose this group, be sure to answer questions **76–80**.

Base your answers to questions 76 through 80 on the *Earth Science Reference Tables,* the diagram below, and your knowledge of Earth science. The diagram represents a cross section of the shoreline of Lake Erie.

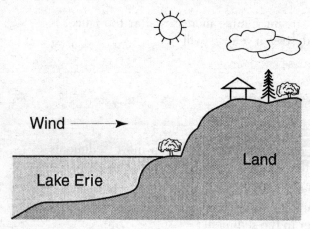

76 From 9 a.m. to 1 p.m. each day, the land surface temperature will usually

  1 rise, then fall   3 rise steadily
  2 fall, then rise   4 fall steadily   76 _____

77 Which characteristics of the land surface have the greatest effect on the amount of insolation the land surface absorbs?

  1 hardness and age
  2 density and hardness
  3 age and roughness
  4 roughness and color   77 _____

78 Most water vapor enters the atmosphere by the processes of

    1 conduction and convection
    2 radiation and condensation
    3 absorption and infiltration
    4 evaporation and transpiration   78 _____

79 Compared with the change in temperature of the water surface, the change in temperature of the land surface will be

    1 faster, because the land has a lower specific heat
    2 faster, because the land has a higher specific heat
    3 slower, because the land has a lower specific heat
    4 slower, because the land has a higher specific heat   79 _____

80 The direction of the wind shown in this diagram is probably due to air moving from areas of

    1 low air pressure to areas of high air pressure
    2 high air pressure to areas of low air pressure
    3 low air humidity to areas of high air humidity
    4 high air humidity to areas of low air humidity   80 _____

GROUP 6

If you choose this group, be sure to answer questions **81–85**.

    Base your answers to questions 81 through 85 on the *Earth Science Reference Tables,* the diagram below, and your knowledge of Earth science. The diagram represents a planet, *P*, in an elliptical orbit around a star located at F_1. The foci of the elliptical orbit are F_1 and F_2. Orbital locations are represented by P_1 through P_6.

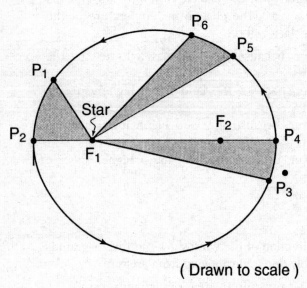

(Drawn to scale)

81 The gravitational attraction between planet *P* and the star is greatest when the planet is located at position

(1) P_1 (3) P_3

(2) P_2 (4) P_4

81 _____

82 When observed from the planet, the star would have its greatest apparent angular diameter when the planet is located at position

   (1) P_1 (3) P_3

   (2) P_2 (4) P_4 82 _____

83 What is the approximate eccentricity of planet P's orbit?

   (1) 0.52   (3) 2.11

   (2) 0.83   (4) 4.47   83 _____

84 If the shaded portions of the orbital plane are equal in area, the time period between P_1 and P_2 will be equal to the time period between

   (1) P_2 and P_3 (3) P_4 and P_5

   (2) P_3 and P_4 (4) P_6 and P_1 84 _____

85 If the mass of planet P were tripled, the gravitational force between the star and planet P would

   1 remain the same

   2 be two times greater

   3 be three times greater

   4 be nine times greater   85 _____

GROUP 7

If you choose this group, be sure to answer questions **86–90**.

Base your answers to questions 86 through 90 on the *Earth Science Reference Tables* and on the diagrams below. The diagrams represent 500-milliliter containers that are open at the top and the bottom and filled with well-sorted, loosely packed particles of uniform size. A piece of screening placed at the bottom of each container prevents the particles from falling out.

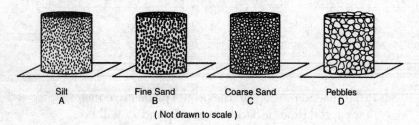

Silt   Fine Sand   Coarse Sand   Pebbles
A   B   C   D
(Not drawn to scale)

86 Container *A* is filled with particles that could have a diameter of

(1) 0.0001 cm   (3) 0.01 cm

(2) 0.001 cm   (4) 0.1 cm   86 _____

87 The sample in which container would have the greatest capillarity when placed in water?

(1) *A*   (3) *C*

(2) *B*   (4) *D*   87 _____

88 Assume that the samples in each container were taken from surface soil in different locations. Which location would produce the *least* amount of runoff during a heavy rainfall?

(1) *A*   (3) *C*
(2) *B*   (4) *D*   88 _____

89 The sample in which container would retain the most water on the particles after 500 milliliters of water is poured through the sample?

(1) *A*   (3) *C*
(2) *B*   (4) *D*   89 _____

90 Which graph best represents the rate of permeability of the samples?

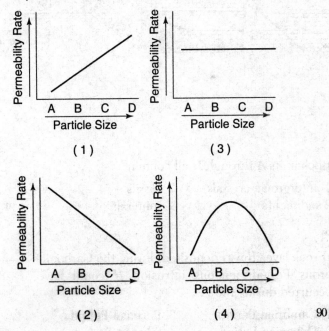

(1)   (3)

(2)   (4)   90 _____

GROUP 8

*If you choose this group, be sure to answer questions **91–95**.*

Base your answers to questions 91 through 95 on the *Earth Science Reference Tables*, the cross section below, and your knowledge of Earth science. Letters *A* through *J* represent rock units. An unconformity is shown at letter *X*. A fault is shown at letter *Y*.

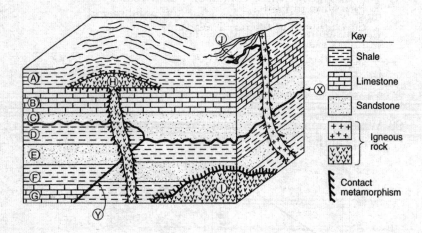

91 Rock units *A* through *H* all contain

   1 intergrown crystals   3 fossils
   2 sediments   4 minerals   91 _____

92 If rock layer *B* was deposited during the Carboniferous Period, igneous intrusion *H* could have occurred during the

   1 Cambrian Period   3 Permian Period
   2 Devonian Period   4 Silurian Period   92 _____

93 Rock *I* was formed deep underground and is composed of 70% pyroxene, 15% plagioclase, 10% olivine, and 5% hornblende. Rock *I* is classified as

1 granite   3 rhyolite
2 gabbro   4 basalt   93 ____

94 Which process occurred most recently?

1 formation of fault *Y*
2 development of unconformity *X*
3 formation of intrusion *H*
4 deposition of rock layers *D, E,* and *F*   94 ____

95 Rock *J* in the diagram represents a lava flow that has cooled rapidly at the surface of Earth. Which diagram and description best represents rock *J*?

Bands of alternating
light and dark
minerals

(1)

Glassy black rock
that breaks with a
shell-shape fracture

(3)

Easily split layers of
0.0001-cm-diameter
particles cemented
together

(2)

Interlocking
0.5-cm-diameter
crystals of
various colors

(4)   95 ____

GROUP 9

If you choose this group, be sure to answer questions **96–100**.

Base your answers to questions 96 through 100 on the *Earth Science Reference Tables,* the diagram below, and your knowledge of Earth science. The diagram shows a cross section of the bedrock where the Niagara River flows over Niagara Falls.

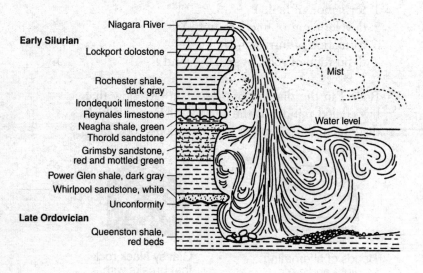

96 The Rochester shale and the Queenston shale are similar in that both

  1 contain the same index fossils
  2 have the same color
  3 are the same age
  4 contain the same size sediment

96 _____

97 What is the most probable age of the Irondequoit limestone?

  1 Silurian   3 Devonian

  2 Cambrian   4 Permian   97 _____

98 The unconformity above the Queenston shale was more likely caused by

  1 faulting   3 erosion

  2 folding   4 volcanism   98 _____

99 The sediment carried by the Niagara River will have its greatest potential energy at the level of the

  1 Lockport dolostone

  2 Rochester shale

  3 Thorold sandstone

  4 Queenston shale   99 _____

100 Which kind of rock in this formation appears to be *least* resistant to weathering?

  1 dolomite   3 limestone

  2 shale   4 sandstone   100 _____

GROUP 10

If you choose this group, be sure to answer questions **101–105**.

101 Which station model correctly shows the weather conditions of a thunderstorm with heavy rain?

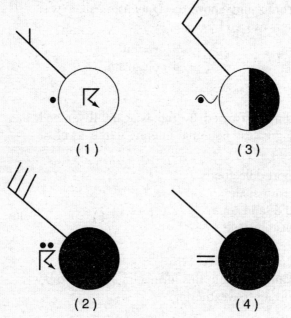

(1)   (3)

(2)   (4)   101 _____

102 Which type of radiation has the shortest wavelength?

1 radar   3 ultraviolet
2 visible light   4 infrared   102 _____

103 What is the dewpoint temperature when the dry-bulb temperature is 12°C and the wet-bulb temperature is 7°C?

(1) 1°C   (2) 6°C
(2) –5°C   (3) 4°C   103 _____

104 Which diagram most accurately represents the diameter of the Moon and the diameter of Earth?

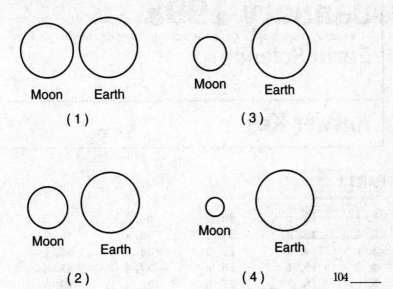

Moon   Earth

(1)

Moon   Earth

(3)

Moon   Earth

(2)

Moon   Earth

(4)   104 _____

105 The end product of the weathering of gabbro or basalt rocks is a solution of dissolved material that most likely would contain high amounts of

1 iron and magnesium
2 magnesium and potassium
3 aluminum and iron
4 aluminum and potassium   105 _____

Answers
January 1998
Earth Science

Answer Key

PART I

1. 4	12. 1	23. 2	34. 3	45. 2
2. 4	13. 2	24. 1	35. 1	46. 2
3. 1	14. 3	25. 2	36. 2	47. 1
4. 2	15. 4	26. 4	37. 2	48. 4
5. 1	16. 3	27. 3	38. 1	49. 3
6. 1	17. 2	28. 4	39. 4	50. 4
7. 3	18. 3	29. 1	40. 3	51. 1
8. 4	19. 4	30. 4	41. 1	52. 3
9. 3	20. 1	31. 2	42. 3	53. 3
10. 2	21. 3	32. 1	43. 4	54. 2
11. 4	22. 1	33. 4	44. 4	55. 1

PART II

56. 1	66. 2	76. 3	86. 2	96. 4
57. 2	67. 4	77. 4	87. 1	97. 1
58. 4	68. 3	78. 4	88. 4	98. 3
59. 1	69. 4	79. 1	89. 1	99. 1
60. 2	70. 1	80. 2	90. 1	100. 2
61. 3	71. 4	81. 2	91. 4	101. 2
62. 3	72. 1	82. 2	92. 3	102. 3
63. 4	73. 2	83. 1	93. 2	103. 1
64. 3	74. 2	84. 2	94. 3	104. 4
65. 1	75. 4	85. 3	95. 3	105. 1

Answers Explained

PART I

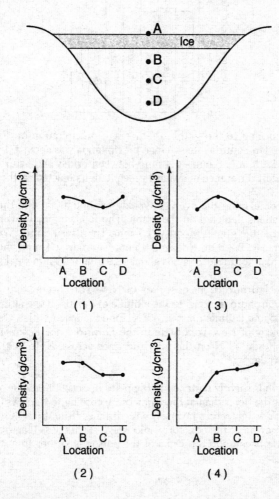

1. **4** The surface of the lake at location *A* is frozen; therefore the temperature is 0°C or colder. At location *D* the temperature is 4°C. Since water becomes denser as the temperature rises between 0°C and 4°C, the water is densest at 4°C. Graph 4, which shows the density increasing from *A* to *D*, best represents the relationship between location and density of the ice or water.

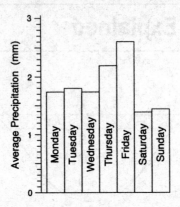

2.   **4** According to the graph, the average precipitation on Friday was about 2.6 mm. On Saturday the average precipitation was about 1.4 mm. This difference of 1.2 mm (2.6 mm − 1.4 mm) between Friday and Saturday represents the greatest difference in average precipitation depicted on the graph.

3.   **1** To determine the average weekly precipitation, sum the average daily precipitations depicted on the graph. The following daily averages may be inferred from the graph: Monday: 1.7 mm, Tuesday: 1.8 mm, Wednesday: 1.7 mm, Thursday: 2.2 mm, Friday: 2.6 mm, Saturday: 1.4 mm, and Sunday: 1.5 mm. The total for these 7 days is about 12.9 mm or 13 mm/week.

4.   **2** The latitude of an observer in New York State, that is, in the Northern Hemisphere, is equal to the altitude of Polaris. If the altitude of the observer is 43°, the latitude must be 43°N. Find the map entitled Generalized Bedrock Geology of New York State in the *Earth Science Reference Tables*. Locate the latitude 43°N on the map and trace across. Note that Utica lies along this latitude.

5.   **1** Earth is slightly flattened at the poles because the radius of Earth is slightly less at the poles than at the Equator. According to the laws of gravitation, a person's weight varies inversely with distance from the center of Earth. Therefore, a person at the equator would weigh slightly less than he would at the North Pole because at the Equator the person is farther from the center of Earth.

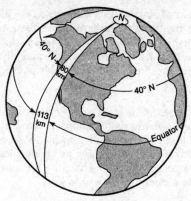

(Not drawn to scale)

6. **1** Find the city of Massena on the map entitled Generalized Bedrock Geology of New York State in the *Earth Science Reference Tables*. Note that the latitude of Massena is about 45°N. In the diagram one degree of longitude at 40°N represents a distance of 80 km. This amount decreases as you move northward. It may therefore be inferred that at 45°N the distance represented by one degree of longitude would be slightly less or about 78 km. (Note: All the other choices are greater, not less, than 80 km.)

7. **3** Find in the *Earth Science Reference Tables* the graph entitled Selected Properties of the Earth's Atmosphere. The graph at the left shows that almost all the water vapor in the atmosphere is found in the troposphere, the layer closest to Earth's surface.

8. **4** The eccentricity of an orbit is its degree of out-of-roundness; an eccentricity of zero represents a perfect circle. Find in the *Earth Science Reference Tables* the Solar System Data chart. Note in the column headed "Eccentricity of Orbit" that the eccentricity of Pluto's orbit is 0.250. The fact that this is the largest value in the column means that Pluto has the most eccentric orbit.

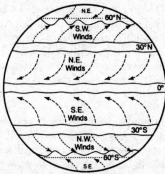

9. **3** The rotation of Earth on its axis causes the planetary winds to be deflected toward the right in the Northern Hemisphere and toward the left in the Southern Hemisphere. Note on the diagram that the arrows in the Northern Hemisphere bend toward the right and that the arrows in the Southern Hemisphere bend toward the left. This phenomenon is referred to as the Coriolis effect.

10. **2** The Sun's apparent daily path through the sky is caused by the rotation of Earth. As Earth rotates, the Sun appears to travel in an arc across the sky. In New York State the Sun typically appears to rise near the horizon and to travel in an easterly direction, reaching its highest point around noon and then setting in a westerly direction. Thus, the apparent daily path is best described as an arc that extends from east to west.

WRONG CHOICES EXPLAINED:
(1) To an observer in the Northern Hemisphere, the stars appear to travel at night in a circular path around the North Star.
(3) The Sun's path is not a straight line; it is curved because Earth's surface is curved.
(4) The apparent path of the Sun varies with the season in a cyclic manner; it is not a random motion. It changes because the inclination of Earth's axis changes relative to the plane of Earth's orbit around the Sun. For example, in the summer, when the axis is pointed toward the Sun, the Sun's path is highest in the sky.

11. **4** The Moon's cycle of phases can be observed from Earth because the Moon revolves around Earth. The Moon completes one cycle of revolution around Earth in a little less than a month. As the relative positions of Earth, the Sun, and the Moon change during this cycle, varying amounts of the Moon's surface appear illuminated to an observer on Earth.

WRONG CHOICES EXPLAINED:
(1) The Moon *is* smaller than Earth, but this fact has no effect on the visibility from Earth of the Moon's cycle of phases.
(2) Any tilt in the Moon's axis would not affect the phases of the moon because it would not change the amount of the Moon's surface that was illuminated; also, it would not affect the ability to observe these phases.
(3) The Moon makes one complete rotation on its axis at the same time that it is completing one revolution. As a result the same side of the Moon is always facing the Earth. The rotation of the Moon does not affect the cycle of phase.

12. **1** In the case of transverse waves, the motion within the waves is perpendicular to the direction of motion of the wave itself. The movement of water in ocean waves illustrates this type of motion. All forms of electromagnetic energy, including visible light, have transverse wave properties. By contrast, sound waves are compressional waves in which the motion is back and forth.

WRONG CHOICES EXPLAINED:

(2), (3) The temperature of electromagnetic waves varies with different wavelengths and depends on the temperature of the source. The surface temperature of the Sun is higher than the surface temperature of Earth. As a result electromagnetic waves radiated from the surface of the Sun have shorter wavelengths than waves emitted by Earth.

(4) The half-life of a radioactive substance is the amount of time it takes for half the substance to decay. Half-life is unrelated to electromagnetic energy.

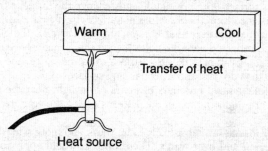

13. **2** Conduction is the primary method of heat transfer in solids such as an iron bar. The molecules at one end of the bar speed up when the bar is heated and strike nearby molecules, causing them to accelerate in turn. This process of heat transfer by conduction continues throughout the length of the bar.

WRONG CHOICES EXPLAINED:

(1) Convection is the primary method of heat transfer within fluids. When a liquid or gas is heated, it expands, rises, and is then replaced by cooler and denser material. The result is a pattern of circular energy transfer referred to as convection.

(3) Water being soaked up by a sponge is an example of absorption. Absorption occurs because there is an attraction between the molecules of different composition.

(4) Advection involves the horizontal movement of air.

14. **3** The fact that dark-colored surfaces are good absorbers of electromagnetic energy explains why a dark-colored surface gets hotter during the summer than a light-colored surface. Similarly, an object that is a good absorber of electromagnetic energy is most likely also a good radiator of energy.

WRONG CHOICES EXPLAINED:

(1) A convector is a material that transfers energy by expanding and rising when it is heated. Convection is the primary method of heat transfer in fluids.

(2) A reflector is a material whose surface causes energy to "bounce off." Light-colored surfaces are good reflectors.

(4) A refractor is a material, such as water, that causes light to bend as it passes through. Refraction is illustrated by the bent appearance of a stick in a pool of water.

15. **4** The higher the Sun is in the sky, the more concentrated is the insolation. For this reason the hottest temperatures occur near the Equator, which receives the most nearly perpendicular insolation. Near the poles, on the other hand, the Sun's rays strike the surface at low angles. Solar energy is spread out over a large area, resulting in the coldest temperatures.

WRONG CHOICES EXPLAINED:

(1) The Sun is so far from Earth that traveling from one region to another along Earth's surface does not significantly change the distance from the Sun. Therefore distance from the Sun does not affect climate on Earth.

(2) The more insolation that is reflected, the *cooler* is the climate. More solar energy is reflected near the poles than at the Equator because of the low angle at which insolation strikes the surface. Also, snow, ice, and water reflect more energy than do dark and irregular land surfaces.

(3) The total annual numbers of hours of daylight near the poles and at the Equator are similar although the patterns during the year are different.

16. **3** Air masses are most likely to be warm and moist when they form in a tropical region and over water. These criteria would apply to an air mass over the Gulf of Mexico.

WRONG CHOICES EXPLAINED:

(1) An air mass over Central Canada would be cold and dry because it formed in a northerly region and over land.

(2) An air mass over Central Mexico would be warm and dry because it formed in a tropical region and over land.

(4) An air mass over the North Pacific Ocean would be cold and moist because it formed in a northerly region and over the ocean.

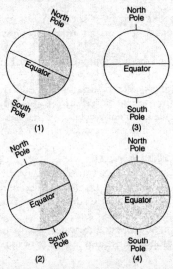

17. **2** Diagram (2) depicts Earth's tilt on its axis as it would appear near June 21, when the Sun's apparent path in the sky is longest in New York State. The longest path produces the longest duration of insolation.

18. **3** If Earth's tilt were to increase from 23½° to 33½°, in the Northern Hemisphere the angle of insolation during the winter months would be smaller resulting in colder temperatures. Also, during the summer months, the angle of insolation would be greater, producing warmer temperatures.

WRONG CHOICES EXPLAINED:
(1) The lengths of day and night at the Equator would remain the same.
(2) The difference between summer and winter temperatures in New York State would be *greater* because summer temperatures would be warmer and winter temperatures would be colder.
(4) The total amount of solar radiation received by Earth would remain the same. What would change is the pattern at which the radiation was distributed.

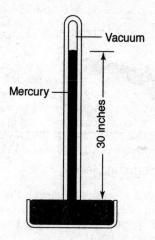

19. **4** The diagram illustrates a mercury barometer, which is designed to measure air pressure. The height to which the mercury rises in the column depends on the air pressure. The greater the air pressure, the higher the mercury will rise.

20. **1** When the height of the mercury is much greater than 30 inches, the air is likely to be cold and dry and the skies clear. Colder air exerts more pressure than warmer air because colder air is denser. When air is cooled, it contracts. Since the volume is less but the mass remains the same, the density of the air increases. When the air is cold and dry, as indicated by an air pressure exceeding 30 inches, the sky will be clear. As moisture is added to the air, the air pressure normally decreases. Clouds form when the moisture content of the air is high.

WRONG CHOICES EXPLAINED:

(2) Warm, moist air would most likely produce lower air pressure.

(3) Wind direction is not directly related to air pressure.

(4) A violent storm suggests highly moist air, which would be indicated by lower air pressure.

21. **3** The transfer of energy to the oceans by prevailing winds produces ocean currents. This is an example of the transfer of energy from the atmosphere (i.e., the prevailing winds) to the hydrosphere (i.e., the ocean currents.)

22. **1** Find the maps entitled Generalized Bedrock Geology of New York State in the *Earth Science Reference Tables*, and locate Utica, New York. The prevailing wind direction in New York State is from west to east. Accordingly, to find the city where the weather conditions were most likely the same a few hours earlier, travel due west from Utica to Syracuse.

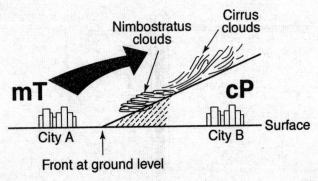

23. **2** In the diagram the warm air mass on the left side is pushing the cold air mass on the right side as the warm air moves up and over the cold air. As a result the front at ground level is moving toward city *B*. The type of weather front shown is a warm front.

24. **1** Warm air is cooled as it rises and expands along the frontal surface. The excess moisture in the air begins to condense and form clouds. If there is too much moisture in the air as it condenses, the excess moisture falls as precipitation.

25. **2** Find the map entitled Surface Ocean Currents in the *Earth Science Reference Tables*. Note the presence of the Florida Current along the eastern coast of the United States. According to the arrows on the map, this current flows northward, bringing warmer water to coastal regions of the eastern United States.

26. **4** Typically the highest rainfall in New York State occurs in the spring and causes high stream discharge. During the summer the amount of precipitation is less. Also, by late summer the higher temperatures have caused more evaporation which also removes water from the streams. As a result of these two factors—less precipitation and more evaporation—stream discharge is typically lowest in late summer. At that time the local water budget most likely shows a deficit because more water is returning to the atmosphere by evaporation and transpiration than is being replaced by precipitation.

WRONG CHOICES EXPLAINED:

(1), (2) Potential evapotranspiration is typically *greater* in late summer than in spring because temperatures are highest and the greatest amount of moisture returns to the atmosphere through transpiration by plants.

(3) The local water budget is more likely to show a surplus in the spring, when precipitation is at its highest and evaporation and transpiration are relatively low.

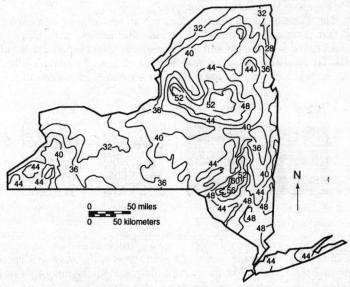

27. **3** Find the map entitled Generalized Bedrock Geology of New York State in the *Earth Science Reference Tables*, and note the locations of Watertown and Plattsburgh. Now find the Generalized Landscape Regions of New York State map in the *Earth Science Reference Tables*. It is obvious that Watertown and Plattsburgh lie on opposite sides of the Adirondack Mountains. Since the prevailing wind direction in New York State is from west to east, you would expect more precipitation to occur at Watertown because that city is located on the windward side. As the air rises over the mountains (i.e., the elevation increases), moisture condenses, causing the air at Plattsburgh to be drier with less precipitation.

28. **4** There is more space between the particles in a well-sorted soil. When soil is poorly sorted, some of the smaller particles fill the spaces between the larger particles, resulting in less total space between all particles. A soil with particles that are well sorted and loosely packed has the greatest porosity.

29. **1** Climate has the greatest influence on the development of soil. Rock material weathers faster when the temperature is warmer and much moisture is present. Therefore, soil forms at the fastest rate in a warm, moist climate, and at the slowest rate in a cold, dry climate.

WRONG CHOICES EXPLAINED:

(2) Longitude does not directly affect the rate of soil formation. Soil may form at markedly different rates at two locations that have the same longitude if the climates are different.

(3) Actually, soil formation occurs faster if the sediment is angular because more surface area is available for weathering.

(4) The slope of the landscape does not affect soil development to the extent that climate does, although slope can affect erosion rates. During erosion, material is carried away and new material becomes exposed to weathering. Hence, local soil formation may be slowed down by heavy erosion. In general, however, climate has the most influence on soil formation.

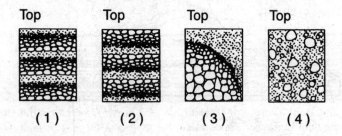

30. **4** A glacier can carry a wider range of particle sizes than can any other agent of erosion. When the water in a glacier melts, the material deposited directly will tend to look like the pattern illustrated in diagram (4). This diagram shows a wide range of particle sizes that are unsorted.

WRONG CHOICES EXPLAINED:

(1), (2), (3) All these diagrams show a range of particle sizes that are well sorted. Sorting of particles is a common characteristic of erosion by running water. For example, as the water in a river slows down, the largest particles are deposited first. Then, as the velocity continues to decrease, smaller and smaller particles are deposited. The result is a series of layers of deposits containing particles of different sizes.

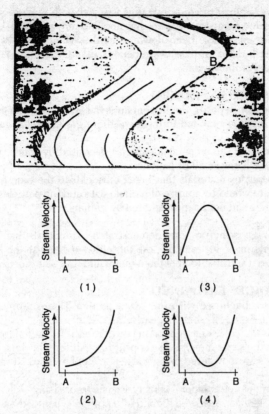

31. **2** The velocity of the water in a stream increases along the outside of a curve and slows down along the inside of a curve. Therefore, the stream velocity would be slower at point *A* in the map and faster at point *B*. Graph (2) shows the velocity of the water increasing across the stream from *A* and *B*.

(Not actual size)

32. **1** The cobbles are rounded, suggesting that they were weathered while being carried by a stream. Also, since cobbles are relatively large particles, the stream must have been flowing rapidly. When the velocity of the stream decreased, these particles were deposited and subsequently collected from the channel of the stream.

WRONG CHOICES EXPLAINED:

(2) Volcanic ash deposits would contain smaller and more angular particles. Particles become rounded through weathering while being carried by a stream.

(3) A desert sand dune contains particles deposited by wind. Winds normally can carry only the smallest particles.

(4) Gravity carries materials that break off a cliff to the base of the cliff. Most likely, there would be a range of particle sizes and the particles would be angular since they had not been weathered by running water.

33. **4** Cleavage is the property demonstrated by minerals that break into layers. Muscovite mica, for example, exhibits cleavage because the layering of the atoms makes it possible to peel the mineral into thin, flat sheets.

WRONG CHOICES EXPLAINED:

(1) Luster is concerned with the way a mineral shines. Some common types of luster are termed "shiny," "pearly," and "dull."

(2) Streak is the color of a mineral in powder form. In natural form, different varieties of the same mineral may appear to be different colors. When the mineral samples are powdered using a streak plate, however, all samples are the same color.

(3) The hardness of a mineral is its resistance to breaking. Talc, for example, is very soft and can easily be ground into powder. Diamonds, at the other extreme, are very hard and cannot easily be scratched.

34. **3** Fossil limestone is a sedimentary rock that may form as a result of biologic processes. The shells of animals settle to the bottom of the ocean and become buried by overlying sediment. Pressure from these sediments and cementing by minerals dissolved in the water can eventually convert the shells into the rock limestone. The characteristics of breccia can be found in the scheme for sedimentary rock identification in the *Earth Science Reference Tables*.

WRONG CHOICES EXPLAINED:

(1) Shale forms when particles of clay become cemented together under pressure. Shale is not formed as a result of biologic processes.

(2) Siltstone forms in a manner similar to shale except that the particles are larger.

(4) Breccia forms in a manner similar to shale and siltstone. However, breccia contains a wide range of particle sizes.

35. **1** Rock salt is most likely to be monomineralic; it normally contains only the mineral halite. The other rock samples all contain two or more minerals. The mineral compositions of these other rocks can be found in the schemes for rock identification in the *Earth Science Reference Tables*.

36. **2** The sediments illustrated in the diagram range widely in size, and they are rounded. The characteristics of the various types of sedimentary rocks are given in the Scheme for Sedimentary Rock Identification in the *Earth Science Reference Tables*. The chart indicates that in conglomerate the particle size is mixed and the fragments are rounded.

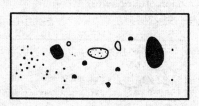

WRONG CHOICES EXPLAINED:

(1) Chemical limestone is a sedimentary rock that forms by evaporation and therefore does not contain rounded fragments.

(3) Granite is an igneous, not a sedimentary, rock.

(4) Sandstone is a clastic sedimentary rock (i.e., it is made up of fragments), but the fragments are the size of sand grains. It contains no larger particles like those in the diagram.

37. **2** Metamorphic rocks form by the recrystallization of unmelted rock material under conditions of high temperature and pressure. Find the Scheme for Metamorphic Rock Identification in the *Earth Science Reference Tables*. Note that gneiss appears on the chart and is therefore a metamorphic rock formed under the conditions stated.

WRONG CHOICES EXPLAINED:

(1), (3), and (4) Granite is an igneous rock, and rock gypsum and bituminous coal are sedimentary rocks. These types of rocks are not formed under the condition stated in the question.

38. **1** Find the map entitled Tectonic Plates in the *Earth Science Reference Tables*. Note the location of the African and Antarctic plates. The arrows between these two plates point outward from a line, indicating that seafloor spreading is occurring at the boundary between these plates.

WRONG CHOICES EXPLAINED:

(2) Find the Nazca and South American plates on the map. Since they are on opposite sides of South America, no seafloor spreading is occurring between them.

(3), (4) No arrows on the map indicate that seafloor spreading is occurring between the China and Philippine plates or between the Australian and Eurasian plates.

39. **4** Find the Inferred Properties of Earth's Interior diagram in the *Earth Science Reference Tables*. On the pressure graph find the depth where the pressure is 2.5 million atmospheres. According to the diagram, this location lies within the outer core.

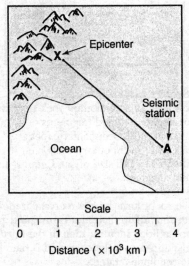

40. **3** Use the scale in the question to determine that the distance between the earthquake epicenter and the seismic station is about 3.5×10^3 km. Then find the graph entitled Earthquake P-wave and S-wave Travel Time in the *Earth Science Reference Tables*, and locate an epicenter distance of 3.5×10^3 km on the horizontal axis. Trace up to the *P*-wave curve and then across to the vertical axis. The travel time is about 6 min and 20 sec.

41. **1** *S*-waves travel more slowly than *P*-waves. Since the two waves leave the epicenter at the same time, the initial *S*-wave will reach the seismic station later than the *P*-wave.

42. **3** Find the Inferred Properties of Earth's Interior diagram in the *Earth Science Reference Tables*. Note that a single line represents the crust, whereas the mantle appears as a much thicker zone. According to the dia-

gram, the density of the crust varies between 2.7 and 3.0 g/cm³, while the density of the mantle varies between 3.4 and 4.7 g/cm³. The crust is therefore thinner and less dense than the mantle.

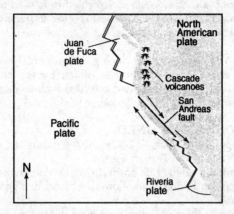

43. **4** The presence of the plates and the fault line along the west coast of North America suggests that this is a zone of frequent crustal movement. The arrows along the fault line indicate the directions of movement caused by the shifting of the plates.

WRONG CHOICES EXPLAINED:

(1) The San Andreas fault lies along the eastern boundary of the Pacific plate.

(2) The presence of the San Andreas fault suggests that the Pacific plate is still moving.

(3) No evidence is presented regarding the relative ages of the continental rocks.

44. **4** Faults that have broken the rock must have occurred after the sandstone rock layer formed and therefore are usually the youngest feature.

WRONG CHOICES EXPLAINED:

(1), (2) Sand grains in a sandstone rock layer and the cement binding the rock layer together must have existed before the rock layer formed and are therefore older than the rock layer.

(3) If fossils are present in the rock, they must have existed before the rock formed.

45. **2** Fossils are found almost exclusively in sedimentary rocks. Occasionally fossils are found in metamorphic rocks that formed from sedimentary rocks. Since shale is a sedimentary rock, the Cambrian Shale bedrock would be most likely to contain fossils.

WRONG CHOICES EXPLAINED:

(1), (3) Granite and basalt are igneous rocks. Since they are formed from molten material, they cannot contain fossils.

(4) Quartzite is a metamorphic rock formed from sandstone, but few life-forms were preserved from the Middle-Proterozoic period. Therefore, this bedrock is unlikely to contain fossils.

46. **2** Find the chart entitled Geologic History of New York State at a Glance in the *Earth Science Reference Tables*. Locate the column headed "Important Fossils of New York," and note that both ammonoids and Naples trees existed during the Devonian Period.

WRONG CHOICES EXPLAINED:

(1) According to the chart, trilobites existed during the Cambrian Period and mastodonts during the Tertiary Period.

(3) The column headed "Life on Earth" states that armored fish appeared during the Devonian Period, while flowering plants first appeared during the Cretaceous Period.

(4) According to the chart, dinosaurs lived throughout the Mesozoic Era, while the earliest humans did not appear until the Cenozoic Era.

50,000 B.C.: Gak invents the first and last
silent mammoth whistle.

47. **1** Find the chart entitled Geologic History of New York State at a Glance in the *Earth Science Reference Tables*. The time scale at the left shows that 50,000 B.C. would fall within the most recent geologic time period. The column headed "Life on Earth" indicates that the dominant life-forms during this period were all mammals.

48. **4** Find the map entitled Generalized Bedrock Geology of New York State in the *Earth Science Reference Tables*. The 45° N latitude line lies along the northern border of New York State. The code of symbols below the map indicates that the bedrock along this line ranges in age from Ordovician on the left to Cambrian on the right.

49. **3** Find the Physical Constants chart in the *Earth Science Reference Tables*. Note in the section headed "Radioactive Decay Data" that the half-life of carbon-14 is 5.7×10^3 years. This means that in 5700 years half of the carbon-14 in the original sample will have decayed, and therefore half or 50% is still present in the fossil.

50. **4** Scientists have estimated that the age of Earth is about 4.6×10^9 (4.6 billion) years. To obtain this value, find the chart entitled Geologic History of New York State at a Glance in the *Earth Science Reference Tables*. On the left side, near the bottom, note "Estimated Time of Origin of Earth and Solar System" at 4600 million (4.6 billion or 4.6×10^9) years.

51. **1** Glaciers form U-shaped valleys when they erode a landscape. The valley tends to develop this shape because while the ice moves over the landscape it carves out the sides of the valley as well as its base. Thus, as the ice moves over the land, it also polishes and creates grooves in the bedrock.

WRONG CHOICES EXPLAINED:

(2) As an agent of erosion, wind is limited to localized effects and does not have the power to carve out valleys.

(3) Wave action can produce erosion along coastlines; it breaks off material from cliffs. But because of the locations where wave action occurs, it does not produce valleys.

(4) Running water tends to produce V-shaped valleys because most of the erosion occurs at the base of the valley where the water is flowing. Over time, water flowing down the sides may widen a valley, but the valley will still retain its V-shape.

52. **3** Surface features, such as level plains, mountains, and plateaus, are used to distinguish landscape regions. Another distinguishing feature is bedrock structure, which affects how a landscape develops. For example, plateaus form when horizontal bedrock is uplifted and erosion carves out valleys.

WRONG CHOICES EXPLAINED:

(1) Bedrock fossils and depositional patterns are features found in some landscapes, but are too localized to be used to define a landscape region.

(2) Rainfall and temperature changes help to alter a local landscape, but they do not produce the unique effects that would make them useful for classifying landscapes.

(4) The boundaries of the drainage basins of major rivers are characteristics of certain landscapes but are not useful in classifying them.

53. **3** Find Albany on the map entitled Generalized Bedrock Geology of New York State in the *Earth Science Reference Tables*. Now find the corresponding area on the Generalized Landscape Regions of New York State map in the *Earth Science Reference Tables*. Note that Albany lies within the region designated as the Hudson-Mohawk Lowlands.

54. **2** Find Binghamton on the map entitled Generalized Bedrock Geology of New York State in the *Earth Science Reference Tables*. The code at the bottom of the map indicates that the bedrock is from the Devonian Period and consists of limestones, shales, sandstones, and conglomerates. The code also tells you that these rocks are dominantly of sedimentary origin.

55. **1** Find the map entitled Generalized Bedrock Geology Map of New York State in the *Earth Science Reference Tables*. Note that the bedrock in the southeastern portion of the state around New York City is composed extensively of rock that the code at the bottom of the map identifies as metamorphic. According to the Generalized Landscape Regions of New York State map in the *Earth Science Reference Tables*, this region is the Manhattan Prong.

PART II

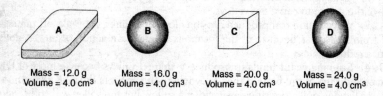

GROUP 1

56. **1** The simplest way to find the volume of each sample is to immerse it in a graduated cylinder containing water. The difference in the water level before and after the samples are added is the volume of the sample.

57. **2** For a uniform sample of material, the mass and volume are proportional; if the volume is doubled, the mass doubles. Increasing the mass of a sample of mineral A from 12.0 g to 48.0 g increases the mass by a factor of 4 (48.0 / 12.0 = 4.0). The volume is then also increased by a factor of 4 to 16.0 cm³ (4.0 cm³ × 4 = 16.0 cm³).

58. **4** Find in the Equations and Proportions chart of the *Earth Science Reference Tables* the equation for percent deviation from accepted value and substitute the given values:

$$\text{deviation (\%)} = \frac{\text{difference from accepted value}}{\text{accepted value}} \times 100$$

$$= \frac{17.5\,\text{g} - 16.0\,\text{g}}{16.0\,\text{g}} \times 100 = \frac{1.5\,\text{g}}{16.0\,\text{g}} \times 100 = 9.4\%$$

59. **1** The more nearly round the sample, the faster it will settle in a quiet body of water. Thus, sample *B* would settle fastest because of its round shape. Sample *A*, because of its flat shape, would most likely settle at the slowest rate.

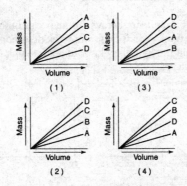

60. **2** Note that all the samples have the same volume (4.0 cm³) but the masses range upward from 12.0g from sample *A* to 24g for sample *D*. Therefore, graph (2) best represents the density of each sample because the masses of the samples are shown to follow the correct order: *A* < *B* < *C* < *D*.

GROUP 2

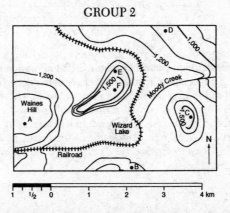

61. **3** Note on the topographic map the 1,000-m and 1,200-m contour lines in the upper right corner. There is one contour line between these two lines. This line would have a value of 1,100-m, making the contour interval 100 m.

62. **3** Note that the contour lines point toward the southwest as they cross Moody Creek. Contour lines point upstream as they cross the creek because here the elevation is the same as on the surrounding banks. If the upstream direction is southwesterly, then the creek must be flowing toward the northeast.

63. **4** From the contour lines on the map, it may be inferred that the lowest elevations occur in the northeastern part of the map, where point D is located. Its elevation of between 1,000 m and 1,100 m makes point D the lowest of the locations designated on the map.

64. **3** The scale that appears below the map can be used to estimate the length of the railroad track even though the track does not lie in a straight line. The scale indicates that the length is closer to 8 km than it is to 15, 12, or 4 km.

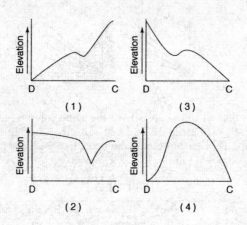

65. **1** The elevation at point D is about 1,050 m, and at point C it is more than 1,500 km. The values of the contour lines between points D and C indicate that the elevation is generally rising from D to C. This profile is illustrated in diagram (1).

GROUP 3

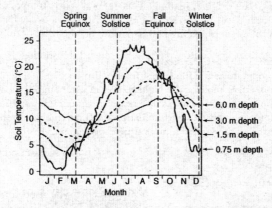

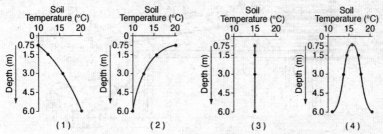

66. **2** Note the approximate location of June 21 (the summer solstice) along the horizontal axis of the graph. At this time the soil temperature is highest at the 0.75-m depth and lowest at the 6.0-m depth. Therefore, as the depth below the surface is decreasing (i.e., in moving closer to the surface) the soil temperature is rising. This pattern is illustrated in graph (2).

67. **4** According to the graph, at a depth of 0.75 m the maximum soil temperature occurred during July and August. At a depth of 1.5 m the maximum soil temperature occurred during August. At a depth of 3.0 m the maximum temperature occurred during September. Therefore, the maximum temperature at depths of less than 5 m occurred between July and September, or between the summer solstice (about June 21) and the fall equinox (about September 21).

68. **3** Note the line on the graph that represents the 3-m depth. During the month of November this line rises higher than any other line. Therefore, on November 1, the temperature at this depth was higher than the temperature at any of the other three depths.

69. **4** Each line on the graph shows a pattern of decreasing soil temperature, then increasing soil temperature, and finally decreasing soil temperature
with a return to the original reading. This general yearly temperature pattern
of change at each depth in the soil is best described as a cyclic change, that is
a change that follows a repeating pattern over a fixed period of time.

70. **1** Note that, over a period of 1 year, the greatest difference between
the highest temperature and the lowest temperature occurs at a depth of 0.75 m
(i.e., the 0.75-m curve reaches the highest level in winter and the lowest level in
summer). Conversely, the smallest difference in annual temperatures occurs at
a depth of 6.0 m (i.e., this curve is the flattest). Overall, the group shows that, as
the depth increases, the annual temperature range decreases.

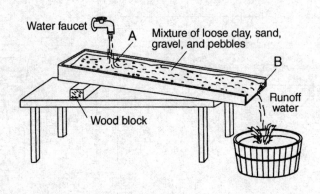

GROUP 4

71. **4** Find in the Equations and Proportions chart of the *Earth Science
Reference Tables* the equation for gradient:

$$\text{gradient} = \frac{\text{change in field value}}{\text{change in distance}}$$

In this case the change in field value is represented by the difference in elevation (in centimeters) between points *A* and *B*. The change in distance is represented by the distance (in meters) between points *A* and *B*. Equation (4)
should be used to determine the gradient of the stream.

72. **1** The smaller the particles, the more easily they are transported by
running water. Of the particles listed, clay is the smallest. The size of each of
these particle types can be found in the Scheme for Sedimentary Rock
Identification in the *Earth Science Reference Tables*.

73. **2** The greater the volume of water flowing in a stream, the greater will be the velocity of the water. After periods of heavy rainfall, it may be noticed that the velocity of the water flowing in rivers and streams has greatly increased because of the larger volume of water. Similarly, when stream volume increases after the faucet is opened, stream velocity will increase.

74. **2** The water flowing in the stream contains kinetic energy or energy of motion. As the water flows over sediments in the stream channel, some of this energy is transferred from the water to the sediments. As a result, smaller sediments are carried along by the water and larger sediments are pushed along the bed of the stream.

75. **4** Materials that are soluble and dissolve in the water are carried in solution by the stream. Also, if particles are small enough, the stream may be able to carry them in suspension. Finally, streams can often move large particles by causing them to roll along the bed of the stream. Thus, streams transport sediments in solution, in suspension, and by rolling.

GROUP 5

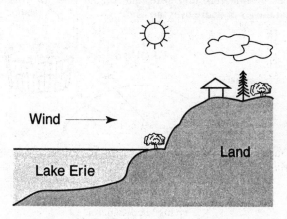

76. **3** Between 9 a.m. and noon, the sun is rising in the sky. The increasing angle of insolation will normally cause the land surface temperature to rise. For a short time after noon, say until 1 p.m., the surface temperature of the land may continue to rise because the air temperature is still warmer than the land temperature.

77. **4** When the surface of the land is rough, more surface area is exposed to insolation, thereby normally increasing the amount of insolation that is absorbed. In addition, dark-colored surfaces absorb energy more readily than light-colored surfaces, a fact that helps to explain why a blacktop driveway

gets hotter on a summer day than nearby concrete pavement. These two characteristics—roughness or color—have the greatest effect on the amount of insolation the land surface absorbs.

78. **4** Water enters the atmosphere when it evaporates from the surfaces of bodies of water such as oceans, rivers, and lakes. Additional water is returned to the atmosphere when plants transpire.

79. **1** Find the Physical Constants chart in the *Earth Science Reference Tables*. Note in the section entitled "Specific Heats of Common Materials" that the specific heat of liquid water is 1.0 cal/g•C°. This is the highest specific heat of any common Earth material. The specific heat of a material is the amount of heat necessary to raise the temperature of 1 g of that material 1 C°. Compared with the change in temperature of the water surface, the change in temperature of the land surface will be faster because the land has a lower specific heat.

80. **2** Air moves from areas of high air pressure to areas of low air pressure. The arrow in the diagram shows the wind blowing from the lake to the land. It may be inferred, therefore, that the air pressure over the lake is greater than the air pressure over the land.

GROUP 6

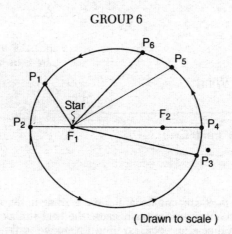

(Drawn to scale)

81. **2** Find in the Equations and Proportions chart of the *Earth Science Reference Tables* the equation for the universal law of gravitation:

$$\text{force} = \frac{\text{mass}_1 \times \text{mass}_2}{(\text{distance between their centers})^2}$$

It states that the force of gravitational attraction between two objects is inversely proportional to the square of the distance between them. The distance between planet P and the star is least when the planet is at position P_2. Therefore, the force of gravitational attraction is greatest when the planet is located at this position. The masses of the star and planet remain the same at all positions.

82. **2** The apparent angular diameter is a relative measure of how a large object appears to an observer. As the object moves closer to the observer, it appears bigger (i.e., its angular diameter becomes greater). Therefore, when observed from the planet, the apparent angular diameter of the star would be greatest at position P_2, where the planet and the star are closest together.

83. **1** Find in the Equations and Proportions chart of the *Earth Science Reference Tables* the equation for the eccentricity of an ellipse:

$$\text{eccentricity} = \frac{\text{distance between foci}}{\text{length of major axis}}$$

Now measure the distance between foci F_1 and F_2 (distance between foci) and the distance between points P_2 and P_4 (length of major axis.) To make these measurements, use a ruler or the measuring scale in the *Earth Science Reference Tables*. Dividing these two values should produce a value of about 0.52.

84. **2** Equal areas carve out equal periods of time. As a result the time period between P_1 and P_2 is equal to the time period between P_3 and P_4 because these periods represent equal areas.

85. **3** Find in the Equations and Proportions chart of the *Earth Science Reference Tables* the equation for the universal law of gravitation:

$$\text{force} = \frac{\text{mass}_1 \times \text{mass}_2}{(\text{distance between their centers})^2}$$

It states that the force of gravitational attraction between two objects is directly proportional to the masses of the two objects. Therefore, if the mass of planet P were tripled, the gravitational force between planet P and the star would be three times greater. This inference assumes that the distance between the planet and the star remains the same.

GROUP 7

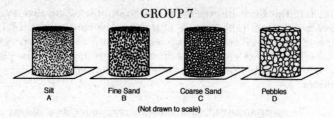

Silt
A

Fine Sand
B

Coarse Sand
C

Pebbles
D

(Not drawn to scale)

86. **2** Find the Scheme for Sedimentary Rock Identification in the *Earth Science Reference Tables*. Locate silt in the column headed "Grain Size," and note that silt contains particles ranging in size from 0.0004 to 0.006 cm. Particles with a diameter of 0.001 cm fall within this range.

87. **1** Capillarity is the process in which water rises because of attraction to another material. An example occurs when a strip of paper is dipped into a beaker of water and the strip becomes moist above the water level in the beaker. Capillarity increases as the amount of surface area increases. Therefore, the sample in container *A* would have the greatest capillarity when placed in water because it contains the smallest particles. These particles have the greatest amount of surface area, resulting in the most attraction between the water and the particles.

88. **4** The pebbles in container *D* provide the most open space because of their large size and irregular shape. As a result they can retain the most water. The more water that is retained, the less water will run off during a heavy rainfall.

89. **1** After 500 milliliters of water is poured through a sample, some water will be retained on the surface of the particles because of the attraction between the water and the particles. The silt particles in container *A* contain the most exposed surface area because of their small size. Therefore, these particles will retain the most water.

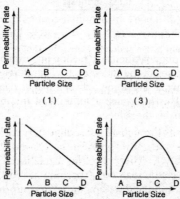

90. **1** Permeability is the rate at which water filters through a material. Permeability increases as particle size increases. Therefore, permeability will be fastest through the pebbles (*D*) because they have the largest particles, and slowest through the silt (*A*) because it has the smallest particles. Graph (1) shows permeability rate increasing as particle size increases.

GROUP 8

91. **4** All rocks are composed of minerals. The cross section in the questions contains igneous, sedimentary, and metamorphic rocks. Sediments are the original materials from which some sedimentary rocks, such as sandstone, limestone, and shale, are formed. These sedimentary rocks would not contain intergrown crystals. Fossils are not found in igneous rocks.

92. **3** Since igneous intrusion *H* cuts through rock layer *B*, it must be younger than rock layer *B*, which was deposited during the Carboniferous Period. Find the Geologic History of New York State at a Glance chart in the *Earth Science Reference Tables,* and locate the Carboniferous Period in the time line. Of the periods listed as choices, only the Permian Period occurred after the Carboniferous. Igneous intrusion *H* must have occurred during this period because it is younger than rock layer *B*.

93. **2** The key indicates that rock *I* is igneous rock. Igneous rocks that form deep underground contain large grains because the molten rock material cooled slowly. Find the Scheme for Igneous Rock Identification in the *Earth Science Reference Tables.* A rock containing 70% pyroxene, 15% plagioclase, 10% olivine, and 5% hornblende is diagramed toward the right side of the chart. Since the rock also contains coarse grains, it would be classified as gabbro.

94. **3** Intrusion *H* cuts through all the sedimentary rock layers in the cross section and is therefore younger than any of these layers. It also cuts through fault *Y* and unconformity *X*. Therefore, the formation of intrusion *H* must have happened more recently than the deposition of rock layers *D, E,* and *F,* the formation of fault *Y,* and the development of unconformity *X.*

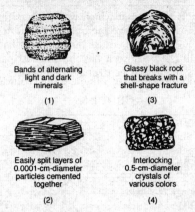

Bands of alternating
light and dark
minerals

(1)

Glassy black rock
that breaks with a
shell-shape fracture

(3)

Easily split layers of
0.0001-cm-diameter
particles cemented
together

(2)

Interlocking
0.5-cm-diameter
crystals of
various colors

(4)

95. **3** When molten rock material cools very rapidly at the surface, the resulting rock often has a glassy texture. Since cooling has occurred too quickly to allow mineral crystals to grow, these igneous rocks look and break like glass. Therefore, diagram (3), showing a glassy black rock that breaks with a shell-shape fracture, best represents rock V. None of the other rocks represents a lava flow that cooled rapidly at the surface of Earth.

GROUP 9

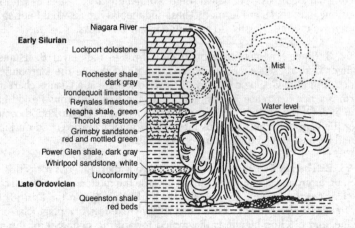

Niagara River

Early Silurian

Lockport dolostone

Mist

Rochester shale
dark gray

Irondequoit limestone

Reynales limestone

Neagha shale, green

Thorold sandstone

Grimsby sandstone
red and mottled green

Water level

Power Glen shale, dark gray

Whirlpool sandstone, white

Unconformity

Late Ordovician

Queenston shale
red beds

96. **4** Since both layers are composed of shale, they will contain the same size sediment. Since index fossils are unique to a particular time period, two layers of different ages could not contain the same index fossils. According to the diagram, the two shale layers are different in color. The Rochester shale lies above the Queenston shale, so it must be a younger rock layer.

97. **1** Find the Geologic History of New York State at a Glance chart in the *Earth Science Reference Tables*. Note that the Silurian Period directly follows the Ordovician Period. This means that the Irondequoit limestone must have formed during one of these two periods because it lies within the zone ranging from Late Ordovician to Early Silurian. The Ordovician Period is not given as a choice.

98. **3** The unconformity at the top of the Queenston shale consists of missing rock material. Since it may be inferred that the shale was level when it was first formed, erosion, in which rock material was removed, is the most likely cause of this unconformity.

99. **1** Potential energy is greatest where the elevation is highest. As the water in the Niagara River falls, some of the potential energy in the sediment carried by the river is converted to kinetic energy and the amount of potential energy continues to decrease. Therefore, the sediment will have its greatest potential energy at the level of the Lockport Dolostone.

100. **2** Note that the greatest indentation along the edges of the rock layers occurs in the shale layers. The reason is that shale is the *least* resistant to weathering of the rock types present. Along the exposed surface of the shale layer, material is more easily worn away.

GROUP 10

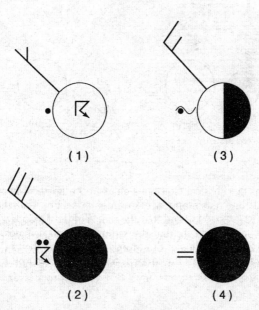

(1)   (3)

(2)   (4)

101. **2** Find the Weather Map Information chart in the *Earth Science Reference Tables*. Note in the "Present Weather Symbols" section the symbol for a thunderstorm, which appears in choice (2). Also note the two dark dots that appear above the symbol. According to the chart, these dots represent heavy rain.

102. **3** Find the Electromagnetic Spectrum diagram in the *Earth Science Reference Tables*. Note that the wavelength of ultraviolet light (10^{-7} m) is shorter than the wavelength of radar (10^{-1} m), visible light (between 10^{-7} and 10^{-6} m) or infrared (10^{-4} m) radiation.

103. **1** Find the Dewpoint Temperatures chart in the *Earth Science Reference Tables*. First determine the difference between the dry and wet bulb temperatures (i.e., $12°C - 7°C = 5°C$) given in the question. Then find a difference of $5°C$ along the horizontal axis and a dry-bulb temperature of $12°C$ along the vertical axis. The intersection of these two readings yields a dewpoint temperature of $1°C$.

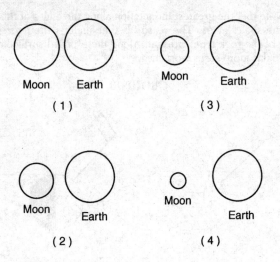

104. **4** Find the Physical Constants chart in the *Earth Science Reference Tables*. Note in the "Astronomy Measurements" section that the radius of Earth is 6370 (6.37×10^3) km and that the radius of the Moon is 1740 (1.74×10^3) km. These values indicate that the radius of Earth is almost four times (6370 km/1740 km) the radius of the Moon. Diagram (4) shows this relationship. In all the other diagrams the Moon is too large relative to Earth.

105. **1** Find the Scheme for Igneous Rock Identification in the *Earth Science Reference Tables*. Note that gabbro and basalt appear on the right side of the chart and contain mafic minerals (i.e., minerals with high iron and magnesium content). From this fact, it may be inferred that, when these rocks weather, the end product will be a solution of dissolved material that most likely would contain high amounts of iron and magnesium.

Topic	Question Numbers (total)	Wrong Answers (x)	Grade
Observation/ Measurement of the Environment; The Changing Environment	1, 56–58, 60 (5)		$\dfrac{100\,(5-x)}{5} = \%$
Measuring the Earth	4, 6, 7, 104 (4)		$\dfrac{100\,(4-x)}{4} = \%$
Earth Motions	5, 8, 10, 11, 17, 61–65, 81–85 (15)		$\dfrac{100\,(15-x)}{15} = \%$
Energy in Earth's Processes	12–14, 79, 99, 102 (6)		$\dfrac{100\,(6-x)}{6} = \%$
Insolation and the Earth's Surface	15, 18, 76, 77 (4)		$\dfrac{100\,(4-x)}{4} = \%$
Energy Exchanges in the Atmosphere	16, 19–21, 23, 24, 78, 80, 101, 103 (10)		$\dfrac{100\,(10-x)}{10} = \%$
Moisture and Energy Budgets	2, 3, 9, 22, 25–28, 66–70, 87–90, (17)		$\dfrac{100\,(17-x)}{17} = \%$
Processes of Erosion	29, 31, 32, 51, 71–75, 100 (10)		$\dfrac{100\,(10-x)}{10} = \%$
Processes of Deposition	30, 59, 86 (3)		$\dfrac{100\,(3-x)}{3} = \%$
The Formation of Rocks	33–37, 91, 93, 95, 96, 105 (10)		$\dfrac{100\,(10-x)}{10} = \%$
The Dynamic Crust	38–43 (6)		$\dfrac{100\,(6-x)}{6} = \%$
Interpreting Geologic History	44–50, 54, 92, 94, 97, 98 (12)		$\dfrac{100\,(12-x)}{12} = \%$
Landscape Development	52, 53, 55 (3)		$\dfrac{100\,(3-x)}{3} = \%$

To further pinpoint your weak areas, use the Topic Outline in the front of the book.

Examination
June 1995
Earth Science —
Program Modification Edition

PART I

Answer all 40 questions in this part.

Directions (1–40): For *each* statement or question, select the word or expression that, of those given, best completes the statement or answers the question. Record your answers in the spaces provided. [40]

1 The diagrams below represent samples of five different minerals found in the rocks of the Earth's crust.

Which physical property of minerals is represented by the flat surfaces in the diagrams?

1 magnetism
2 hardness
3 cleavage
4 crystal size

1 _____

2 The best evidence that the Earth has a spherical shape would be provided by
  1 the prevailing wind direction at many locations on the Earth's surface
  2 the change in the time of sunrise and sunset at a single location during 1 year
  3 the time the Earth takes to rotate on its axis at different times of the year
  4 photographs of the Earth taken from space   2 _____

3 A contour map shows two locations, X and Y, 5 kilometers apart. The elevation at location X is 800 meters and the elevation at location Y is 600 meters. What is the gradient between the two locations? [Refer to the *Earth Science Reference Tables*.]
  (1) 12 m/km   (3) 120 m/km
  (2) 40 m/km   (4) 160 m/km   3 _____

4 Nine rock samples were classified into three groups as shown in the table below.

Group *A*	Group *B*	Group *C*
Granite	Shale	Marble
Rhyolite	Sandstone	Schist
Gabbro	Conglomerate	Gneiss

This classification system was most likely based on the
  1 age of the minerals in the rock
  2 size of the crystals in the rock
  3 way in which the rock formed
  4 color of the rock   4 _____

Base your answers to questions 5 and 6 on the diagram below which represents the formation of a sedimentary rock. [Sediments are drawn actual size.]

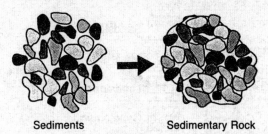

Sediments   Sedimentary Rock

5 The formation of which sedimentary rock is shown in the diagram?

  1 conglomerate   3 siltstone
  2 sandstone   4 shale   5 _____

6 Which two processes formed this rock?

  1 folding and faulting
  2 melting and solidification
  3 compaction and cementation
  4 heating and application of pressure   6 _____

7 The geologic cross section below represents an igneous intrusion into layers of rock strata.

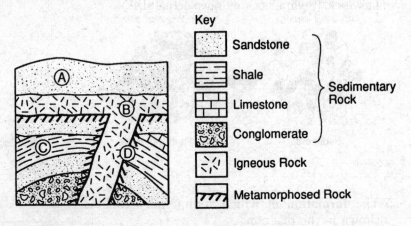

Key

Sandstone

Shale   Sedimentary Rock

Limestone

Conglomerate

Igneous Rock

Metamorphosed Rock

Which letter indicates a point where unmelted rock changed as a result of increased temperature and pressure?

1 A   3 C
2 B   4 D   7 _____

8 Rock X and rock Y are igneous rocks with identical mineral composition. Rock X has no visible crystals and rock Y has large, visible crystals. What can be inferred about rock Y?

1 It cooled at the Earth's surface, more slowly than rock X.
2 It cooled beneath the Earth's surface, more slowly than rock X.
3 It cooled at the Earth's surface, more quickly than rock X.
4 It cooled beneath the Earth's surface, more quickly than rock X.   8 _____

9 In which type of climate does the greatest amount of chemical weathering of rock occur?

1 cold and dry   3 warm and dry
2 cold and moist   4 warm and moist   9 _____

10 The diagram below represents a cross section of the Earth's crust in which no overturning has occurred.

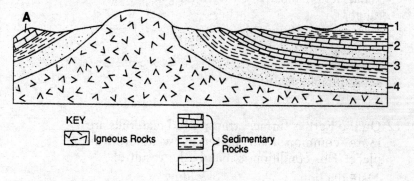

KEY

Igneous Rocks

Sedimentary Rocks

Which numbered rock layer on the right is most likely the same age as layer *A* on the left?

(1) 1   (3) 3
(2) 2   (4) 4   10 _____

11 The block diagram below represents a geologic cross section of a mountain range.

What action most likely formed this mountain range?

1 contact metamorphism
2 glacial erosion
3 volcanic eruptions
4 earthquake faulting

11 _____

12 On the Earth's surface, transported materials are more common than materials weathered in place. This condition is mainly the result of

1 subduction   3 folding
2 erosion   4 recrystallization

12 _____

13 How are dissolved materials carried in a river?

1 in solution
2 in suspension
3 by precipitation
4 by bouncing and rolling

13 _____

14 According to the *Earth Science Reference Tables*, during which period were North America, Africa, and South America closest?

1 Tertiary   3 Triassic
2 Cretaceous   4 Ordovician

14 _____

15 The cartoon below represents the time of the last dinosaurs and the earliest mammals.

According to the *Earth Science Reference Tables*, the cartoon could represent the boundary between which two units of geologic history?

1 Archean and Proterozoic
2 Precambrian and Paleozoic
3 Ordovician and Silurian
4 Mesozoic and Cenozoic

15 _____

16 The geologic cross section below shows an unconformity between gneiss and the Cambrian-age Potsdam sandstone in northern New York State.

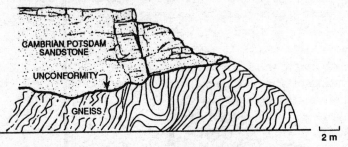

According to the *Earth Science Reference Tables*, what is the most probable age of the gneiss at this location?

1 Precambrian   3 Ordovician

2 Silurian   4 Cretaceous   16 _____

17 The diagram below represents the surface topography of a mountain valley.

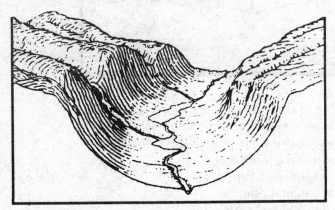

Which agent of erosion most likely created the shape of the valley shown in the diagram?

1 wind   3 ocean waves

2 glaciers   4 running water   17 _____

18 What is the approximate dewpoint temperature when the dry-bulb temperature is 2°C and the wet-bulb temperature is 0°C?

(1) –1°C   (3) –3°C
(2) –2°C   (4) –6°C   18 _____

19 According to the *Earth Science Reference Tables*, if atmospheric pressure measurements were taken at regular intervals from sea level to the stratopause, the measurements would most likely show that the pressure

1 decreases, only
2 increases, only
3 remains the same
4 decreases, then increases   19 _____

20 Which conditions are most likely to develop over a land area next to an ocean during a hot, sunny afternoon?

1 The air temperature over the land is lower than the air temperature over the ocean, and a breeze blows from the land.
2 The air temperature over the land is higher than the air temperature over the ocean, and a breeze blows from the land.
3 The air pressure over the land is higher than the air pressure over the ocean, and a breeze blows from the ocean.
4 The air pressure over the land is lower than the air pressure over the ocean, and a breeze blows from the ocean.   20 _____

Base your answers to questions 21 through 23 on the *Earth Science Reference Tables*, the diagram below, and your knowledge of Earth science. The diagram represents a cross section of a hillside in eastern New York State. The names of the rock layers and the geologic time periods in which they were formed are given. Fossil corals, brachiopods, and ammonoids are shown in the layers in which they could be found. Layer thicknesses are shown in feet.

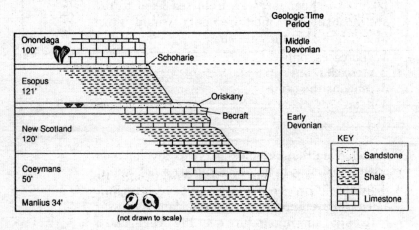

21 Which rock layer is the youngest?

1 Manlius   3 New Scotland
2 Coeymans   4 Onondaga   21 _____

22 Which rock layer appears to be most resistant to erosion?

1 Esopus   3 Coeymans
2 New Scotland   4 Manlius   22 _____

23 The fossil evidence suggests that these rock layers were most likely formed in a

1 desert   3 volcanic region
2 sea   4 tropical jungle   23 _____

24 The data table below shows average daily air temperature, windspeed, and relative humidity for 4 days at a single location.

Day	Air Temperature (°F)	Windspeed (mph)	Relative Humidity (%)
Monday	40	15	60
Tuesday	65	10	75
Wednesday	80	20	30
Thursday	85	0	95

On which day was the air closest to being saturated with water vapor?

(1) Monday   (3) Wednesday

(2) Tuesday   (4) Thursday   24 _____

25 On a July afternoon in New York State, the barometric pressure is 29.85 inches and falling. This reading most likely indicates

1 an approaching storm
2 rapidly clearing skies
3 continuing fair weather
4 gradually improving conditions   25 _____

26 Which type of surface absorbs the greatest amount of electromagnetic energy from the Sun?

1 smooth, shiny, and dark in color
2 rough, dull, and dark in color
3 smooth, shiny, and light in color
4 rough, dull, and light in color   26 _____

27 The curving path of planetary winds is caused by
   1 the Earth's revolution
   2 the Earth's rotation
   3 ocean currents
   4 weather fronts   27 _____

28 Which event usually occurs when air is cooled to
   its dewpoint temperature?
   1 freezing   3 condensation
   2 evaporation   4 transpiration   28 _____

29 A weather map of New York State shows isobars
   that are close together, indicating a steep
   pressure gradient. Which weather condition is
   most likely present?
   1 dry air   3 low temperatures
   2 strong winds   4 low visibility   29 _____

30 The diagram below shows the noontime shadow
   cast by a vertical post located in New York State.

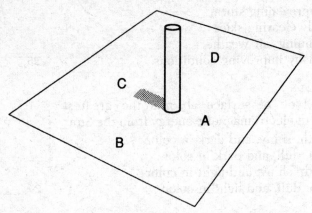

   Which letter indicates a location south of the
   post?
   (1) A   (2) B   (3) C   (4) D   30 _____

31 Which statement best explains the apparent daily motion of the stars around Polaris?

1 The Earth's orbit is an ellipse.
2 The Earth has the shape of an oblate spheroid.
3 The Earth rotates on its axis.
4 The Earth revolves around the Sun.   31 _____

32 During which month does the Sun rise north of due east in New York State?

1 February   3 October
2 July   4 January   32 _____

33 A student drew the phase of the Moon observed from one location on the Earth on each of the dates shown below.

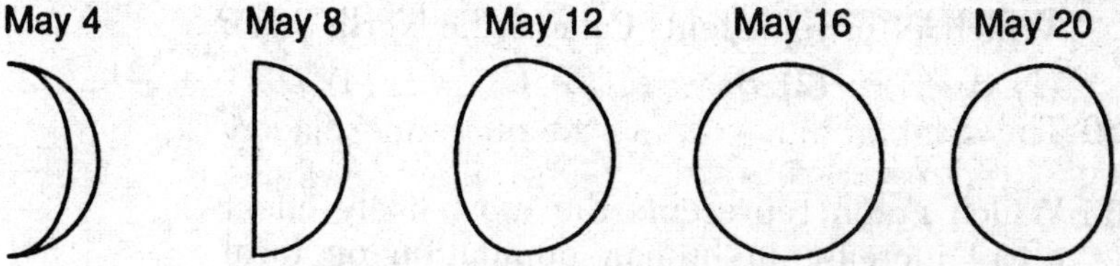

May 4   May 8   May 12   May 16   May 20

Which diagram best shows the Moon's phase on May 24?

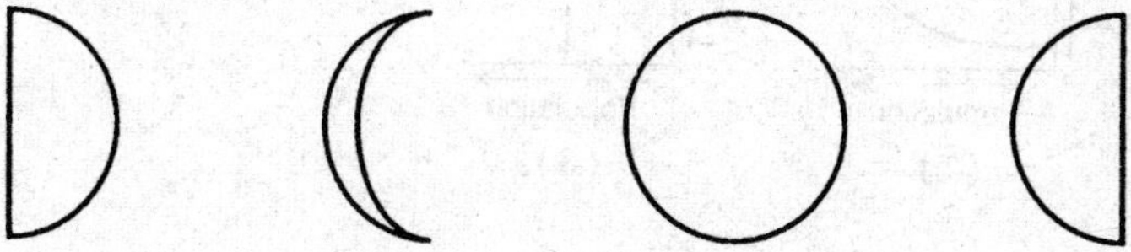

(1)   (2)   (3)   (4) 33 _____

34 The Big Dipper, part of the star constellation Ursa Major, is shown below.

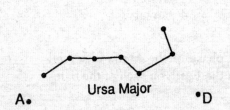

Ursa Major

Which letter represents Polaris (the North Star)?
(1) *A*   (2) *B*   (3) *C*   (4) *D*   34 _____

35 Which graph represents the most likely effect of an increase in human population on local environmental pollution?

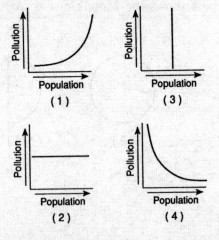

35 _____

Base your answers to questions 36 through 38 on the *Earth Science Reference Tables*, the diagram below, and your knowledge of Earth science. The diagram represents the orbits of three planets, *X*, *Y*, and *Z*, around star *A*. Star *A* is located at one focus and point *B* is the other focus. Numbers 1 through 7 represent different positions of the three planets. The arrows show the direction of revolution.

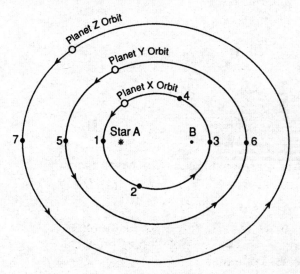

36 The orbital paths of these planets around star *A* can best be described as having

    1 the same period of rotation
    2 major axes of the same length
    3 an elliptical shape, with star *A* at one focus
    4 a circular shape, with star *A* at one focus   36 _____

37 Which number indicates the position at which a planet would have the greatest gravitational attraction to star *A*? [Assume that all three planets have the same mass.]

    (1) 7   (2) 6   (3) 3   (4) 5   37 _____

38 At which position does planet X have the greatest orbital velocity?

(1) 1   (3) 3

(2) 2   (4) 4   38 _____

39 The diagram below represents the Milky Way Galaxy.

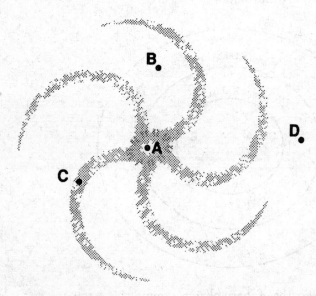

Which letter best represents the location of the Earth's solar system?

(1) *A*   (3) *C*

(2) *B*   (4) *D*   39 _____

40 In which parts of the Earth's interior would melted or partially melted material be found?

1 stiffer mantle and inner core

2 stiffer mantle and outer core

3 crust and inner core

4 asthenosphere and outer core   40 _____

PART II

This part consists of six groups, each containing ten questions. Choose any one of these six groups. Be sure that you answer all ten questions in the single group chosen. Record the answers to these questions in the spaces provided. [10]

GROUP A — **Rocks and Minerals**

If you choose this group, be sure to answer questions 41–50.

Base your answers to questions 41 and 42 on the diagram below which shows the grade of metamorphism of felsic rocks and its relationship to temperature and depth. [Refer to the Earth Science Reference Tables.]

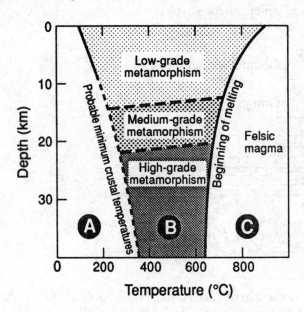

41 Which rock could be formed at a temperature of 500°C and a depth of 25 kilometers?

1 gneiss   3 granite
2 schist   4 slate   41 _____

42 Which igneous rock will most likely be formed if magma represented in region *C* is forced to the surface and cooled quickly?

1 gabbro   3 rhyolite
2 basalt   4 diorite   42 _____

Base your answers to questions 43 and 44 on the diagram below which represents a cross section of a portion of the Earth's crust. Letters *A* through *D* indicate layers of different rock types.

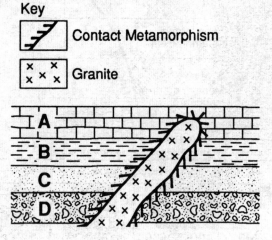

43 According to the *Earth Science Reference Tables*, which rock layer is most likely to have the greatest range of particle sizes?

(1) *A*   (2) *B*   (3) *C*   (4) *D*   43 _____

44 Which rock would most likely result from the metamorphism of the rock in layer *C*?

1 quartzite   3 slate

2 marble   4 metaconglomerate   44 _____

Base your answers to questions 45 through 47 on the *Earth Science Reference Tables*, the diagrams below, and your knowledge of Earth science. Diagram I shows the order in which silicate minerals crystallize from magma. Diagram II shows the minerals that form as the number of linked tetrahedra increases.

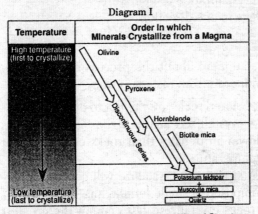

Diagram I

Temperature	Order in which Minerals Crystallize from a Magma
High temperature (first to crystallize)	Olivine
	Pyroxene
	Hornblende
	Biotite mica
	Potassium feldspar + Muscovite mica + Quartz
Low temperature (last to crystallize)	

Discontinuous Series

Diagram II **Types of Sharing in Silicate Mineral Structures**

Increased Sharing of Oxygen →

Minerals	Olivine	Hornblende	Mica	Quartz
SiO₄ Patterns	Single Tetrahedron	Chains of Tetrahedra	Sheets of Tetrahedra	Networks of Tetrahedra
Chemical Composition	$(Fe, Mg)_2SiO_4$	$Ca_2(Mg, Fe)_5(Si_8O_{22})(OH)_2$	$K(Mg, Fe)_3(AlSi_3O_{10})(OH)_2$	SiO_2

Fe + Mg Decreasing →

45 Which statement about the minerals quartz and olivine must always be true?

  1 They have the same form at the same temperature.
  2 They have the same density.
  3 They contain the elements silicon and oxygen.
  4 They contain the elements iron and magnesium.   45 _____

46 The arrangement of the silicon-oxygen tetrahedra in the mineral mica is responsible for mica's

  1 black color
  2 colorless streak
  3 high crystallization temperature
  4 cleavage in one direction   46 _____

47 Which statement about the process of crystallization of minerals from magma is true?

  1 Dark-colored rocks are the last to crystallize.
  2 Rocks crystallized at the highest temperatures are lower in density than rocks crystallized at low temperatures.
  3 The first rocks to crystallize will contain minerals composed of single tetrahedra.
  4 The rocks crystallized last contain the greatest amounts of iron (Fe) and magnesium (Mg).   47 _____

48 The diagram below shows a sample of rock material that contains coarse-grained intergrown crystals of several minerals. [Mineral crystals are shown actual size.]

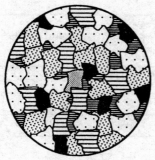

This rock sample should be identified as

1 rhyolite   3 scoria

2 granite   4 basalt   48 _____

Base your answers to questions 49 and 50 on the world map and graph below. The graph shows the relationship between average annual income and average annual energy used per person for several countries.

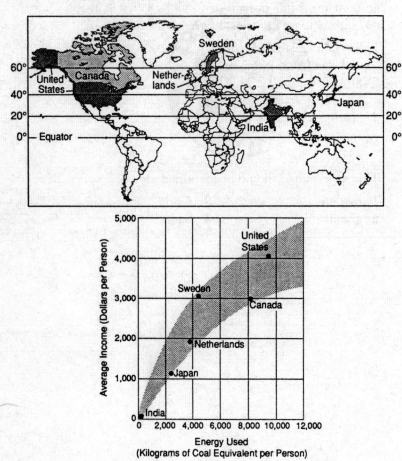

Energy Used
(Kilograms of Coal Equivalent per Person)

49 In which country is the average energy used per person the greatest?

1 India   3 United States
2 Netherlands   4 Japan

49 _____

50 According to the graph, the average income of a person in Sweden is about the same as the average income of a person in Canada, but the average energy used per person in Sweden is about half the amount used per person in Canada. What is the most likely reason for this difference in energy use?

1 Sweden has more daylight per year than Canada and uses less energy for lighting.

2 Since Sweden's climate is much warmer than Canada's, Sweden uses less energy for heating.

3 Homes in Canada are better insulated than homes in Sweden and require less energy for heating.

4 People in Canada travel more and for longer distances by automobile than people in Sweden.   50 _____

GROUP B — **Plate Tectonics**

*If you choose this group, be sure to answer questions **51–60**.*

51 The diagram below represents a site in western New York State. The remains of a mastodont were found in sediments from a swamp that was located above limestone bedrock. Fossils of Paleozoic coral were discovered in the limestone beneath the swamp.

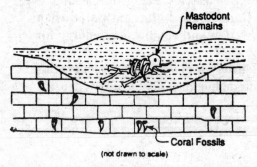

(not drawn to scale)

Observations made at this site provide evidence that

1 the local climate and environment have changed dramatically
2 fossil coral provided food for mastodonts
3 North America has drifted to the east from Europe and Africa
4 mastodonts and coral were alive at the same time   51 _____

52 Which information would be most useful for predicting the occurrence of an earthquake at a particular location?
1 elevation
2 climate
3 seismic history
4 number of nearby seismic stations   52 _____

53 Mesozoic rocks and fossils found in Australia are most likely to match Mesozoic rocks and fossils found in

1 Europe
2 Antarctica
3 the Atlantic Ocean
4 North America

53 _____

54 The Mercalli-scale intensity of an earthquake that occurred before the invention of the seismograph can be inferred by

1 observing the local topography
2 reading historical observations of the event
3 measuring the rate at which heat escapes from the Earth
4 measuring the strength of the Earth's gravitational field

54 _____

55 Evidence of subduction exists at the boundary between the

1 African and South American plates
2 Australian and Antarctic plates
3 Pacific and Antarctic plates
4 Nazca and South American plates

55 _____

Base your answers to questions 56 through 60 on the *Earth Science Reference Tables*, the table below, and your knowledge of Earth science. The table shows some of the data collected at two seismic stations, *A* and *B*. Some data have been omitted.

Station	Arrival Time of P-Wave	Arrival Time of S-Wave	Difference in Arrival Times of P- and S-Waves	Distance to Epicenter
A	6:02:00 p.m.	6:07:30 p.m.	5 min 30 sec	— km
B	— p.m.	6:11:20 p.m.	7 min 20 sec	5,700 km

56 Which seismogram most accurately represents the arrival of the *P*- and *S*-waves at station *A*?

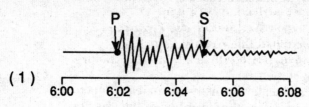

(1)

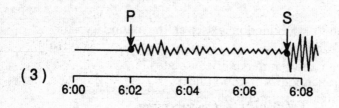

(2)

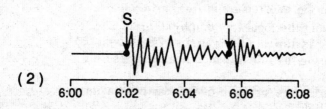

(3)

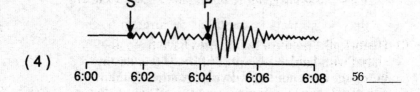

(4)

56 _____

57 What is the approximate distance from the epicenter to station *A*?

(1) 1,400 km   (3) 3,000 km
(2) 1,900 km   (4) 4,000 km   57 _____

58 Which statement best describes the seismic waves received at station *B*?

1 The *P*-wave arrived at 6:12 p.m.
2 The *S*-wave arrived before the *P*-wave.
3 The *P*-wave had the greatest velocity.
4 The *S*-wave passed through a fluid before reaching station *B*.   58 _____

59 What was the origin time of this earthquake?

(1) 5:55:00 p.m.   (3) 6:06:00 p.m.
(2) 6:00:00 p.m.   (4) 6:11:20 p.m.   59 _____

60 What is the minimum number of *additional* stations from which scientists must collect data in order to locate the epicenter of this earthquake?

(1) 1   (3) 3
(2) 2   (4) 0   60 _____

GROUP C — **Oceanography**

If you choose this group, be sure to answer questions 61–70.

61 Historically, political boundaries have been associated with land areas. Most of the ocean basins, however, have not been divided politically. Why are people becoming more concerned about the political status of the ocean basins?

1 Worldwide, rising sea level is changing the shapes of the continents.
2 The salinity of ocean water is changing.
3 The size of the ocean basins is decreasing.
4 Some ocean basins contain important mineral resources.   61 _____

62 At which location is ocean water most likely to be clean and unpolluted?
1 near the west coast of North America
2 near the east coast of South America
3 in the mid-Pacific
4 around Australia   62 _____

63 The table below shows the percentage by mass of dissolved ions in seawater.

Dissolved Ion	Percentage
Chloride (Cl^-)	55.04
Sulfate (SO_4^{2-})	7.68
Bicarbonate (HCO_3^-)	0.41
Bromide (Br^-)	0.19
Sodium (Na^+)	30.61
Magnesium (Mg^{2+})	3.69
Calcium (Ca^{2+})	1.16
Potassium (K^+)	1.10
All others	0.12
	100.00

Which mineral is most abundant in seawater?
1 calcite (calcium carbonate)
2 halite (sodium chloride)
3 quartz (silicon dioxide)
4 magnetite (iron oxide)   63 _____

64 Points *A* through *E* on the map below represent locations in the Atlantic Ocean between the United States and Africa. Assume that a complete core sample of sediment and sedimentary rock could be taken from the ocean bottom at each of these five locations.

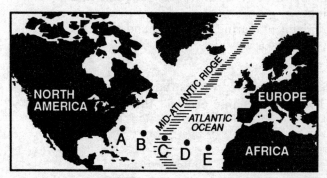

Which diagram of core samples best represents the relative thickness of the sediments at the five locations?

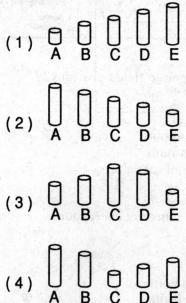

64 _____

65 Most of the sediments deposited on the deep ocean bottoms consist of

1 volcanic dust and materials dissolved by ocean currents
2 windblown particles and meteorites
3 materials directly deposited by streams and glaciers
4 the remains of marine organisms and particles from land areas

65 _____

66 The diagram below represents the magnetic fields of rocks in the oceanic crust at the mid-ocean ridge.

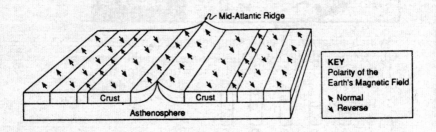

The mapping of these magnetic fields provides information about

1 the origin of ocean basins
2 climatic patterns over the oceans
3 the circulation of ocean currents
4 differences in the climates of coastal areas

66 _____

67 Which is the most common source of energy for surface ocean waves?

1 planetary winds
2 the Earth's rotation
3 disturbances of the ocean bottom
4 heat flow and seafloor spreading

67 _____

68 The usual direction of movement of major surface ocean currents is best described as

1 clockwise in the Northern Hemisphere and
counterclockwise in the Southern Hemisphere
2 clockwise in the Southern Hemisphere and
counterclockwise in the Northern Hemisphere
3 clockwise in both hemispheres
4 counterclockwise in both hemispheres   68 _____

69 Structures such as jetties and breakwaters are
built along the shore of the ocean in order to

1 prevent ice formation along the shore
2 increase the speed of longshore currents
3 absorb and deflect the energy of the water
4 lower the salinity of the water   69 _____

70 Which is a characteristic of water that helps the
oceans to moderate the climates of the Earth?

1 Water is a fluid with a high specific heat.
2 Water can exist as a high-density solid.
3 Water can dissolve and transport minerals.
4 Water can flow into loose sediments to deposit
mineral cements.   70 _____

GROUP D — Glacial Processes

*If you choose this group, be sure to answer questions **71–80**.*

71 One similarity between a valley glacier and a
mountain stream is that they both

1 carry only small particles of sediment
2 form a V-shaped valley
3 transport materials at the same rate
4 move fastest on the steepest slope   71 _____

72 At which location in New York State are unconsolidated gravels, sands, and clays found in the
greatest amounts?
(1) 42°30′ N 76°30′ W
(2) 40°45′ N 73°30′ W
(3) 43°30′ N 74°45′ W
(4) 42°45′ N 78°30′ W   72 _____

73 Which evidence suggests that temperatures in
New York State were much colder in the geologic past than at present?
1 large rock-salt deposits
2 petrified fern-tree trunks
3 dinosaur footprints in sedimentary rock
4 parallel grooves in the surface bedrock   73 _____

    Base your answers to questions 74 through 80 on
the map below and your knowledge of Earth science.
The map shows the inferred position of the continental
ice sheet in New York State approximately 12,000 years
ago.

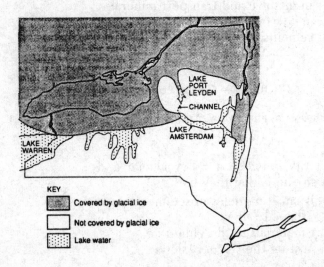

74 Which evidence would have been most useful to geologists for locating the edges of the ice sheet shown on the map?

1 flat, thick deposits of impermeable clay
2 piles of unlayered, unsorted sediment
3 folded layers of bedrock
4 formations of rocks with interlocking crystals   74 _____

75 Which geologic feature is located in the former region of Lake Amsterdam?

1 Taconic Mountains   3 Finger Lakes
2 Atlantic Coastal Plain   4 Mohawk River   75 _____

76 A steep-walled channel was cut by meltwater pouring from Lake Port Leyden into Lake Amsterdam about 12,000 years ago. What condition probably existed in the channel at that time?

1 The channel contained a small volume of water.
2 The channel had a very gentle slope.
3 The water flowed at a high velocity in the channel.
4 The water in the channel was very warm.   76 _____

77 Which map best represents the inferred position of the ice sheet at the time of glacial deposition on Long Island? [The shaded portion represents the areas covered by glacial ice.]

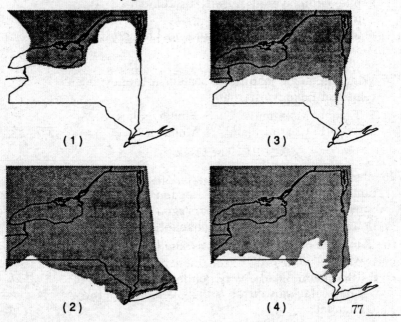

(1)   (3)

(2)   (4)   77 _____

78 Which fossil found in sediments in New York State supports the inference that these sediments were deposited during the Pleistocene Epoch?

1 eurypterid fossils
2 stromatolite mounds
3 mastodont bones
4 coelophysis footprints   78 _____

79 Which radioactive isotope is commonly used to date organic matter buried during the Pleistocene ice age?

1 carbon-14   3 rubidium-87
2 potassium-40   4 uranium-238   79 _____

80 Which evidence best supports the inference that the movement of the ice sheet was generally from north to south over New York State during the Pleistocene glaciation?

1 pieces of anorthositic rock from the Adirondacks found near Albany
2 striations aligned east to west on bedrock near Utica
3 rocks of Devonian age found near Elmira
4 the direction of flow of the Niagara River at Niagara Falls   80 _____

GROUP E — Atmospheric Energy

If you choose this group, be sure to answer questions 81–90.

Base your answers to questions 81 through 86 on the *Earth Science Reference Tables*, the diagram below, and your knowledge of Earth science. The diagram shows the conditions necessary for the development of a typical lake-effect snowstorm along the shore of Lake Erie in western New York State.

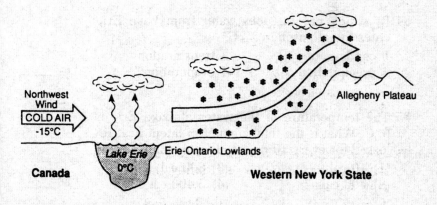

81 The air temperature at the surface of Lake Erie is
0°C and the dewpoint temperature is –6°C. To
which altitude must the air rise to form a cloud?
(1) 1.0 km   (3) 0.7 km
(2) 1.2 km   (4) 0.5 km   81 _____

82 How should the air mass that formed over
central Canada be labeled?
(1) cP   (3) cT
(2) mP   (4) mT   82 _____

83 What is the correct order of events that occur as
air travels from the Erie-Ontario Lowlands to the
Allegheny Plateau as shown in the diagram?
1 the air rises → the water vapor condenses →
   the air expands and cools
2 the air rises → the air expands and cools →
   the water vapor condenses
3 the air expands and cools → the air rises →
   the water vapor condenses
4 the water vapor condenses → the air rises →
   the air expands and cools   83 _____

84 By which process does water from Lake Erie
enter the atmosphere?
1 condensation   3 transpiration
2 precipitation   4 evaporation   84 _____

85 The temperature of the water of Lake Erie is
0°C. What is the total amount of latent heat released for every 10 grams of water that freezes?
(1) 10 cal   (3) 800 cal
(2) 400 cal   (4) 5,400 cal   85 _____

Note that question 86 has only three choices.

86 As Lake Erie freezes over, the chance of having a
lake-effect snowstorm will

1 decrease
2 increase
3 remain the same   86 _____

87 The weather map below shows two air-pressure
systems covering a large geographic area. Points *A*
through *D* identify four locations on the map.

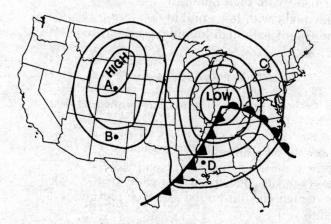

At which location will the greatest weather change
occur during the next 6 hours?

(1) *A*   (3) *C*
(2) *B*   (4) *D*   87 _____

88 Which planetary wind pattern is present in areas
of great rainfall?

1 Winds diverge and air rises.
2 Winds converge and air rises.
3 Winds diverge and air sinks.
4 Winds converge and air sinks.   88 _____

89 The most difficult type of weather event to forecast several hours in advance is the development of a

1 thunderstorm along a cold front
2 hurricane in the Atlantic Ocean
3 tornado along a cold front
4 snowstorm along Lake Erie   89 _____

90 Under which conditions will the greatest amount of cooling by terrestrial reradiation occur?

1 a clear night with low humidity
2 a clear night with high humidity
3 a cloudy night with low humidity
4 a cloudy night with high humidity   90 _____

GROUP F — Astronomy

If you choose this group, be sure to answer questions 91–100.

Note that question 91 has only three choices.

91 If the average distance from a satellite to the Earth is decreased, the period of revolution of the satellite will

1 decrease
2 increase
3 remain the same   91 _____

92 On which planet do large amounts of water exist in all three states of matter?

1 Venus   3 Mars
2 Earth   4 Saturn   92 _____

93 Which three planets are known as terrestrial planets because of their high density and rocky composition?

  1 Venus, Neptune, and Pluto
  2 Venus, Saturn, and Neptune
  3 Jupiter, Saturn, and Uranus
  4 Mercury, Mars, and Venus   93 _____

94 Why do rock samples brought back from the Moon show no signs of chemical weathering?

  1 The Moon has no gravity.
  2 The Moon has no atmosphere.
  3 Temperatures on the Moon are very cold.
  4 Temperatures on the Moon are very hot.   94 _____

95 Diagram I below represents a group of celestial objects observed in the night sky on October 25. Diagram II represents the same objects 1 month later.

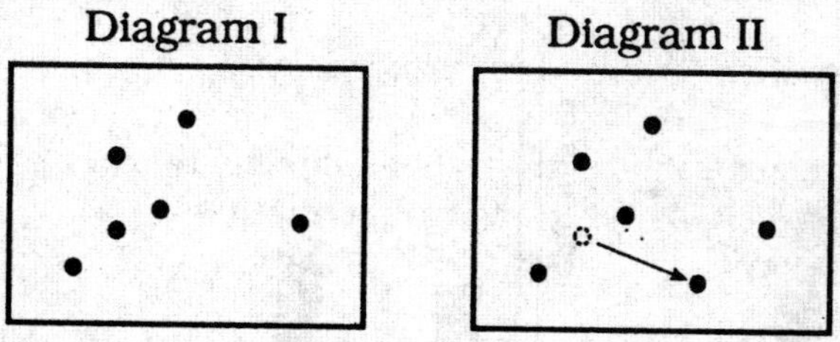

Diagram I   **Diagram II**

The object in diagram II that moved was most likely a

  1 galaxy   2 meteor   3 planet   4 star   95 _____

96 The presence of which atmospheric gas causes the high temperature of Venus?

  1 carbon dioxide (CO_2)
  2 nitrogen (N_2)
  3 oxygen (O_2)
  4 hydrogen (H_2)   96 _____

97 Planetary temperatures were recorded by the Voyager I space probe as it traveled from the Earth past the outer planets Jupiter, Saturn, and Uranus before leaving the solar system. Which graph best represents the relationship between the temperature of the planets and their distance from the Earth?

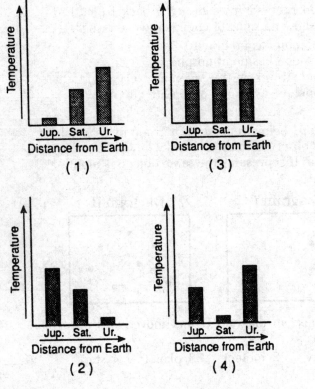

97 _____

98 In which group are the parts listed in order from oldest to youngest?

1 solar system, Milky Way, universe
2 Milky Way, solar system, universe
3 universe, solar system, Milky Way
4 universe, Milky Way, solar system

98 _____

99 The diagram below shows the standard dark-line spectrum for an element.

VIOLET	BLUE	GREEN	YELLOW	ORANGE	RED

The spectral lines of the same element are observed in light from four distant galaxies. Which spectral lines most likely represent the galaxy farthest from the Earth?

(1) VIOLET .... RED

(2) VIOLET .... RED

(3) VIOLET .... RED

(4) VIOLET .... RED

99 _____

100 Which diagram best represents a geocentric model of the solar system? [Diagrams are not drawn to scale. Key: *E* = Earth; *P* = Planet; *S* = Sun]

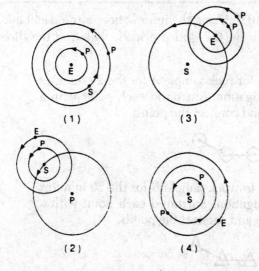

(1)   (3)

(2)   (4)

100 _____

PART III

This part consists of 14 questions. Be sure that you answer all questions in this part. Record your answers in the spaces provided. Some questions may require the use of the *Earth Science Reference Tables.*   [25]

Base your answers to questions 101 through 106 on the data table below. Samples of three different rock materials, *A, B,* and *C,* were placed in three containers of water and shaken vigorously for 20 minutes. At 5-minute intervals, the contents of each container were strained through a sieve. The mass of the materials remaining in the sieve was measured and recorded as shown in the data table below.

Mass of Material Remaining in Sieve

Shaking Time (minutes)	Rock Material A (grams)	Rock Material B (grams)	Rock Material C (grams)
0	25.0	25.0	25.0
5	24.5	20.0	17.5
10	24.0	18.5	12.5
15	23.5	17.0	7.5
20	23.5	12.5	5.0

Directions (101–103): Using the information in the data table, construct a line graph on the grid provided, following the directions below.

101 Plot the data for rock sample *A* for the 20 minutes of the investigation. Surround each point with a small circle and connect the points.   [1]

Example:

102 Plot the data for rock sample *B* for the 20 minutes of the investigation. Surround each point with a small triangle and connect the points.   [1]

Example:

103 Plot the data for rock sample *C* for the 20 minutes of the investigation. Surround each point with a small square and connect the points. [1]

Example:

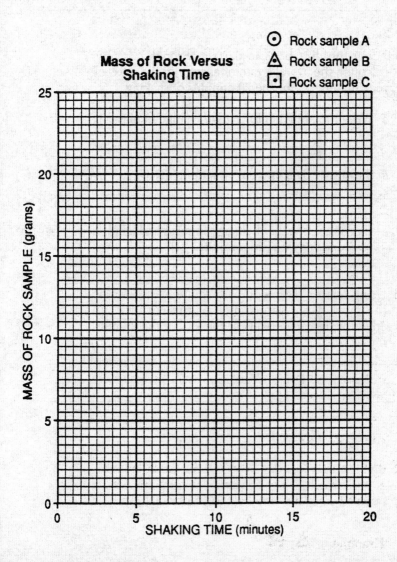

Mass of Rock Versus Shaking Time

⊙ Rock sample A
△ Rock sample B
⊡ Rock sample C

104 Using one or more complete sentences, state the most likely reason for the differences in the weathering rate of the three rock materials. [2]

_____

_____

_____

105 Using the directions in parts *a* through *c* below, calculate the average rate of change in the mass of rock material *C* for the 20 minutes of shaking.

   a Write the equation for rate of change in mass. [1]

   b Substitute data into the equation. [1]

   c Calculate the rate of change in mass and label your answer with proper units. [1]

106 Using one or more complete sentences, describe the most likely appearance of the corners and edges of rock material *C* at the end of the 20 minutes. [2]

_____

_____

_____

Base your answers to questions 107 through 110 on the diagram below and your knowledge of Earth science. The diagram represents a profile view of a rock outcrop. The layers are labeled *A* through *H*.

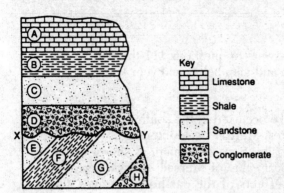

Key
- Limestone
- Shale
- Sandstone
- Conglomerate

107 State the range of particle sizes of the sediment that formed rock layer *C*. [1]

_____

108 Using one or more complete sentences, briefly describe the geologic process that resulted in the boundary represented by the line *XY*. [2]

_____

_____

_____

109 State two ways in which the composition of rock layer *A* differs from the composition of rock layer *B*. [2]

_____

_____

_____

110 None of the layers has been overturned. Layer *D* is 505 million years old and layer *B* is 438 million years old. State the geologic period during which layer *C* could have formed. [1]

Base your answers to questions 111 through 114 on the information and map below and your knowledge of Earth Science.

An earthquake occurred in the southwestern part of the United States. Mercalli-scale intensities were plotted for selected locations on a map, as shown below. (As the numerical value of Mercalli ratings increases, the damaging effects of the earthquake waves also increase.)

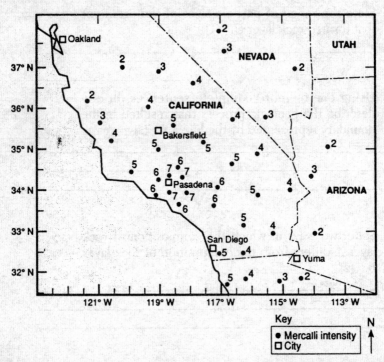

111 Using an interval of 2 Mercalli units and starting with an isoline representing 2 Mercalli units, draw an accurate isoline map of earthquake intensity. Record your answer on the map provided. [4]

112 State the name of the city that is closest to the earthquake epicenter. [1]

_____

113 State the latitude and longitude of Bakersfield. [2]

_____

114 Using one or more complete sentences, identify the most likely cause of earthquakes that occur in the area shown on the map. [2]

_____
_____
_____

Answers
June 1995

Earth Science —
Program Modification Edition

Answer Key

PART I

1. 3	**8.** 2	**15.** 4	**22.** 3	**29.** 2	**36.** 3
2. 4	**9.** 4	**16.** 1	**23.** 2	**30.** 1	**37.** 4
3. 2	**10.** 3	**17.** 2	**24.** 4	**31.** 3	**38.** 1
4. 3	**11.** 4	**18.** 3	**25.** 1	**32.** 2	**39.** 3
5. 1	**12.** 2	**19.** 1	**26.** 2	**33.** 4	**40.** 4
6. 3	**13.** 1	**20.** 4	**27.** 2	**34.** 2	
7. 4	**14.** 3	**21.** 4	**28.** 3	**35.** 1	

PART II

Group A	Group B	Group C	Group D	Group E	Group F
41. 1	**51.** 1	**61.** 4	**71.** 4	**81.** 3	**91.** 1
42. 3	**52.** 3	**62.** 3	**72.** 2	**82.** 1	**92.** 2
43. 4	**53.** 2	**63.** 2	**73.** 4	**83.** 2	**93.** 4
44. 1	**54.** 2	**64.** 4	**74.** 2	**84.** 4	**94.** 2
45. 3	**55.** 4	**65.** 4	**75.** 4	**85.** 3	**95.** 3
46. 4	**56.** 3	**66.** 1	**76.** 3	**86.** 1	**96.** 1
47. 3	**57.** 4	**67.** 1	**77.** 2	**87.** 4	**97.** 2
48. 2	**58.** 3	**68.** 1	**78.** 3	**88.** 2	**98.** 4
49. 3	**59.** 1	**69.** 3	**79.** 1	**89.** 3	**99.** 3
50. 4	**60.** 1	**70.** 1	**80.** 1	**90.** 1	**100.** 1

PART III See answers explained.

Answers Explained

PART I

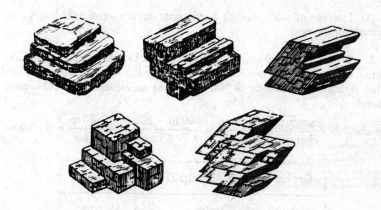

1. **3** Cleavage is a property of some minerals that break along flat surfaces. Cleavage occurs because of the arrangement of the atoms in a mineral. Mica is an example of a mineral in which the atoms are arranged in layers. For example, when mica cleaves, it breaks in sheets that represent the layers of atoms in the mineral. All of the samples shown in the diagrams have flat surfaces, indicating that the minerals break along straight lines between layers of atoms.

WRONG CHOICES EXPLAINED:
   (1) A few minerals, such as magnetite, that contain iron display magnetic properties.
   (2) The hardness of a mineral is its resistance to being scratched. Diamond, the hardest commonly found mineral, can scratch any other mineral. Talc is a very soft mineral that cannot scratch any of the other commonly found minerals.
   (4) The size of the crystals in a mineral depends upon the rate at which they formed. When cooling or evaporation takes place slowly, large crystals form.

2. **4** Photographs of the Earth taken from space show the Earth's true shape. From these photographs scientists have determined that the Earth is very nearly spherical. A slight flattening at the poles is too small to be noticed by the naked eye and is revealed only when measurements are made.

WRONG CHOICES EXPLAINED:

(1) Variations in the prevailing wind direction at the Earth's surface are caused by the rotation of the Earth.

(2) Changes in the time of sunrise and sunset at a single location during the year are caused by the tilt in the Earth's axis as it revolves around the Sun. When the axis is pointed toward the Sun, the Sun's apparent daily path in the sky is higher and longer, causing sunrise to occur earlier and sunset to occur later.

(3) The time of rotation of the Earth on its axis is a constant 24 hours a day throughout the year.

3. **2** Find the equation for gradient in the Equations and Proportions section of the *Earth Science Reference Tables*. In this case the change in field value is from an elevation of 800 meters to an elevation of 600 meters. Therefore:

$$\text{gradient} = \frac{\text{change in field value}}{\text{change in distance}} = \frac{800 \text{ m} - 600 \text{ m}}{5 \text{ km}} = \frac{200 \text{ m}}{5 \text{ km}} = 40 \text{ m/km}$$

Group A	Group B	Group C
Granite	Shale	Marble
Rhyolite	Sandstone	Schist
Gabbro	Conglomerate	Gneiss

4. **3** The rocks in group A are all igneous. The rocks in group B are sedimentary, and the rocks in group C are metamorphic. This represents a classification system based on how the rocks formed. Igneous rocks, for example, form when molten rock material solidifies, while sedimentary rocks form when sediments are compacted or when water evaporates. Information about rock types can be found in the *Earth Science Reference Tables*.

WRONG CHOICES EXPLAINED:

(1) All the minerals in an igneous rock are of the same age since igneous rocks form when molten material solidifies. The minerals in a sedimentary rock, however, could be of different ages since most sedimentary rocks form from sediments that come from other rocks. The age of the minerals would therefore not be a practical method of classifying them.

(2) For igneous rocks only, the size of the crystals depends upon the rate at which the rock cooled. The slower the cooling rate, the larger will be the crystal size. Grain size is used to classify only some sedimentary rocks. For example, sandstone can be distinguished from shale by the difference in grain size. Metamorphic rocks are classified based on the rocks from which they formed, not on crystal size. Crystal size therefore cannot be used to classify all rocks.

(4) The color of a rock sample depends upon the minerals that are present. Small variations in mineral composition can sometimes produce marked differences in color, and therefore color is not useful in classifying rocks.

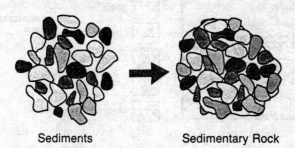

Sediments   Sedimentary Rock

5. **1** Find the Scheme for Sedimentary Rock Identification in the *Earth Science Reference Tables*. Note that, when the grain size is larger than 0.2 cm, the rock formed is conglomerate. Many of the sediments shown in the diagram are larger than 0.2 cm. Note also that the sediments are rounded, not angular. The sedimentary rock formed from these rounded sediments will be conglomerate. If the particles were angular, the rock formed would be breccia.

WRONG CHOICES EXPLAINED:
(2) Sandstone forms when the sediments range in size from 0.006 to 0.2 cm.
(3) Siltstone forms when the sediments range in size from 0.0004 to 0.006 cm.
(4) Shale forms when the sediments are less than 0.0006 cm in size.

6. **3** Sedimentary rocks can form when sediments are compacted under pressure (e.g., buried by overlying sediments) and cemented together. Cementing can occur when water filters through the sediments and evaporates or runs off leaving dissolved material or "cement" behind.

WRONG CHOICES EXPLAINED:
(1) Folding and faulting are processes that alter the shape of rock layers near the Earth's surface. Most rock layers form horizontally. When rock layers appear angular, it may be inferred that folding or faulting has occurred.
(2) Melting and solidification are processes involved in the formation of igneous and metamorphic, but not sedimentary, rock.
(4) Heating and application of pressure can be part of the formation of any rock type, but either cementation or evaporation must be present when sedimentary rocks form.

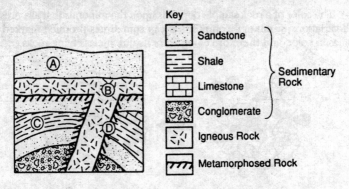

Key
- Sandstone
- Shale
- Limestone
- Conglomerate

} Sedimentary Rock

- Igneous Rock
- Metamorphosed Rock

7. **4** According to the key, there is metamorphosed rock at point D. When molten rock material comes in contact with existing rock layers, the rock material undergoes metamorphism because of the heat and pressure applied. The diagram shows that the existing rock layers were altered by metamorphism at point D.

8. **2** When molten rock material cools slowly, large crystals form, like those in rock Y. This slower rate of cooling usually occurs deep below the Earth's surface. On the other hand, material that cools quickly near the surface will have small crystals. The igneous rock that forms around volcanoes is fine-grained because the molten material solidifies rapidly.

9. **4** Chemical weathering occurs more rapidly in a moist climate because chemicals dissolved in water react with rock materials. Warmer temperatures also accelerate the process of chemical weathering. As a result, the greatest amount of chemical weathering occurs in a climate that is warm and moist.

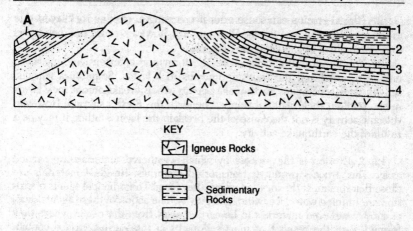

KEY

⬛ Igneous Rocks

⬛ ⬛ ⬛ } Sedimentary Rocks

10. **3** Rock layer A is the third layer above the igneous rocks on the left side of the diagram. Note that the same sequence of layers appears on the right side. Rock layer 3 has the same limestone composition as rock A, according to the symbols in the diagram, and is also the third layer above the igneous rocks.

11. **4** The rock layers appear to be offset across a fault line near the center of the diagram. This suggests that a major earthquake occurred at some time in the past, causing the break across rock layers. The rock layers to the left of the fault appear to have moved upward relative to the layers on the right.

WRONG CHOICES EXPLAINED:

(1) Contact metamorphism occurs when molten rock material comes in contact with existing rock layers. The heat and pressure from the molten material cause the existing rock to undergo metamorphism.

(2) Glacial erosion can cause extensive carving of existing rock layers due to the abrasive action of the ice. It would not, however, cause the rock layers to become offset as shown in the diagram.

(3) Volcanic eruptions can cause the formation of extensive deposits of molten material and the resulting igneous rock layers. Sometimes volcanic activity does accompany earthquake activity when breaks across rock layers release molten rock material trapped beneath the Earth's surface. However, volcanic activity is not the cause of the break in the layers; rather, it may be a result of the earthquake activity.

12. **2** Erosion is the process by which weathered materials are carried away. This process produces transported materials. Residual materials are those that remain at the location where they form. The principal agents of erosion are running water, ice, wind, and gravity. The action of these agents causes most weathered materials to be carried away from the places where they formed, with the result that most sediments at the Earth's surface contain transported material.

13. **1** When material is dissolved in water, a solution forms. Some weathered rock material becomes dissolved and is carried in solution by rivers and streams. The amount that dissolves depends upon mineral composition because some minerals are more soluble in water than others.

14. **3** Find in the *Earth Science Reference Tables* the chart entitled Geologic History of New York State at a Glance. The column at the far right shows the inferred position of the Earth's landmasses during past geologic periods. According to the sequence of diagrams, starting with the Ordovician Period, the landmasses representing North America, Africa, and South America drifted together and were closest during the Triassic Period. After the Triassic Period they began to drift apart again.

15. **4** Find in the *Earth Science Reference Tables* the chart entitled Geologic History of New York State at a Glance. The column labeled "Life on Earth" indicates that the boundary between the Mesozoic Era and the Cenozoic Era was characterized by the last of the dinosaurs and the first placental mammals.

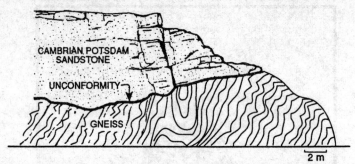

16. **1** Find in the *Earth Science Reference Tables* the chart entitled Geologic History of New York State at a Glance. Since in the diagram the Cambrian sandstone overlays the gneiss, it may be inferred that the gneiss is older. According to the geologic history chart, the time before the Cambrian Period was the Precambrian Era.

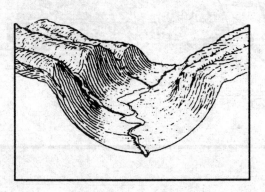

17. **2** The diagram shows the typical U-shaped valley that is formed by glacial erosion. Glaciers tend to form U-shaped valleys because the ice carves out the walls of the valley as well as the base.

WRONG CHOICES EXPLAINED:

(1) As an agent of erosion, wind is not powerful enough to carve out a valley. Wind erosion is most noticeable in deserts, where particles of sand scour other particles and low-lying rock formations.

(3) Erosion by ocean waves is most noticeable along coastal areas, where beach sand is removed or rock layers are worn away.

(4) Running water tends to form V-shaped valleys because the water erosion occurs primarily at the bed of the river or stream.

18. **3** Find in the *Earth Science Reference Tables* the chart entitled Dewpoint Temperatures. The difference between the dry-bulb and wet-bulb temperatures is 2°C–0°C, or 2°C. Find, at the top of the chart, the 2°C difference column, and then the row that shows a 2°C dry-bulb temperature. Where the two values meet is the dewpoint temperature of –3°C.

19. **1** Find in the *Earth Science Reference Tables* the graphs entitled Selected Properties of Earth's Atmosphere. Note on the graph of atmospheric pressure that, as the elevation above the surface increases, the pressure decreases.

20. **4** On a hot, sunny afternoon the land near the ocean will be warmer than the water. The land heats up faster than the water because it has a lower specific heat. The air over the land becomes warmer and the air pressure is lower than the air pressure over the ocean. This warmer air rises and is replaced by cooler air that blows in from the ocean.

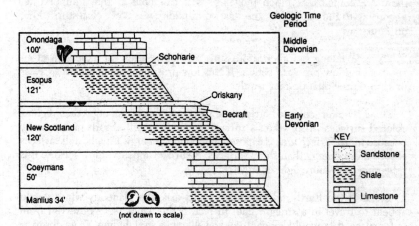

21. **4** In rock layers that have not been overturned, the youngest layer is on top. According to the chart entitled Geologic History of New York State at a Glance in the *Earth Science Reference Tables*, the sequence of geologic time periods along the right edge of the diagram range from oldest on the bottom to youngest on the top, so the rock layers have not been over turned. The top rock layer in the diagram, and the youngest, is the Onondaga.

22. **3** Notice in the diagram how the Oriskany, Becraft, and Coeymans layers appear to protrude. The reason is that they are more resistant to erosion than the layers above and below them. The layer that appears to protrude the farthest is the Coeymans layer, so it appears to be the most resistant to erosion.

23. **2** Notice in the diagram that the Manlius, Oriskany, and Onondaga layers contain fossils. Refer to the chart entitled Geologic History of New York State at a Glance in the *Earth Science Reference Tables*. This chart shows that the fossils in the Onondaga are coral heads, the fossils in the Oriskany are brachiopods, and the fossils in the Manlius are ammonoids. These are all fossils of marine organisms, so the fossil evidence suggests that these rock layers were most likely formed in a sea.

Day	Air Temperature (°F)	Windspeed (mph)	Relative Humidity (%)
Monday	40	15	60
Tuesday	65	10	75
Wednesday	80	20	30
Thursday	85	0	95

24. **4** A relative humidity of 100% means that the air is saturated and therefore contains as much moisture as it can hold at that temperature. According to the data table, the relative humidity was 95%, closest to 100%, on Thursday.

25. **1** A falling barometric pressure reading is often an indication of an approaching low-pressure center. If this low-pressure center contains excess moisture, precipitation may result.

26. **2** Darker colored surfaces absorb more solar energy than lighter colored surfaces. Also, when a surface is rough, it will absorb more energy because more surface area is exposed to incoming energy. Finally, dull surfaces absorb more energy than shiny surfaces. A surface appears shiny because it is reflecting back more light energy.

27. **2** If the Earth did not rotate on its axis, the planetary winds would appear to travel in a straight line. In fact, if the winds were observed from outer space, they would appear to be traveling in a straight line. To an observer on the Earth's surface, however, the rotation of the Earth causes the winds to appear to travel in a curved path.

28. **3** The dewpoint temperature represents the temperature at which the air is saturated. If air is cooled below the dewpoint temperature, the excess moisture will begin to condense. Condensation is the process by which clouds form in the atmosphere.

29. **2** Winds are caused by pressure differences. They blow from regions of higher pressure to regions of lower pressure. When the isobars on a weather map are closer together, the pressure is dropping more rapidly and the wind speed will be greater, producing stronger winds.

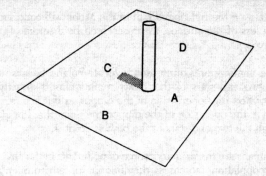

30. **1** The Sun shines in a direction that is opposite to the direction of the shadow. In New York State the Sun shines from a southerly direction. Therefore, at noon the Sun is shining from direction *A*, and direction *A* must be south of the post.

31. **3** Polaris, the North Star, is located directly over the North Pole. As the Earth rotates on its axis, Polaris appears to be stationary in the nighttime sky, and the other stars appear to move in a circular orbit around Polaris.

32. **2** Of the months listed, July is the month when the Sun's path is longest and highest in the sky in New York State. At this time the Sun rises and sets north of due east. Between September and March, a period that includes the other three choices, the Sun's apparent path in the sky is shorter, and the Sun rises south of due east in New York State.

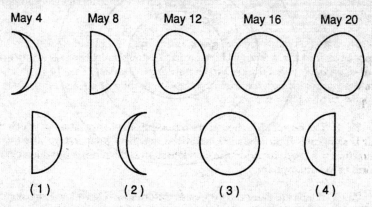

33. **4** Notice in the diagram that between May 16 and May 20 the amount of the Moon's surface that appears illuminated is decreasing. This process of

less illumination on the right-hand half of the Moon will continue, so that on May 24 even less of the surface, as represented by diagram (4), will appear illuminated.

34. **2** The Big Dipper is often used to locate Polaris, because a line drawn through the two end stars of the "bowl" points almost directly at Polaris. Polaris is always on the "open" side of the dipper, as if it is being poured out of the bowl. On the open side of the dipper, the point that lies along the line drawn through the two end stars of the bowl is point *B*.

35. **1** Humans are the primary source of pollutants in the environment. As the human population increases, the amount of environmental pollution increases. The graph that shows this relationship is choice (1).

WRONG CHOICES EXPLAINED:
(2) This graph shows no change in pollution as population increases.
(3) This graph shows pollution occurring only at a specific population. If this graph were correct, there would be no pollution at a lower population, and none at a higher population. In fact, pollution occurs at all levels of population.
(4) This graph shows pollution decreasing as population increases. This trend is the opposite of what actually occurs.

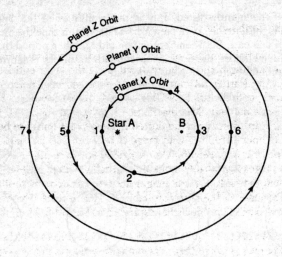

36. **3** According to the diagram, the shape of the orbits of the planets is elliptical. Star *A* and point *B* are located at the two foci of these orbits.

37. **4** According to the list of Equations and Proportions in the *Earth Science Reference Tables*, the force of gravitational attraction between any two objects depends upon the product of their masses and the square of the distance between their centers. Since all three planets have the same mass, they can be considered equivalent and gravitational attraction will depend solely upon the distance between planet and star. The gravitational attraction between any planet and star *A* is greatest when they are closest together. Of the choices given, a planet in position 5 is the closest to star *A*. The gravitational attraction between star and planet will be greatest when Planet *Y* is in position 5.

38. **1** As planet *X* orbits star *A*, the gravitational attraction changes as the distance between the planet and the star changes. The gravitational attraction is greatest when the distance between them is smallest. When the gravitational attraction is the greatest, planet *X* moves fastest. Thus, the gravitational force and the orbital velocity are greatest when the planet is closest to the star, that is, when planet *X* is at position 1.

39. **3** A system containing billions of stars is called a galaxy. The Earth's solar system is located in a galaxy called the Milky Way. It was given this name because, when we look towards the galactic core, the vast number of stars appears as a band of milky light in the sky. The Milky Way galaxy is a spiral galaxy, and the Earth's solar system is located in one of the spiral arms far from the center. The location of the Earth's solar system is best represented by point *C*.

40. **4** If material is melted or partially melted, then its temperature must be above the melting point of the material. Find the chart entitled Inferred Properties of the Earth's Interior in the *Earth Science Reference Tables*. On the lower graph, note the solid bold line labeled "Actual Temperature" and the two dotted lines marked "Melting Point." Notice that there are two depth ranges in the Earth for which the dotted melting-point lines are beneath the bold actual temperature line. One range is between 100 km and 600 km in depth. The other is between 3000 km and 4800 km. in depth. The material in these depth ranges is probably molten or partially melted. Follow the dashed lines that mark the ends of these ranges up to the cross-section of the Earth, on which the names of the layers are labeled. The layers in which melting or partial melting occurs are the asthenosphere and the outer core.

PART II

GROUP A—Rocks and Minerals

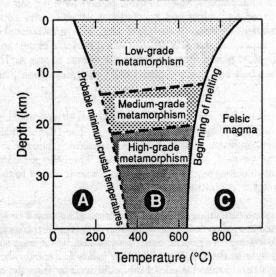

41. **1** Note that, in the diagram, a temperature of 500°C and a depth of 25 km correspond to a point in the "High-grade metamorphism" section of region *B*. This point indicates that the metamorphic rock formed under these conditions has undergone high-grade metamorphism. Find the Scheme for Metamorphic Rock Identification in the *Earth Science Reference Tables*. Note from the "Rock Name" column that the only choices that are metamorphic rocks are choice (1)—gneiss, choice (2)—schist, and choice (4)—slate. Refer to the column labeled "Comments." Gneiss is a metamorphic rock that has undergone high-grade metamorphism.

42. **3** Note that region *C* is labeled "Felsic magma." If a magma is forced to the surface and cools quickly, the magma forms an extrusive rock. Find the Scheme for Igneous Rock Identification in the *Earth Science Reference Tables*. Note that, in the middle of the chart, rocks with a felsic composition are on the left side of the scheme. Note also that, in the upper section of the chart, rhyolite, pumice, and obsidian are listed as extrusive rocks. Of the choices given, only choice (3), rhyolite, is a felsic rock that formed in an extrusive environment.

Key

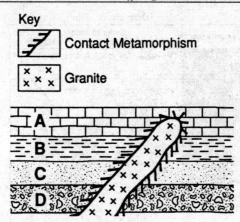

Contact Metamorphism

Granite

43. **4** Note the symbols for the four rock layers, *A*, *B*, *C*, and *D*. Find the Scheme for Sedimentary Rock Identification in the *Earth Science Reference Tables*. Refer to the columns labeled "Rock Name" and "Map Symbol." Note that layer *A* in the question is limestone, layer *B* is shale, layer *C* is sandstone, and layer *D* is conglomerate. Now refer to the column labeled "Grain Size" in the Scheme for Sedimentary Rock Identification. Conglomerate, which can contain grains ranging from silt 0.001 cm in diameter to boulders, has the greatest range of particle sizes. Although the "Grain Size" column indicates that limestone can have coarse to fine grains, the grains in limestone do not approach the size of a boulder.

44. **1** Find the Scheme for Sedimentary Rock Identification in the *Earth Science Reference Tables*. Refer to the columns labeled "Rock Name" and "Map Symbol." Note that layer *C* is sandstone. Now find the Scheme for Metamorphic Rock Identification. Look in the column labeled "Comments." "Metamorphism of sandstone" is the comment listed for the rock named quartzite. Thus, quartzite is a rock formed by the metamorphism of the rock in layer *C*.

WRONG CHOICES EXPLAINED:
(2) Marble forms by the metamorphism of limestone or dolostone, not sandstone.
(3) Slate forms by the metamorphism of shale, not sandstone.
(4) Metaconglomerate forms by the metamorphism of conglomerate, not sandstone.

Diagram I

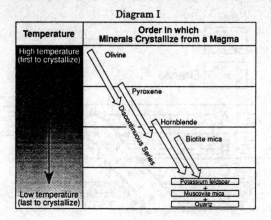

Diagram II
Types of Sharing in Silicate Mineral Structures

Minerals	Increased Sharing of Oxygen →			
	Olivine	Hornblende	Mica	Quartz
SiO₄ Patterns	Single Tetrahedron	Chains of Tetrahedra	Sheets of Tetrahedra	Networks of Tetrahedra
Chemical Composition	(Fe, Mg)₂SiO₄	Ca₂(Mg, Fe)₅(Si₈O₂₂)(OH)₂	K(Mg, Fe)₃(AlSi₃O₁₀)(OH)₂	SiO₂
	Fe + Mg Decreasing →			

45. **3** Olivine and quartz are both silicate minerals, which, by definition, contain silicon and oxygen.

WRONG CHOICES EXPLAINED:
(1) In diagram I, note that olivine forms at a high temperature, whereas quartz forms at a low temperature. Therefore, olivine may have a crystalline form while quartz is still a liquid.

(2) From diagram II, note that olivine and quartz have different compositions. These minerals are therefore expected to have different densities.

(4) In diagram II, find the row labeled "Chemical Composition." Compare the formulas for olivine and quartz. Note that quartz contains no iron and no magnesium.

46. **4** The bonding between tetrahedra in a silicate mineral is responsible for the strengths and weaknesses in its internal structure. When a mineral breaks, it generally does so along planes of weakness. The sheet structure of mica is directly responsible for its pattern of cleavage in one direction.

WRONG CHOICES EXPLAINED:

(1) A mineral is colored if certain wavelengths of light are absorbed when light is reflected from the mineral or passes through it. The absorption has more to do with the composition of the mineral than with the arrangement of the tetrahedra. For example, biotite and muscovite are two varieties of mica. Both are composed of sheets of tetrahedra, but biotite is black in color and muscovite is colorless. Iron, found in biotite but not in muscovite, is often associated with coloring in minerals and is chiefly responsible for the black color of biotite.

(2) The *streak* is the color of a powdered mineral. In a powder, the mineral has been broken down into very small particles. As described in wrong choice (1), above, the color has more to do with the composition of the mineral than with the arrangement of the tetrahedra.

(3) From diagram I, it can be seen that mica has a relatively *low* crystallization temperature.

47. **3** From diagram I, note that the first mineral to crystallize from a magma is olivine. In diagram II, find the row labeled "SiO_4 Patterns" and note that olivine is composed of single tetrahedra. Thus, the first rocks to crystallize would contain olivine, which is composed of single tetrahedra.

WRONG CHOICES EXPLAINED:

(1) From diagram I, the last minerals to crystallize are potassium feldspar, which is pink to white in color, muscovite mica, which is colorless, and quartz, which is white to colorless. So, the last rocks to crystallize are light-colored, not dark-colored.

(2) From diagram I, note that olivine and pyroxene crystallize at high temperatures, whereas potassium feldspar and quartz crystallize at low temperatures. Find the Scheme for Igneous Rock Identification in the *Earth Science Reference Tables*. On the density scale, note that olivine and pyroxene have high densities, whereas potassium feldspar and quartz have low densities.

(4) Note that, from diagram I, quartz is the last to form. In diagram II, the scale labeled "Fe + Mg Decreasing" shows that quartz contains the least iron and magnesium. So, the last rock to crystallize contains the smallest amounts of iron and magnesium, not the greatest amounts.

48. **2** Note that all of the choices are igneous rocks. Find the Scheme for Igneous Rock Identification in the *Earth Science Reference Tables*. Refer to the column labeled "Texture." Note that all of the choices are fine grained, except for granite. The rock sample should be identified as granite.

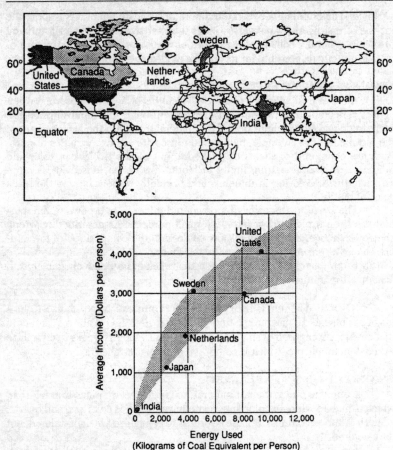

Energy Used
(Kilograms of Coal Equivalent per Person)

49. **3** The horizontal axis on the graph is labeled "Energy Used (Kilograms of Coal Equivalent per Person)." Note that the scale increases to the right. The country farthest to the right on the horizontal scale is the United States.

50. **4** Sweden is a much smaller country than Canada, and Sweden's population tends to be clustered where the inhabitants both live and work. Canada's population is concentrated in larger cities, with people living in suburbs and working in the city centers. This pattern results in the need for more travel by automobile in Canada, and a higher per capita energy use.

WRONG CHOICES EXPLAINED:

(1) Sweden, which is located farther north than most of the major cities in Canada, has less daylight per year, not more.

(2) Although Sweden lies astride the Arctic Circle, its climate is moderated by the North Atlantic Drift, a warm ocean current that brings warm, moist air across most of the Scandinavian Peninsula. However, its climate is not *much* warmer than Canada's, it is about the same as Canada's. For example, consider the following average daily Fahrenheit temperatures from the 1995 World Almanac and Book of Facts:

	Jan.		July	
	max	min	max	min
Stockholm	31	23	70	55
Toronto	30	16	79	59

Such small differences in temperature alone would not explain why Sweden uses *one-half* as much energy per person as Canada.

(3) If homes in Canada were better insulated than homes in Sweden and required less energy for heating, energy consumption in Canada would be *lower*, which isn't the case.

GROUP B—Plate Tectonics

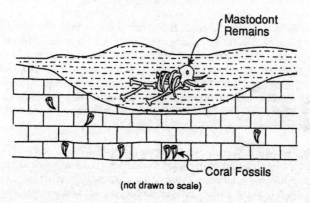

(not drawn to scale)

51. **1** Mastodonts were terrestrial organisms that preferred cool, temperate climates. Corals are marine organisms found in tropical climates. The juxtaposition of these two organisms indicates a change in the local environment (marine to terrestrial) and in the local climate (tropical to temperate).

WRONG CHOICES EXPLAINED:

(2) Find the "Important Fossils of New York" column in the chart entitled Geologic History of New York State at a Glance in the *Earth Science Reference Tables*. Note that, in New York State, the mastodonts lived during the Cenozoic era. The corals are identified as Paleozoic. The time scale indicates that the Paleozoic coral died more than 400 million years before the arrival of the Cenozoic mastodont, so the coral could not have served as food for the mastodont.

(3) Refer to the column labeled "Inferred Position of Earth's Landmasses" in the chart entitled Geologic History of New York State at a Glance in the *Earth Science Reference Tables*. It can be seen that North America drifted to the *west* from Europe and Africa, not to the east.

(4) As explained in wrong choice (2), the mastodonts existed during the Cenozoic era, whereas the corals existed during the Paleozoic era. These organisms were not alive at the same time.

52. **3** There is no causal relationship between earthquakes and the elevation, the climate, or the number of seismic stations found near a particular location. However, the seismic history of a location indicates how frequently earthquakes have occurred there in the past. From this information, the likelihood of an earthquake occurring again could be inferred and a prediction could be made.

53. **2** In the *Earth Science Reference Tables*, find the chart entitled Geologic History of New York State at a Glance. In the column headed "Era," find "Mesozoic." Follow the row for this era to the right to the last column, labeled "Inferred Positions of Earth's Landmasses." Note that during the Mesozoic, Australia was next to Antarctica and has since drifted away to the northeast. None of the other choices was located near Australia during the Mesozoic. Therefore, you would most likely find Mesozoic rocks and fossils in Australia matching those in Antarctica.

54. **2** The Mercalli scale is a measure of earthquake intensity. The scale is based upon a description of earthquake damage. Long before modern instruments were invented, humans witnessed and recorded the damage caused by earthquakes. Thus, by reading historical observations of an event, we can infer the Mercalli-scale intensity of an earthquake.

55. **4** Locate the map entitled Tectonic Plates in the *Earth Science Reference Tables*. Subduction is the process by which the edge of one tectonic plate is pushed beneath the edge of an adjacent plate. The downward plunge of a plate edge produces a trench. So, you are looking for converging plates and a trench. Of the choices, only the boundary between the Nazca and South American plates matches these criteria.

Station	Arrival Time of P-Wave	Arrival Time of S-Wave	Difference in Arrival Times of P- and S-Waves	Distance to Epicenter
A	6:02:00 p.m.	6:07:30 p.m.	5 min 30 sec	— km
B	— p.m.	6:11:20 p.m.	7 min 20 sec	5,700 km

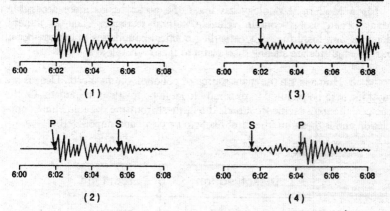

56. **3** According to the chart, the P-wave arrived at station A at 6:02:00 p.m. and the S-wave arrived at 6:07:30. These arrival times are illustrated in seismogram (3).

57. **4** Find the Earthquake P-wave and S-wave Travel Time graph in the *Earth Science Reference Tables*. According to the chart in the question, the difference in arrival times at station A is 5 min 30 sec. On the graph locate the point where the two curves are 5 min 30 sec apart. Trace this position down the graph to obtain an epicenter distance of about 4,000 km.

58. **3** According to the chart in the question, the S-wave arrived at station B at 6:11:20 p.m. while the P-wave had arrived 7 min 20 sec earlier. Since the P-wave arrived first, it may be inferred that it traveled at a greater velocity because it covered the same distance in less time.

59. **1** Find the Earthquake P-wave and S-wave Travel Time graph in the *Earth Science Reference Tables*. For station B the distance to the epicenter is 5,700 km. Locate this distance along the horizontal axis of the graph and trace upward to the S-wave curve. Note that the S-wave travel time for this distance is about 16 min. For station B the S-wave arrival time was 6:11:20 p.m. The origin time for the earthquake was about 16 minutes earlier, or 5:55:00 p.m.

60. **1** A minimum of three stations is required to determine the location of the epicenter. If circles are drawn on a map with centers at the three stations, in this case *A, B,* and one other, and diameters equal to their respective distances from the epicenter, they will intersect at one point. This point is the location of the epicenter.

GROUP C—Oceanography

61. **4** Modern technology has made the ocean basins more accessible. Many ocean basins contain valuable natural resources such as mineral deposits, and fossil fuels. By extending political boundaries to include ocean basins, countries are staking their claim to these valuable resources.

62. **3** Humans are the main source of pollution on the Earth. Oceans are most likely to be polluted in proximity to large populations of humans. Of the choices, the mid-Pacific location is the farthest from any sizeable human population and is therefore the most likely to be clean and unpolluted.

Dissolved Ion	Percentage
Chloride (Cl^-)	55.04
Sulfate (SO_4^{2-})	7.68
Bicarbonate (HCO_3^-)	0.41
Bromide (Br^-)	0.19
Sodium (Na^+)	30.61
Magnesium (Mg^{2+})	3.69
Calcium (Ca^{2+})	1.16
Potassium (K^+)	1.10
All others	0.12
	100.00

63. **2** From the table, the most abundant dissolved ions are the chloride ion and the sodium ion. Of the choices, only halite contains these ions, so halite is the most abundant mineral in seawater.

64. **4** The longer an ocean bottom exists, the more time there is for sediments to accumulate on it. The Mid-Atlantic ridge is a site of ocean floor formation. The farther from the ridge, the older the rock of the ocean bottom and the more sediments will have built up. Therefore, you would find thicker sediments to the east and west of the ridge, and thinner sediments near the ridge, as shown in choice (4).

65. **4** Sediments that reach the deep ocean bottom have been carried far from land. They are generally fine-grained clays, plant or animal skeletal materials, or a mixture of clays and skeletal materials. The fine-grained clays are sediments eroded from the land. They reach the deep ocean bottom because they settle so slowly that there is time for them to be carried far from land. The skeletal materials are derived from material dissolved in seawater. This material was originally eroded from the land. Thus, most deep ocean sediments consist of particles from land areas and the remains of marine organisms.

WRONG CHOICES EXPLAINED:

(1) Materials dissolved in seawater do not settle out. Volcanic dust is sometimes found in deep ocean sediments, but not as a major component.

(2) Meteorites are rare and do not reach the Earth frequently enough to build up in great thicknesses on deep ocean bottoms. Windblown particles would be deposited in the ocean as soon as the wind speed decreased. Rarely does a wind blow strongly and consistently enough to carry particles out to deep ocean basins.

(3) Streams and glaciers exist too far from the deep ocean bottoms to deposit sediments directly onto them.

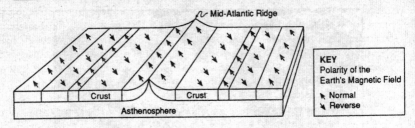

66. **1** As shown in the diagram, the pattern of magnetic field polarity is identical on both sides of the ridge. This discovery suggests that rock forms at the ridge and moves to either side away from the ridge. Thus, the indication is that the ocean basin originated from rock created at the ridge.

WRONG CHOICES EXPLAINED:

(2) The polarity of the minerals in the rock of the ocean floor is the result of the Earth's magnetic field acting upon magnetic minerals in a crystallizing magma. Climatic patterns exert no magnetic force on the minerals in rocks.

(3) Ocean currents involve the movement of water, which is a non-magnetic material. Therefore, ocean currents would not affect the polarity of magnetic minerals in rocks forming on the ocean floor.

(4) As described in wrong choice (2), climatic patterns exert no magnetic force on the minerals in forming rocks.

67. **1** Surface ocean waves are most commonly formed by the transfer of energy from moving air to the water. The most common and most consistent winds that come in contact with the oceans are the planetary winds.

WRONG CHOICES EXPLAINED:

(2) The oceans and the surface of the Earth rotate as a unit. Since there is no relative motion, there is no transfer of energy.

(3) Although disturbances of the ocean floor during an earthquake can produce surface waves called tsunami, these are uncommon occurrences.

(4) Heat flow could cause density differences in ocean water and hence produce an ocean current, but a current is not a wave. Seafloor spreading is very slow and transfers energy to the water near the ocean bottom, not near the surface.

68. **1** Find the chart entitled Surface Ocean Currents in the *Earth Science Reference Tables*. Find the major ocean currents in the Northern Hemisphere, such as the California Current, the Gulf Stream, and the Kuroshio Current.

Note that they all move in a clockwise direction. Find the major ocean currents in the Southern Hemisphere, such as the Peru Current, the Benguela Current, the Brazil Current, and the East Australia Current. Note that they all move in a counterclockwise direction. The direction of major ocean currents can best be described as clockwise in the Northern Hemisphere and counterclockwise in the Southern Hemisphere.

69. **3** Jetties are built perpendicular to the shoreline, and breakwaters are built parallel to the shoreline. They are built to decrease shoreline erosion by absorbing or deflecting the energy of moving water, so that the water is less able to transport sediments.

WRONG CHOICES EXPLAINED:

(1) Ice formation is the result of low air temperatures. Jetties and breakwaters do not change the air temperature.

(2) A *longshore current* is a flow of water parallel to the shore. Jetties block this flow and decrease the speed of a longshore current.

(4) The salinity of seawater depends on the concentrations of dissolved ions. Jetties and breakwaters are made of insoluble materials and do not affect the salinity of the water.

70. **1** Liquid water has an unusually high specific heat. This means that a large change in the heat content of water produces a relatively small change in temperature. Therefore, a large body of water undergoes less drastic changes in temperature than adjacent land bodies and has a moderating effect on their climates. Since water is a fluid, it can move from place to place, which helps it moderate the climates of the Earth. Convection currents and global ocean currents carry water warmed at the Equator north and water cooled near the Poles south. Refer to the "Surface Ocean Currents" diagram in the *Earth Science Reference Tables*.

WRONG CHOICES EXPLAINED:

(2) Water does not exist as a high-density solid. Solid water is, in fact, less dense than liquid water. For this reason, ice floats on water.

(3) A description of the climate of a region gives a general sense of the weather patterns it experiences, including the temperature and rainfall. Anything that directly influences either of them also affects the climate. The ability of water to dissolve and transport minerals does not affect temperature, rainfall, or other aspects of a region's climate.

(4) The ability of water to infiltrate loose sediments and deposit mineral cements does not affect the temperature or rainfall experienced in a region, or any other aspects of its climate.

GROUP D—**Glacial Processes**

71. **4** Under pressure, ice behaves like a fluid. Ice flows in a manner similar to liquid water, but at a much slower rate because ice is so viscous. Both water in a stream and ice in a glacier move fastest on the steepest slope.

WRONG CHOICES EXPLAINED:

(1) Valley glaciers carry sediments ranging from fine clays to huge boulders. Mountain streams, which flow rapidly down steep slopes, often transport coarse sediments, such as pebbles and cobbles.

(2) Streams form V-shaped valleys, but glaciers form U-shaped valleys.

(3) Even the fastest glaciers move downslope much slower than liquid water. It is highly unlikely that both would transport materials at the same rate.

72. **2** Find the Generalized Bedrock Geology of New York State map in the *Earth Science Reference Tables*. In the section labeled "Geological Periods in New York," note the description "unconsolidated gravels, sands, clays (not bedrock)" next to the symbol for Cretaceous, Tertiary, and Pleistocene materials. The largest area shaded with this symbol is Long Island. The location with coordinates 40°45′ N 73°30′ W is the only one of the choices that falls within Long Island.

73. **4** Pebbles and other large particles frozen into glacial ice scratch deep, parallel grooves, called glacial striations, into the surface of the bedrock over which the glacier moves. The presence of such grooves in the bedrock of New York State is evidence of glaciers, which suggests much colder temperatures in the geologic past than at present.

WRONG CHOICES EXPLAINED:

(1) Large rock-salt deposits form when the water in shallow seas evaporates, and the salt in the water crystallizes out. Such large-scale evaporation is associated with warmer temperatures than exist at present.

(2) Petrified fern-tree trunks are evidence of a tropical climate, with temperatures warmer than at present.

(3) Dinosaurs are generally associated with tropical or warm-temperate climates. These climates are warmer than the present climate of New York State.

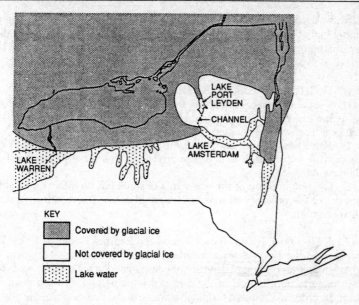

KEY

▨ Covered by glacial ice

☐ Not covered by glacial ice

⋯ Lake water

74. **2** The edges of ice sheets are marked by *terminal moraines*, which are piles of unlayered, unsorted sediments deposited when the ice melted.

WRONG CHOICES EXPLAINED:

(1) Flat, thick deposits of clay generally accumulate in the still waters of a deep lake. While some lakes associated with glaciers develop such layers, lakes fed by streams far from the edge of a glacier can also develop them.

(3) Folded layers of bedrock are the result of crustal movement, not glacial activity. Glaciers transport and deposit sediments, not rock layers.

(4) Formations of rock with interlocking crystals are formed by the crystallization of magma or as a result of evaporation of water from oceans and lakes. Neither of these processes is associated with glacial margins.

75. **4** Find the maps entitled Generalized Bedrock Geology of New York State and Generalized Landscape Regions of New York State in the *Earth Science Reference Tables*. Find the position of Lake Amsterdam in the diagram in the question, and compare this position with the Landscape Regions map. Note that Lake Amsterdam falls within the region labeled "Hudson-Mohawk Lowlands." Now refer to the Bedrock Geology map, and note that the present-day Mohawk River runs through the former site of Lake Amsterdam.

76. **3** A steep-walled channel indicates that the water was eroding the channel bed at a much faster rate than the walls of the channel were eroding. This situation is typical of a channel in which water is flowing at a very high velocity.

WRONG CHOICES EXPLAINED:

(1) A small volume of water has less ability to transport sediment than a large volume of water. Thus, it would have less power to erode and would not produce a steep-walled channel.

(2) Water flows with a low velocity in a channel with a very gentle slope. Slow-moving water has little power to erode and would not produce a steep-walled channel.

(4) The temperature of the water in a channel has no effect on its ability to erode, so the presence of warm water could not be inferred from a steep-walled channel.

77. **2** Find Long Island on the Generalized Bedrock Geology of New York State map in the *Earth Science Reference Tables*. According to the map legend, the materials on Long Island are unconsolidated gravels, sands, and clays. These materials are typical of the terminal moraine and outwash plain of a glacier. In order to deposit a terminal moraine and outwash plain, the edge of the ice would have had to reach Long Island. Only the diagram in choice (2) shows the glacial ice reaching Long Island.

78. **3** Find the chart entitled Geologic History of New York State at a Glance in the *Earth Science Reference Tables*. In the column labeled "Epoch," find the row labeled "Pleistocene." Move horizontally to the right to the column labeled "Important Fossils of New York." Note the picture labeled "Mastodont." Mastodonts lived during the Pleistocene Epoch. Mastodont bones found in sediments in New York State would support the inference that the sediments were deposited during the Pleistocene Epoch.

WRONG CHOICES EXPLAINED:

(1) Locate the "Important Fossils of New York" column of the chart entitled Geologic History of New York State at a Glance in the *Earth Science Reference Tables*. Find the picture labeled "Eurypterid." Follow the row left to the "Epoch" column. Note that eurypterids lived during the Early Silurian and are not associated with Pleistocene sediments.

(2) As described in wrong choice (1), find the picture labeled "Stromatolites." Note that stromatolites lived during the Precambrian and are not associated with Pleistocene sediments.

(4) As described in wrong choice (1), find the picture labeled "Coelophysis." Note that these dinosaurs lived during the Triassic and are not associated with Pleistocene sediments.

79. **1** Carbon-14 is the radioactive isotope commonly used to date organic matter buried during the Pleistocene for two reasons. To understand the first reason, find the Geologic History of New York State at a Glance chart in the *Earth Science Reference Tables*. Find "Pleistocene" in the column labeled "Epoch." Note that the Pleistocene ended 0.01 millions of years ago. (0.01 million years = 10,000 years) Now, find the Radioactive Decay Data in the Physical Constants section of the *Earth Science Reference Tables*. Carbon-14, with a half life of 5.7×10^3 or 5,700 years, could be used to date materials of Pleistocene age. Each of the other listed isotopes has such a long half-life that, in 10,000 years, the quantity that would have decayed would be immeasurably small. The second reason that carbon-14 is used to date organic matter buried during the Pleistocene is that all organic matter contains carbon. Some of this carbon is carbon-14. While an organism is alive, whatever carbon-14 decays is replaced as the organism assimilates air, food, and water from its environment. After an organism dies, the carbon-14 that decays is not replaced. The longer the organism has been dead, the less carbon-14 it contains.

80. **1** Find the maps entitled Generalized Bedrock Geology of New York State and Generalized Landscape Regions of New York State in the *Earth Science Reference Tables*. Note that Albany is south of the Adirondacks. Pieces of anorthositic rock from the Adirondacks found near Albany would support the inference that a glacier moved from north to south over New York State.

WRONG CHOICES EXPLAINED:

(2) Glacial striations align with the direction of glacial movement, so east-west striations would not support an inference that the ice moved from north to south.

(3) The Bedrock Geology map shows that Devonian age rocks are found in all directions around Elmira. Because the rocks could have come from any direction, their presence does not support an inference that the ice moved from north to south.

(4) There is no connection between the river's present-day direction of flow at Niagara Falls and the movement of ice during the Pleistocene glaciation.

GROUP E—Atmospheric Energy

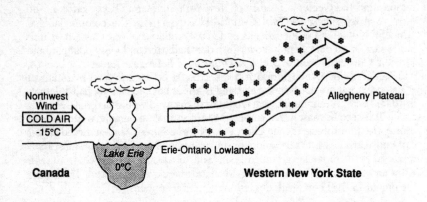

81. **3** As the air rises, both the air temperature and the dewpoint temperature decrease as a result of adiabatic processes. A cloud forms at the altitude at which the air temperature and the dewpoint temperature are the same. To find this altitude, refer to the Lapse Rate graph in the *Earth Science Reference Tables*. Find 0°C on the temperature scale and note the bold black line sloping upward to the left. This line represents the decreasing air temperature with increasing altitude. Now, use the fact that the dashed lines are 2°C apart to find –6°C on the temperature scale. Note the dashed line sloping more steeply upward to the left from this temperature value. This dashed line represents the decreasing dewpoint temperature with increasing altitude. Follow the bold line and the dashed line upward to the left until they intersect. The value of the altitude where they intersect is 0.7 km, so this is the altitude at which a cloud forms.

82. **1** An air mass that forms over central Canada is forming over land in the center of the North American continent. The air mass is a continental air mass. In the diagram, the air coming from Canada is labeled with a temperature of –15°C, making the air mass a polar air mass. Refer to the Weather Map Information in the *Earth Science Reference Tables*. A continental polar air mass should be labeled cP.

83. **2** As it travels from the Erie-Ontario Lowlands to the Allegheny Plateau, the air first rises. As the air rises, the air pressure around it decreases, and the air expands. This expansion causes it to cool. The air cools, until it reaches the dewpoint of the water vapor in it, and the water vapor then condenses.

84. **4** The water in Lake Erie is in the liquid state. The water that enters the atmosphere is in the gaseous state. The change from liquid to gas is called *evaporation*.

WRONG CHOICES EXPLAINED:

(1) *Condensation* is the change from gas to liquid. This change causes water to leave the atmosphere, not enter it.

(2) *Precipitation* is the process in which solid or liquid water falls to the Earth's surface from the atmosphere. This process causes water to leave the atmosphere, not enter it.

(3) *Transpiration* involves the uptake of water by plant roots and its release into the atmosphere through their leaves. Lake Erie is not covered with vegetation, so water from the lake does not enter the atmosphere by transpiration.

85. **3** Lake Erie contains liquid water that is freezing, or changing from solid to liquid. Find the Equations and Proportions section of the *Earth Science Reference Tables*. The equation for finding the latent heat for a change from liquid to solid is listed as $Q = mH_f$. Find the Properties of Water table in the Physical Constants section of the *Earth Science Reference Tables*. The latent heat of fusion (H_f) of water is 80 cal/g. You are given a mass of 10 g in the question. Substituting in the equation for latent heat:

$$Q = mH_f$$
$$Q = (10 \text{ g}) (80 \text{ cal/g}) = 800 \text{ cal}$$

For every 10 g of water that freezes, 800 cal of latent heat is released.

86. **1** A lake-effect snowstorm occurs when water in a lake evaporates into air moving across the lake and the moisture-laden air then moves over a cold land surface adjacent to the lake. Over the land surface, the air cools to its dewpoint temperature, at which the moisture picked up from the lake sublimes to form snow. When the lake is covered with ice, the evaporation of water into the air above the lake is blocked, and the lake cannot act as the source of moisture for the snow. Therefore, as Lake Erie freezes over, the chance of a lake-effect snowstorm decreases.

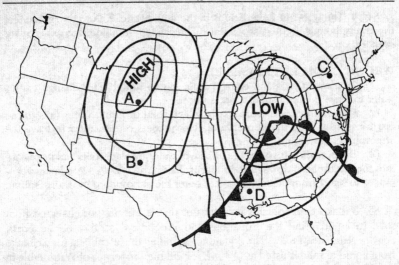

87. **4** An air mass has relatively uniform characteristics throughout. However, an adjacent air mass may have very different characteristics. Changes in weather are associated with the movement of one air mass into another across a boundary called a front. Refer to the front symbols in the Weather Map Information in the *Earth Science Reference Tables*. Note the presence of two fronts on the map in the question. A cold front lies to the west of point *D* and is moving east. A warm front lies to the south of point *C* and is moving north. During the next 6 hours, the greatest weather change will occur as a front approaches and then passes through one of the points. This change is most likely to occur at point *D*, which is closest to an approaching front.

88. **2** Great rainfall is associated with wet moisture belts. Find the diagram entitled Planetary Wind and Moisture Belts in the Troposphere in the *Earth Science Reference Tables*. Note the wet moisture belts at the equator, 60° N, and 60° S. Look carefully at the dashed arrows indicating the planetary winds to the north and south of each region. In each case, the winds are converging. Look at the outside ring of the diagram and note the solid arrows, which show the converging air rising. Thus, in areas of great rainfall, winds converge and air rises.

89. **3** Tornadoes are violent disturbances that develop over land from intense thunderstorms. (Most tornadoes are less than 100 meters in diameter, and last only a few minutes.) Their small size and brief occurrence make tornadoes difficult to predict. Weather stations are too far apart and record data over too large a time interval to be of much use in identifying, tracking, or predicting tornadoes.

WRONG CHOICES EXPLAINED:

(1) Thunderstorms are associated with a region just ahead of a cold front. Once the front has been identified and its motion plotted, the occurrence of thunderstorms associated with the front can be predicted well in advance.

(2) Hurricanes are large-scale disturbances that develop over a period of days or weeks. This slow development makes it possible to track the motion of a hurricane and to make forecasts.

(4) A snowstorm can be predicted by tracking air masses moving toward the lake. Air masses are large enough and move slowly enough to allow forecasts several hours in advance of their arrival.

90. **1** Cooling by terrestrial reradiation involves the escape into space of longwave radiation from the Earth. Both water vapor (humidity) and water droplets (clouds) in the atmosphere block longwave radiation. Therefore, the greatest cooling by terrestrial reradiation occurs on a clear night with low humidity.

GROUP F—Astronomy

91. **1** Find the Equations and Proportions section of the *Earth Science Reference Tables*. Under "Proportions," find Kepler's harmonic law of planetary motion, which states:

$$(\text{period of revolution})^2 \propto (\text{mean radius of orbit})^3$$

From this law, there is a direct relationship between the period of revolution of a satellite and its mean radius of orbit. Thus, if the average distance from a satellite to the Earth is decreased, the period of revolution will decrease.

92. **2** Only the Earth has large amounts of water in all three states of matter. Other planets have some water, but their surfaces are either too hot or too cold for water to exist in all three states. For example, the Venera probes sent to Venus revealed a surface temperature of 750K (890°F). The Mariner probes sent to Mars found some water frozen in the polar ice caps, but the amount was small compared to the amount found on Earth.

93. **4** Locate the chart entitled Solar System Data in the *Earth Science Reference Tables*. Examine the column labeled "Density." Note that only Mercury, Venus, and Mars have high densities, similar to the Earth's density. Because Mercury, Venus, and Mars have a high density and their rocky composition resembles the Earth's, they are called terrestrial planets.

94. **2** Chemical weathering involves reactions between rocks and gases or liquids that were not present in the environment in which the rocks formed. On the Earth, oxygen, water, and carbon dioxide are chiefly responsible for chemical weathering. Because the moon has no atmosphere or liquid water, chemical weathering is non-existent.

WRONG CHOICES EXPLAINED:

(1) The moon does have gravity, though less than the Earth. Moreover, the absence of gravity does not stop chemical reactions from occurring.

(3) Low temperatures slow reaction rates but do not stop them from occurring altogether. Furthermore, the increase in surface temperatures on the moon during the lunar day would increase reaction rates. The reactions fail to take place because there is nothing to react with, not because it is too cold.

(4) High temperatures would increase reaction rates and produce more evidence of chemical weathering, if there were substances on the moon that react with rocks.

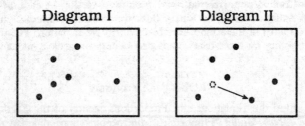

95. **3** From Diagram I to Diagram II, only one object changed position in relation to the others. This movement is typical of planets, which change position in relation to the background of distant stars when observed from the Earth at one-month intervals.

WRONG CHOICES EXPLAINED:

(1) A galaxy is very far from the Earth. So, in order to change position enough in one month for its motion to be detectable from the Earth, the galaxy would have to travel faster than the speed of light, which is impossible.

(2) A *meteor* is an atmospheric phenomenon caused by a piece of solid matter from space falling through the air. Because of friction, the matter heats up until it glows. At most, meteors last for seconds, not months.

(4) Stars are so far from the Earth that they do not appear to change their relative position significantly during the course of a single month.

96. **1** The atmosphere of Venus is mostly carbon dioxide, CO_2, which is a greenhouse gas. When energy from the sun is absorbed by Venus and then reradiated, the carbon dioxide traps the energy. The result is high temperatures. Nitrogen, oxygen, and hydrogen are not greenhouse gases. They are transparent to reradiated longwave radiation and would allow Venus to lose energy and cool.

97. **2** As radiation travels away from the sun, individual waves diverge. The farther they are from the sun, the more they diverge, and the fewer waves will strike a given area of surface. So, the farther a planet is from the sun, the

less solar energy strikes a given area of its surface, and the lower is its surface temperature. The only graph that correctly shows this relationship is choice (2). Please note: Since all three planets are in orbits outside the Earth's, as distance from the Earth increases, distance from the sun also increases.

98. **4** Observations of the universe indicate that it is expanding. Tracing the motions of galaxies back in time suggests a point about 20–30 billion years ago at which they were very close to each other. Thus, we have a model in which the universe started out with all its matter in a very small volume and then expanded in all directions. Because the model suggests motion similar to an explosion, the model is known as the big bang theory. The universe existed first, and the matter then split into large masses that we call galaxies. The galaxies, including the Milky Way galaxy, then coalesced into clumps of matter as a result of gravity. Some of the clumps were large enough that the intense force of gravity caused nuclear reactions to begin, so that a star formed. The formation of the star we refer to as the sun marked the beginning of our solar system. Thus, the correct sequence of formation is universe, Milky Way, solar system.

99. **3** To be the farthest galaxy from the Earth, the galaxy must be traveling away from the Earth at the greatest speed. The light waves from galaxies traveling away from the Earth are shifted toward the red end of the spectrum. The faster the galaxy is moving away, the more its light waves are shifted toward the red. The spectrum in choice (3) is shifted farthest toward the red and represents the galaxy farthest from the Earth.

100. **1** The geocentric model places the Earth at the center of the solar system, with the sun and planets orbiting the Earth. Only choice (1) shows this configuration.

PART III

Mass of Material Remaining in Sieve

Shaking Time (minutes)	Rock Material *A* (grams)	Rock Material *B* (grams)	Rock Material *C* (grams)
0	25.0	25.0	25.0
5	24.5	20.0	17.5
10	24.0	18.5	12.5
15	23.5	17.0	7.5
20	23.5	12.5	5.0

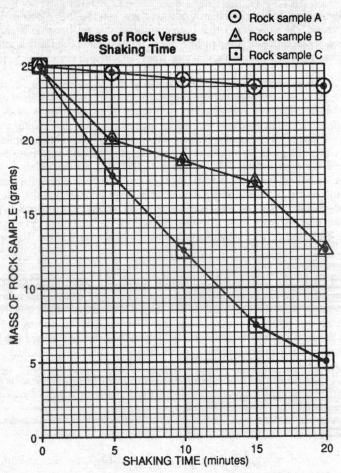

Mass of Rock Versus Shaking Time

- ⊙ Rock sample A
- △ Rock sample B
- ⊡ Rock sample C

MASS OF ROCK SAMPLE (grams) vs. SHAKING TIME (minutes)

101–103. Using the data in the table, mark points for rock material A at the appropriate coordinates, (0,25.0), (5,24.5), (10,24.0), and so on. Surround each point with a circle, as shown in the example. Repeat for rock materials B and C, surrounding the points with triangles and squares, respectively. Then, connect the points as shown in the graph, above.

104. All three rock samples were shaken for the same length of time under the same conditions. However, rock sample A wore away less than rock sample B, and rock sample B wore away less than rock sample C. To wear away less under the same conditions, rock sample A must be harder and more resistant to weathering. This information must be stated in one or more complete sentences.

<u>Example</u>: Sample *A* weathered more slowly than samples *B* and *C* because sample *A* was harder and more resistant to weathering than samples *B* and *C*.

105. (a) Find the Equations and Proportions in the *Earth Science Reference Tables*. The equation for rate of change is:

$$\text{rate of change} = \frac{\text{change in field value}}{\text{change in time}}$$

In this case, the field value is mass. The equation for the rate of change in mass is:

$$\text{rate} = \frac{\text{change in mass}}{\text{change in time}}$$

This equation can be expressed in a number of ways for full credit. Or,

$$r = \frac{\Delta m}{\Delta t}$$

$$\text{rate} = \frac{\text{change in mass}}{\text{change in time}}$$

$$\text{rate} = \Delta \frac{m}{t}$$

(b) Rock sample *C* changed from a mass of 25.0 grams at 0 minutes to a mass of 5.0 grams after 20 minutes of shaking. These data should be substituted as follows:

$$\text{rate} = \frac{25.0 \pm 5.0 \text{ grams}}{20 \text{ minutes}}$$

(c) Calculate the rate of change in mass and label the answer with the correct units.

$$= \frac{20 \text{ grams}}{20 \text{ minutes}} = 1.0 \text{ g/min.}$$

106. During the process of abrasion, any part of a rock surface that juts out, such as an edge or a corner, is more likely to be struck by another rock particle and broken off. The softer and less resistant the material, the more such irregularities will be broken off, and the smoother and more rounded the appearance of the rock. This information must be stated in one or more complete sentences for full credit.

Example: The edges and corners of the rock sample would be rounded and smooth.

You will get one credit for a scientifically correct answer and one additional credit for stating your scientifically correct answer in a complete sentence.

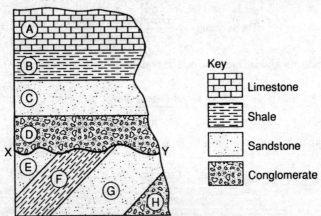

Key
- Limestone
- Shale
- Sandstone
- Conglomerate

107.   Look at the diagram and note the dotted pattern in rock layer C. According to the key in the diagram, rock layer C is sandstone. Refer to the Scheme for Sedimentary Rock Identification in the *Earth Science Reference Tables*. In the column labeled "Rock Name," find "Sandstone." Follow the row that contains "Sandstone" to the left until you reach the column labeled "Grain Size." Note that sandstone is composed of sand, which ranges from 0.006 cm to 0.2 cm in diameter. Also note that in the column labeled "Comments," sand particles are described as fine to coarse. Either answer is acceptable, but stating that the particles are 0.006 cm to 0.2 cm in diameter is the better answer.

108.   Line XY represents a buried erosional surface. Your answer must include a reference to the concept of erosion.

Example: After layers E, F, G, and H were deposited, they were tilted and uplifted. Then, erosion wore down these layers unevenly, forming the surface marked by the line XY.

You will receive one credit for a scientifically correct reason that includes the concept of erosion and one additional credit for writing the answer in a complete sentence.

109.   Look at the diagram and note the brick-like pattern in rock layer A. The key shows that this pattern represents limestone. Next, look at the diagram and note the dashed-line pattern in rock layer B. The key shows that this pattern represents shale. Now, refer to the Scheme for Sedimentary Rock Identification in the *Earth Science Reference Tables*. In the column labeled

"Rock Name," find the rows that contain "Limestone" and "Shale." Read across these rows to find the characteristics of each rock. You will receive one credit for each difference in composition.

Example: *A* is nonclastic; *B* is clastic.
           A is made primarily of calcite; *B* is not.

110. The *principle of superposition* states that, unless they are disturbed, the oldest sedimentary rock layers are at the bottom, and the youngest are at the top. Layer *C* is younger than layer *D* and older than layer *B*. Layer *C* is between 438 and 505 million years old. Refer to the chart entitled Geologic History of New York State at a Glance in the *Earth Science Reference Tables*. To the right of the column labeled "Epoch" is a time scale. Find 438 million years ago and 505 million years ago. Note the name for this period is listed to the left of the time scale as Ordovician. Rock layer *C* could have formed during the Ordovician period.

111. An isoline connects points of equal field value, in this case, equal Mercalli intensities. Begin by connecting all of the points marked 2 with a smooth line. The question instructs you to use an interval of 2 units, so the next isoline should connect points labeled as 2 units higher, that is, 4 Mercalli units. The third isoline connect points labeled as 2 higher than 4, that is, 6 Mercalli units. Your finished map should look like this:

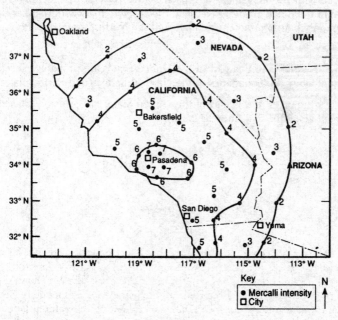

You will receive one credit each for any three isolines you plot correctly and one credit for plotting only the 2, 4, and 6 Mercalli unit isolines.

112. The damage caused by an earthquake is usually greatest at locations closest to the epicenter, because the seismic waves have the most energy there. The higher the Mercalli rating, the greater the damage and the closer the location is to the epicenter. You will receive one credit for stating that Pasadena is closest to the earthquake epicenter.

113. The latitude scale is marked along the left side of the map. Draw a horizontal line through Bakersfield to intersect this scale. The line intersects the scale midway between 35° N and 36° N. Each degree consists of 60', so midway between these latitudes is a latitude of 35°30' N. The longitude scale is on the bottom edge of the map. Draw a vertical line through Bakersfield to intersect this scale. The longitude is closest to the 119° W marking. Bakersfield's latitude is 35°30' N and its longitude is 119° W. You will receive one credit for stating the latitude, correct to within ± 30', and one credit for stating the longitude, correct to within ± 30'.

114. Your answer should contain a scientifically acceptable cause for earthquakes in the California area. Earthquakes are caused by the sudden movement of rock along faults. The force that produces this motion is most often associated with the movement of crustal plates. Look at the Tectonic Plates diagram in the *Earth Science Reference Tables*. Note that California lies along the boundary between the Pacific plate and the North American plate.

(a)   Earthquakes in this part of the United States are probably caused by movement along the boundaries of the North American and Pacific plates.

(b)   Earthquakes in this part of the United States are probably caused by movement along the San Andreas fault.

Unit (1–9) or Optional/Extended Topic (A–F)	Question Numbers (total)	Wrong Answers (x)	Grade
1. Earth Dimensions	2, 3, 11, 34, 113 (5)		$\dfrac{100(5-x)}{5} = \%$
2. Minerals and Rocks	1, 4–8, 40, 109 (8)		$\dfrac{100(8-x)}{8} = \%$
A. Rocks, Minerals, and Resources	41–50 (10)		$\dfrac{100(10-x)}{10} = \%$
3. The Dynamic Crust	11, 54, 112, 114 (4)		$\dfrac{100(4-x)}{4} = \%$
B. Earthquakes and the Earth's Interior	10, 51–60 (11)		$\dfrac{100(11-x)}{11} = \%$
4. Surface Processes and Landscapes	9, 12, 13, 17, 22, 75, 104, 106, 107 (9)		$\dfrac{100(9-x)}{9} = \%$
C. Oceanography	61–70 (10)		$\dfrac{100(10-x)}{10} = \%$
D. Glacial Geology	71–80 (10)		$\dfrac{100(10-x)}{10} = \%$
5. Earth's History	14–16, 21, 23, 79, 108, 110 (8)		$\dfrac{100(8-x)}{8} = \%$
6. Meteorology	18–20, 24, 25, 27–29, 82 (9)		$\dfrac{100(9-x)}{9} = \%$
E. Latent Heat and Atmospheric Energy	81, 83–90 (9)		$\dfrac{100(9-x)}{9} = \%$
7. Water Cycle and Climates	26 (1)		$\dfrac{100(1-x)}{1} = \%$
8. The Earth in Space	30–39, 91 (9)		$\dfrac{100(9-x)}{9} = \%$
F. Astronomy Extensions	91–100 (10)		$\dfrac{100(10-x)}{10} = \%$
9. Environmental Awareness	35, 36 (2)		$\dfrac{100(2-x)}{2} = \%$

To further pinpoint your weak areas, use the Topic Outline in the front of the book.

Examination June 1996

Earth Science — Program Modification Edition

PART I

Answer all 40 questions in this part.

Directions (1–40): For *each* statement or question, select the word or expression that, of those given, best completes the statement or answers the question. Record your answers in the spaces provided. [40]

1 Measurements of the Sun's altitude at the same time from two different Earth locations a known distance apart are often used to determine the

1 circumference of the Earth
2 period of the Earth's revolution
3 length of the major axis of the Earth's orbit
4 eccentricity of the Earth's orbit

1 _____

2 The data table below gives information on mineral hardness.

MINERAL HARDNESS

Moh's Hardness Scale		Approximate Hardness of Common Objects
Talc	1	
Gypsum	2	Fingernail (2.5)
Calcite	3	Copper penny (3.5)
Fluorite	4	Iron nail (4.5)
Apatite	5	Glass (5.5)
Feldspar	6	Steel file (6.5)
Quartz	7	Streak plate (7.0)
Topaz	8	
Corundum	9	
Diamond	10	

Moh's scale would be most useful for

1 identifying a mineral sample
2 finding the mass of a mineral sample
3 finding the density of a mineral sample
4 counting the number of cleavage surfaces of a mineral sample

2 _____

3 Which diagram best represents a sample of the metamorphic rock gneiss? [Diagrams show actual size.]

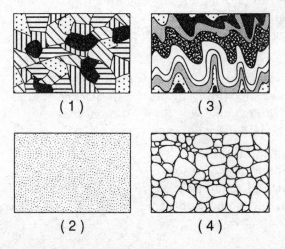

(1)   (3)

(2)   (4)

3 _____

4 Which two igneous rocks could have the same mineral composition?

1 rhyolite and diorite
2 pumice and scoria
3 peridotite and andesite
4 gabbro and basalt   4 _____

5 According to the *Earth Science Reference Tables*, much of the surface bedrock of the Adirondack Mountains consists of

1 slate and dolostone
2 gneiss and quartzite
3 limestone and sandstone
4 conglomerate and red shale   5 _____

6 The theory of plate tectonics suggests that

1 the continents moved due to changes in the Earth's orbital velocity
2 the continents' movements were caused by the Earth's rotation
3 the present-day continents of South America and Africa are moving toward each other
4 the present-day continents of South America and Africa once fit together like puzzle parts   6 _____

7 Contact zones between tectonic plates may produce trenches. According to the *Earth Science Reference Tables*, one of these trenches is located at the boundary between which plates?

1 Australian and Pacific
2 South American and African
3 Australian and Antarctic
4 North American and Eurasian   7 _____

8 Which cross section best represents the general
bedrock structure of New York State's Allegheny
Plateau?

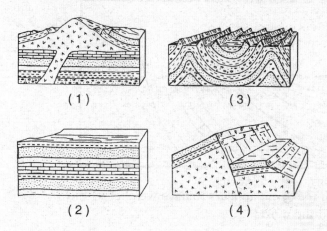

(1)　　　　　　　　　(3)

(2)　　　　　　　　　(4)

8 _____

9 Which statement best describes a stream with a
steep gradient?

1 It flows slowly, producing a V-shaped valley.
2 It flows slowly, producing a U-shaped valley.
3 It flows rapidly, producing a V-shaped valley.
4 It flows rapidly, producing a U-shaped valley.

9 _____

10 The diagram below represents a geologic cross section.

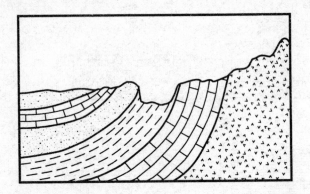

Which rock type appears to have weathered and eroded the most?

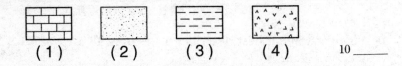

(1)   (2)   (3)   (4)   10 _____

11 In which type of landscape are meandering streams most likely found?

(1) gently sloping plains
(2) regions of waterfalls
(3) steeply sloping hills
(4) V-shaped valleys   11 _____

12 What change will a pebble usually undergo when it is transported a great distance by streams?

1 It will become jagged and its mass will decrease.
2 It will become jagged and its volume will increase.
3 It will become rounded and its mass will increase.
4 It will become rounded and its volume will decrease.

12 _____

13 A large, scratched boulder is found in a mixture of unsorted, smaller sediments forming a hill in central New York State. Which agent of erosion most likely transported and then deposited this boulder?

1 wind   3 a glacier
2 ocean waves   4 running water

13 _____

14 The map below represents a river as it enters a lake.

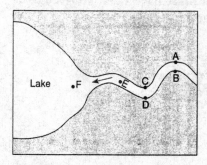

At which locations is the amount of deposition greater than the amount of erosion?

(1) *A*, *C*, and *E*   (3) *B*, *D*, and *F*
(2) *B*, *C*, and *F*   (4) *A*, *D*, and *E*

14 _____

15 The velocity of a stream is decreasing. As the velocity approaches zero, which size particle will most likely remain in suspension?

1 pebble
2 sand
3 clay
4 boulder

15 _____

16 What is the relative age of a fault that cuts across many rock layers?

1 The fault is younger than all the layers it cuts across.
2 The fault is older than all the layers it cuts across.
3 The fault is the same age as the top layer it cuts across.
4 The fault is the same age as the bottom layer it cuts across.

16 _____

17 In order for an organism to be used as an index fossil, the organism must have been geographically widespread and must have

1 lived on land
2 lived in shallow water
3 been preserved by volcanic ash
4 existed for a geologically short time

17 _____

18 The cartoon below is a humorous look at geologic history.

Early Pleistocene mermaids

If Early Pleistocene mermaids had existed, their fossil remains would be the same age as fossils of

1 armored fish   3 trilobites

2 mastodonts   4 dinosaurs   18 _____

19 According to the *Earth Science Reference Tables*, when did the Jurassic Period end?

(1) 66 million years ago

(2) 144 million years ago

(3) 163 million years ago

(4) 190 million years ago   19 _____

20 The table below gives information about the radioactive decay of carbon-14. [Part of the table has been left blank for student use.]

Half-Life	Mass of Original C-14 Remaining (grams)	Number of Years
0	1	0
1	$\frac{1}{2}$	5,700
2	$\frac{1}{4}$	11,400
3	$\frac{1}{8}$	17,100
4		
5		
6		

What is the amount of the original carbon-14 remaining after 34,200 years?

(1) $\frac{1}{8}$ g   (2) $\frac{1}{16}$ g   (3) $\frac{1}{32}$ g   (4) $\frac{1}{64}$ g   20 _____

Note that question 21 has only three choices.

21 If a sample of a radioactive substance is crushed, the half-life of the substance will

1 decrease
2 increase
3 remain the same   21 _____

22 According to the *Earth Science Reference Tables*, which geologic event is associated with the Grenville Orogeny?

  1 the formation of the ancestral Adirondack Mountains

  2 the advance and retreat of the last continental ice sheet

  3 the separation of South America from Africa

  4 the initial opening of the Atlantic Ocean   22 _____

23 In which atmospheric layer is most water vapor found?

  1 troposphere   3 thermosphere

  2 stratosphere   4 mesosphere   23 _____

24 In order for clouds to form, cooling air must be

  1 saturated and have no condensation nuclei

  2 saturated and have condensation nuclei

  3 unsaturated and have no condensation nuclei

  4 unsaturated and have condensation nuclei   24 _____

25 Which graph best represents the relationship between air temperature and air density in the atmosphere?

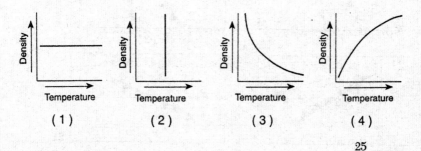

    (1)   (2)   (3)   (4)

25 _____

26 The greatest source of moisture entering the atmosphere is evaporation from the surface of

  1 the land   3 lakes and streams

  2 the oceans   4 ice sheets and glaciers   26 _____

27 Which angle of the Sun above the horizon produces the greatest intensity of sunlight?

  (1) 70°   (2) 60°   (3) 40°   (4) 25°   27 _____

28 The diagram below represents a cross section of air masses and frontal surfaces along line *AB*. The dashed lines represent precipitation.

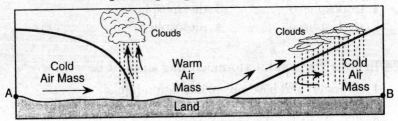

Which weather map best represents this frontal system?

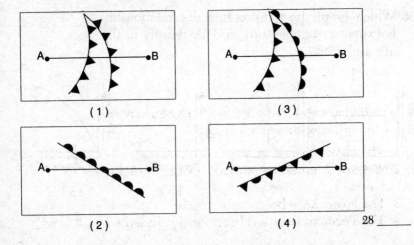

28 _____

29 The cross section below shows several locations in the State of Washington and the annual precipitation at each location. The arrows represent the prevailing wind direction.

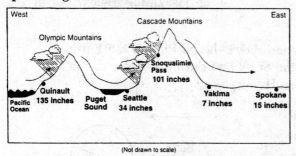

(Not drawn to scale)

Why do the windward sides of these mountain ranges receive more precipitation than the leeward sides?

1 Sinking air compresses and cools.
2 Sinking air expands and cools.
3 Rising air compresses and cools.
4 Rising air expands and cools.   29 _____

30 The diagram below represents a cross section of a series of rock layers of different geologic ages.

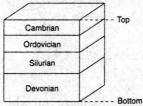

Which statement provides the best explanation for the order of these rock layers?

1 The oldest layer is on the bottom.
2 A buried erosional surface exists between layers.
3 The layers have been overturned.
4 The Permian layer has been totally eroded.   30 _____

31 In New York State, which day has the shortest period of daylight?

1 March 21   3 September 21
2 June 21   4 December 21   31 _____

32 Which diagram shows the position of the Earth relative to the Sun's rays during a winter day in the Northern Hemisphere?

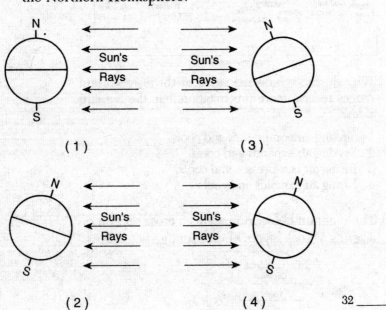

(1)   (3)

(2)   (4)   32 _____

33 To an observer located at the Equator, on which date would the Sun appear to be directly overhead at noon?

1 February 1   3 March 21
2 June 6   4 December 21   33 _____

34 The new-moon phase occurs when the Moon is positioned between the Earth and the Sun. However, these positions do not always cause an eclipse (blocking) of the Sun because the

  1 Moon's orbit is tilted relative to the Earth's orbit
  2 new-moon phase is visible only at night
  3 night side of the Moon faces toward the Earth
  4 apparent diameter of the Moon is greatest during the new-moon phase   34 ____

35 The diagram below shows an instrument made from a drinking straw, protractor, string, and rock.

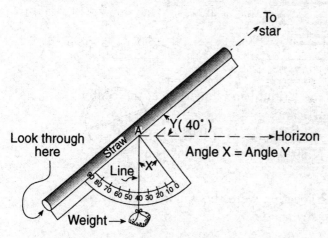

This instrument was most likely used to measure the

  1 distance to a star
  2 altitude of a star
  3 mass of the Earth
  4 mass of the suspended weight   35 ____

Base your answers to questions 36 through 40 on the *Earth Science Reference Tables*, the weather map below, and your knowledge of Earth science. The weather map shows a hurricane that was located over southern Florida. The isobars show air pressure in inches of mercury. Letters *A* through *D* represent four widely separated locations.

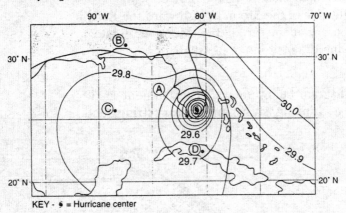

KEY - 🌀 = Hurricane center

36 What is the latitude and longitude at the center of the hurricane?

   (1) 26° N 81° W   (3) 34° N 81° W

   (2) 26° N 89° W   (4) 34° N 89° W   36 _____

37 At which location were the winds of this hurricane the strongest?

   (1) *A*   (2) *B*   (3) *C*   (4) *D*   37 _____

38 What was the direction of movement of surface winds associated with this hurricane?

   1 counterclockwise and away from the center

   2 counterclockwise and toward the center

   3 clockwise and away from the center

   4 clockwise and toward the center   38 _____

39 Which station model best represents some of the atmospheric conditions at location *A*?

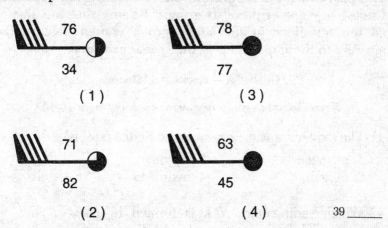

76
34
(1)

78
77
(3)

71
82
(2)

63
45
(4)

39 _____

40 Which map best shows the most likely track of this hurricane?

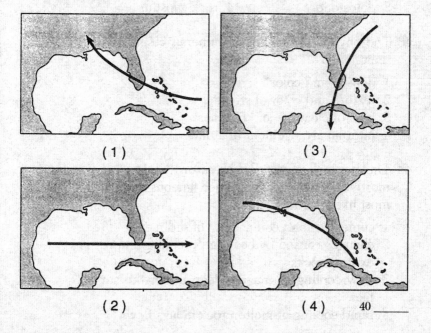

(1)

(3)

(2)

(4)   40 _____

PART II

This part consists of six groups, each containing ten questions. Choose any one of these six groups. Be sure that you answer all ten questions in the single group chosen. Record the answers to these questions in the spaces provided. [10]

GROUP A — Rocks and Minerals

If you choose this group, be sure to answer questions 41–50.

41 The most abundant element in the Earth's crust is

    1 nitrogen   3 silicon
    2 oxygen   4 hydrogen   41 _____

42 Which sedimentary rock is formed by compaction and cementation of land-derived sediments?

    1 siltstone   3 rock salt
    2 dolostone   4 rock gypsum   42 _____

43 The physical properties of minerals result from their

    1 density and color
    2 texture and color of streak
    3 type of cleavage and hardness
    4 internal arrangement of atoms   43 _____

44 The bedrock of the flat areas on the Moon is mostly basalt. This fine-grained igneous rock was most likely formed by the

    1 cementing and compacting of sediments
    2 changes caused by heat and pressure on pre-existing rocks
    3 slow cooling of magma deep under the surface
    4 rapid cooling of molten rock in lava flows   44 _____

45 Heat and pressure due to magma intrusions may result in

1 vertical sorting
2 graded bedding
3 contact metamorphism
4 chemical evaporites

45 _____

46 In the diagram below, each angle of the triangle represents a 100 percent composition of the mineral named at that angle. The percentage of the mineral decreases toward 0 percent as either of the other angles of the triangle is approached. Letter A represents the mineral composition of an igneous rock.

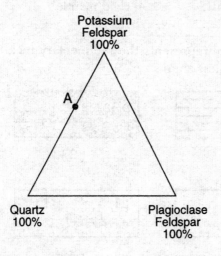

Rock A is a coarse-grained igneous rock that can best be identified as

1 rhyolite   3 granite
2 pumice   4 gabbro

46 _____

47 What is one difference between the metamorphic rocks quartzite and hornfels?

  1 Hornfels is foliated; quartzite is nonfoliated.
  2 Hornfels contains plagioclase; quartzite does *not* contain plagioclase.
  3 Hornfels is produced by regional metamorphism; quartzite is produced by contact metamorphism.
  4 Hornfels is medium grained; quartzite is fine grained.

47 _____

48 In which part of the Earth are felsic rocks most likely to be found?

  1 continental crust   3 plastic mantle
  2 oceanic crust   4 rigid mantle

48 _____

49 Which symbol represents the sedimentary rock with the smallest grain size?

(1 )   (3)

(2)   (4)

49 _____

50 Which diagram best represents the silicon-oxygen tetrahedron of which talc, feldspar, and quartz are composed?

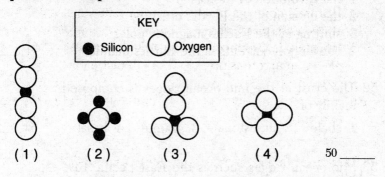

KEY

● Silicon   ○ Oxygen

(1)   (2)   (3)   (4)   50 _____

GROUP B — Plate Tectonics

If you choose this group, be sure to answer questions **51–60**.

Base your answers to questions 51 through 55 on the *Earth Science Reference Tables*, the map below, and your knowledge of Earth science. The map shows mid-ocean ridges and trenches in the Pacific Ocean. Specific areas *A, B, C,* and *D* are indicated by shaded rectangles.

51 Movement of the crustal plates shown in the diagram is most likely caused by

    1 the revolution of the Earth
    2 the erosion of the Earth's crust
    3 shifting of the Earth's magnetic poles
    4 convection currents in the Earth's mantle   51 _____

52 The crust at the mid-ocean ridges is composed mainly of

    1 shale   2 limestone   3 granite   4 basalt   52 _____

53 Mid-ocean ridges such as the East Pacific Rise and the Oceanic Ridge are best described as

    1 mountains containing folded sedimentary rocks
    2 mountains containing fossils of present-day marine life
    3 sections of the ocean floor that contain the youngest oceanic crust
    4 sections of the ocean floor that are the remains of a submerged continent   53 _____

54 Which map best shows the direction of movement of the oceanic crustal plates in the vicinity of the East Pacific Rise (ridge)?

(1)   (2)   (3)   (4)   54 _____

55 The cross section below represents an area of the Earth's crust within the map region.

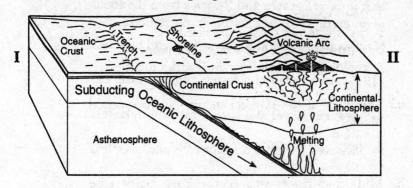

Which shaded rectangular area on the map does this cross section represent?

1 Area *A*   2 Area *B*   3 Area *C*   4 Area *D*   55 _____

56 A seismogram recorded at a seismic station is shown below.

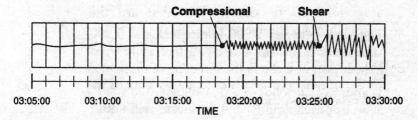

Which information can be determined by using this seismogram?

1 depth of the earthquake's focus
2 direction to the earthquake's focus
3 location of the earthquake's epicenter
4 distance to the earthquake's epicenter   56 _____

57 Which statement best describes the relationship between the travel rates and travel times of earthquake *P*-waves and *S*-waves from the focus of an earthquake to a seismograph station?

(1) *P*-waves travel at a slower rate and take less time.

(2) *P*-waves travel at a faster rate and take less time.

(3) *S*-waves travel at a slower rate and take less time.

(4) *S*-waves travel at a faster rate and take less time.   57 _____

58 An earthquake's *P*-wave traveled 4,800 kilometers and arrived at a seismic station at 5:10 p.m. At approximately what time did the earthquake occur?

(1) 5:02 p.m.   (3) 5:10 p.m.

(2) 5:08 p.m.   (4) 5:18 p.m.   58 _____

59 The rock between 2,900 kilometers and 5,200 kilometers below the Earth's surface is inferred to be

1 an iron-rich solid   3 a silicate-rich solid

2 an iron-rich liquid   4 a silicate-rich liquid   59 _____

60 Where is the thickest part of the Earth's crust?

1 at mid-ocean ridges

2 at transform faults

3 under continental mountain ranges

4 under volcanic islands   60 _____

GROUP C — **Oceanography**

If you choose this group, be sure to answer questions **61–70**.

Base your answers to questions 61 through 63 on the map and profile shown below. The map shows the major areas of the North Atlantic Ocean. Letters *A*, *B*, *C*, and *D* represent locations on the ocean floor. The profile represents the ocean bottom from point *X* in North America along the dashed line to point *Y* in Africa. Note that the profile is vertically exaggerated.

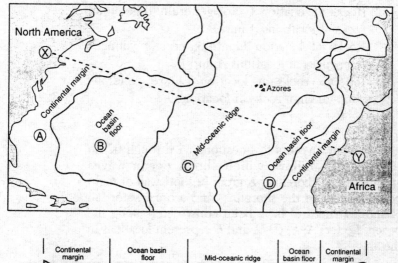

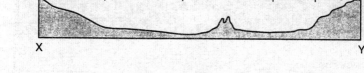

61 Classification of the ocean bottom into the areas shown is based on the

1 distance from continental landmasses
2 topography of the ocean floor
3 age of ocean-bottom rocks
4 type of ocean-bottom sediments

61 _____

62 At which location would land-derived sediments most likely be accumulating on the ocean bottom?

(1) A   (3) C

(2) B   (4) D

62 _____

63 Which statement about the age of ocean-floor rocks is correct?

1 All ocean-floor rocks are generally the same age.

2 Rocks at location C are generally older than rocks at locations A and B.

3 Rocks at location C are generally younger than rocks at locations A and B.

4 Igneous rocks at location D are generally younger than rocks at location C.

63 _____

Base your answers to questions 64 through 67 on the diagram below, which shows ocean waves approaching a shoreline. A groin (a short wall of rocks perpendicular to the shoreline) and a breakwater (an offshore structure) have been constructed along the beach. Letters A, B, C, D, and E represent locations in the area.

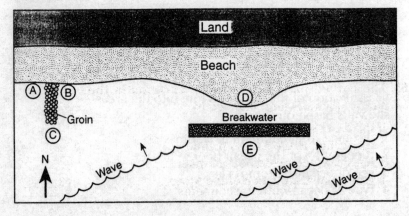

64 What is the most common cause of the approaching waves?

1 underwater earthquakes
2 variations in ocean-water density
3 the gravitational effect of the Moon
4 winds at the ocean surface   64 _____

65 This shoreline is located along the east coast of North America. Which ocean current would most likely modify the climate of this shoreline?

1 Florida Current   3 Brazil Current
2 Canaries Current   4 California Current   65 _____

66 At which location will the beach first begin to widen due to sand deposition?

(1) A   (3) C
(2) B   (4) E   66 _____

Note that question 67 has only three choices.

67 The size of the bulge in the beach at position D will

1 decrease
2 increase
3 remain the same   67 _____

68 The curved pattern of surface currents in the North Atlantic is most affected by

1 the Earth's rotation
2 convection in the Earth's mantle
3 density differences of ocean water
4 gravitational attraction of the Moon   68 _____

69 What is the source of most dissolved minerals in seawater?

1 weathering of seafloor rocks
2 weathering and erosion of continental rocks
3 deep-ocean organic-matter sediments
4 gases from underwater volcanic eruptions   69 _____

70 When the seafloor moves as a result of an underwater earthquake and a large tsunami develops, what will most likely occur?

1 Deep-ocean sediments will be transported over great distances.
2 No destruction will occur near the origin of the earthquake.
3 The direction of the tsunami will be determined by the magnitude of the earthquake.
4 Severe destruction will occur in coastal areas.   70 _____

GROUP D — Glacial Processes

*If you choose this group, be sure to answer questions **71–80**.*

Base your answers to questions 71 through 74 on the diagram below, which represents a landscape in which sediments were deposited by a continental glacier. Letters *A*, *B*, *C*, *D*, and *E* represent locations in this area.

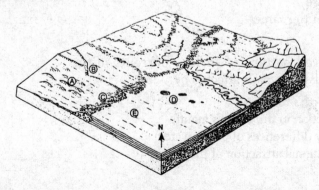

71 The sediments deposited at location *E* are best described as

1 sorted and unlayered
2 sorted and layered
3 unsorted and unlayered
4 unsorted and layered   71 _____

72 A terminal moraine that marks the farthest advance of the glacier is found at location

(1) *A*   (3) *C*
(2) *B*   (4) *D*   72 _____

73 Features formed by the melting of isolated ice blocks surrounded by sediment are found near which location?

(1) *A*   (3) *C*
(2) *B*   (4) *D*   73 _____

74 Which arrow best represents the direction of ice movement that formed the deposits shown in the diagram?

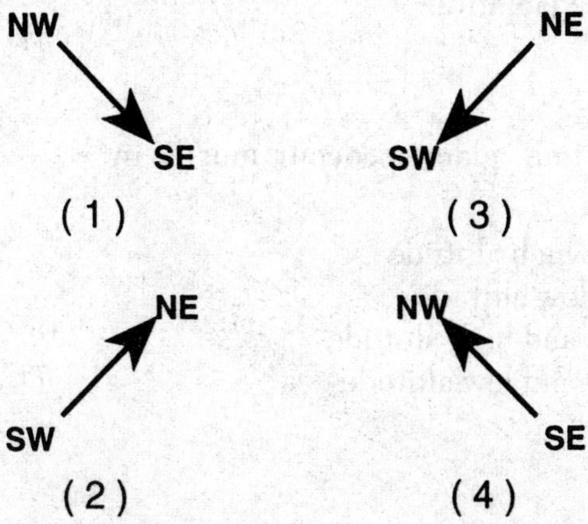

NW
   SE
(1)

NE
SW
(3)

NE
SW
(2)

NW
   SE
(4)   74 _____

75 Which statement identifies a result of glaciation that has had a positive effect on the economy of New York State?

  1 Large amounts of oil and natural gas were formed.
  2 The number of usable water reservoirs was reduced.
  3 Many deposits of sand and gravel were formed.
  4 Deposits of fertile soil were removed.   75 _____

76 The record of glaciation in New York State provides a source of information about

  1 changes in bedrock composition
  2 changes in the global climate
  3 movements of crustal plates
  4 plants and animals from the Paleozoic Era   76 _____

77 Because of glaciation, New York State presently has soils that are best described as

  1 deep and residual
  2 rich in gemstone minerals
  3 unchanged by glaciation
  4 thin and rocky   77 _____

78 At the present time, glaciers occur mostly in areas of

  1 high latitude or high altitude
  2 low latitude or low altitude
  3 middle latitude and high altitude
  4 middle latitude and low altitude   78 _____

79 Wooden stakes were placed on a glacier in a straight line as represented by A–A′ in the diagram below. The same stakes were observed later in the positions represented by B–B′.

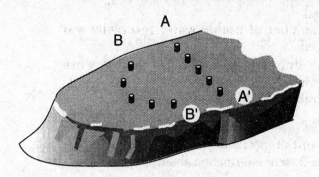

The pattern of movement of the stakes provides evidence that

1 glacial ice does not move
2 glacial ice is melting faster than it accumulates
3 the glacier is moving faster in the center than on the sides
4 friction is less along the sides of the glacier than in the center   79 _____

80 Which landscape region is a flat plain consisting mainly of unsorted clays, gravels, sands, scratched pebbles, boulders, and cobbles?

1 the Adirondacks
2 Long Island
3 the Catskills
4 the Taconics   80 _____

GROUP E — Atmospheric Energy

If you choose this group, be sure to answer questions 81–90.

81 Which is the major source of energy for most Earth processes?

   1 radioactive decay within the Earth's interior
   2 convection currents in the Earth's mantle
   3 radiation received from the Sun
   4 earthquakes along fault zones   81 _____

82 A map of the United States is shown below.

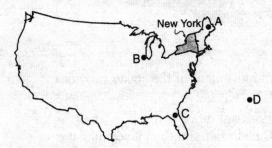

Weather conditions in which location would be of most interest to a person predicting the next day's weather for New York State?

(1) *A*   (2) *B*   (3) *C*   (4) *D*   82 _____

83 Which statement about electromagnetic energy is correct?

   (1) Violet light has a longer wavelength than red light.
   (2) X rays have a longer wavelength than infrared waves.
   (3) Radar waves have a shorter wavelength than ultraviolet rays.
   (4) Gamma rays have a shorter wavelength than visible light.   83 _____

84 The graph below shows the air temperature and air pressure recorded over a 30-day period at one location.

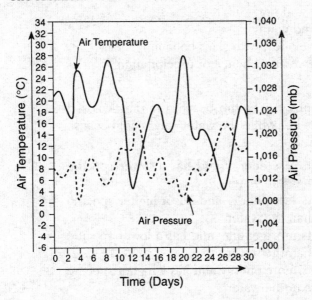

Time (Days)

What were the approximate air temperature and air pressure readings on day 6?

(1) 10°C and 1,024 mb
(2) 19°C and 1,016 mb
(3) 10°C and 1,016 mb
(4) 19°C and 1,024 mb

84 _____

85 Which planetary wind pattern is present in many areas of little rainfall?

1 Winds converge and air sinks.
2 Winds converge and air rises.
3 Winds diverge and air sinks.
4 Winds diverge and air rises.

85 _____

86 Which weather conditions are most probable when the moisture content of the air increases, resulting in a lower atmospheric pressure?

1 sunny and fair
2 cold and windy
3 partly cloudy, with skies becoming clear
4 cloudy, with a chance of precipitation   86 _____

87 On a sunny day at the beach, the dark-colored sand gets hot while the water stays cool because the sand

1 reflects less energy and has a lower specific heat than the water
2 reflects less energy and has a higher specific heat than the water
3 reflects more energy and has a lower specific heat than the water
4 reflects more energy and has a higher specific heat than the water   87 _____

88 During which phase change does water absorb the most heat?

1 freezing
2 melting
3 condensation
4 evaporation   88 _____

89 What is the wet-bulb temperature when the air temperature is 16°C and the relative humidity is 71%?

(1) 11°C   (3) 3°C
(2) 13°C   (4) 19°C   89 _____

90 The diagram below represents the percentage of total incoming solar radiation that is affected by clouds.

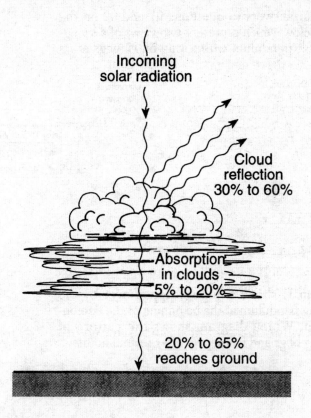

What percentage of incoming solar radiation is reflected or absorbed on cloudy days?

(1) 100%
(2) 35% to 80%
(3) 5% to 30%
(4) 0%

90 _____

If you choose this group, be sure to answer questions **91–100**.

Base your answers to questions 91 and 92 on the diagrams below, which represent two views of a swinging Foucault pendulum with a ring of 12 pegs at its base.

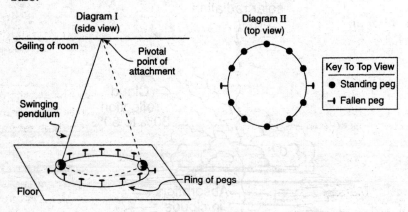

91 Diagram II shows two pegs tipped over by the swinging pendulum at the beginning of the demonstration. Which diagram shows the pattern of standing pegs and fallen pegs after several hours?

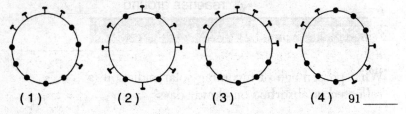

(1)   (2)   (3)   (4)   91 _____

92 The predictable change in the direction of swing of a Foucault pendulum provides evidence that the

  1 Sun rotates on its axis
  2 Sun revolves around the Earth
  3 Earth rotates on its axis
  4 Earth revolves around the Sun   92 _____

93 Which planet's orbital shape would be most similar to Jupiter's orbital shape?

1 Uranus   2 Pluto   3 Venus   4 Mercury   93 _____

94 The diagram below shows several planets at various positions in their orbits at a particular time.

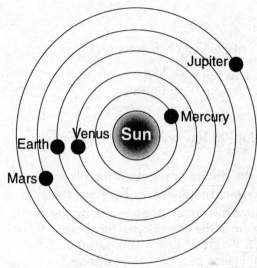

(Not drawn to scale)

Which planet would be visible from the Earth at night for the longest period of time when the planets are in these positions?

1 Mercury   2 Venus   3 Mars   4 Jupiter   94 _____

95 According to current data, the Earth is apparently the only planet in our solar system that has

1 an orbiting moon
2 an axis of rotation
3 atmospheric gases
4 liquid water on its surface   95 _____

96 Which statement best describes the geocentric model of our solar system?

    1 The Earth is located at the center of the model.
    2 All planets revolve around the Sun.
    3 The Sun is located at the center of the model.
    4 All planets *except* the Earth revolve around the Sun.

96 _____

97 Which planet's day is longer than its year?

    1 Mercury   3 Mars
    2 Venus   4 Jupiter

97 _____

Note that question 98 has only three choices.

98 Compared to the distances between the planets of our solar system, the distances between stars are usually

    1 much less
    2 much greater
    3 about the same

98 _____

99 In what way are the planets Mars, Mercury, and Earth similar?

    1 They have the same period of revolution.
    2 They are perfect spheres.
    3 They exert the same gravitational force on each other.
    4 They have elliptical orbits with the Sun at one focus.

99 _____

Base your answers to questions 105 through 108 on the information below and on your knowledge of Earth science.

The climate of an area is affected by many variables such as elevation, latitude, and distance to a large body of water. The effect of these variables on average surface temperature and temperature range can be represented by graphs on grids that have axes labeled as shown below.

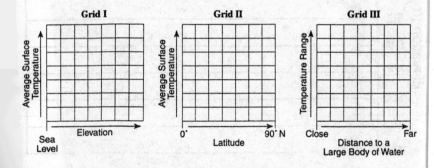

105 On Grid I, draw a line to show the relationship between elevation and average surface temperature. [1]

06 On Grid II, draw a line to show the relationship between latitude and average surface temperature. [1]

7 On Grid III, draw a line to show the relationship between distance to a large body of water and temperature range.   [1]

Barogram for Miami, Florida

100 The symbols below represent the Milky Way galaxy, the solar system, the Sun, and the universe.

◯ = Milky Way Galaxy

⬭ = Solar System

• = Sun

▢ = Universe

Which arrangement of symbols is most accurate?

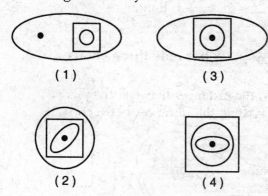

(1)   (3)

(2)   (4)   100 _____

PART III

This part consists of questions 101 through 115. Be sure that you answer all questions in this part. Record your answers in the spaces provided. Some questions may require the use of the *Earth Science Reference Tables.*   [25]

Base your answers to questions 101 through 104 on the topographic map of Cottonwood, Colorado, below. Points A, B, X, and Y are marked for reference.

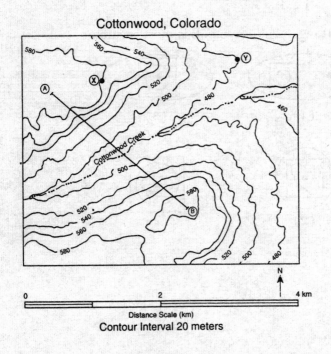

Cottonwood, Colorado

Distance Scale (km)
Contour Interval 20 meters

101 State the general direction in which Cottonwood Creek is flowing. [1]

102 State the highest possible elevation, to the *nearest meter*, for point *B* on the topographic map. [1]

_____ meters

103 On the grid provided, draw a profile of the topography along line *AB* shown on the map. [3]

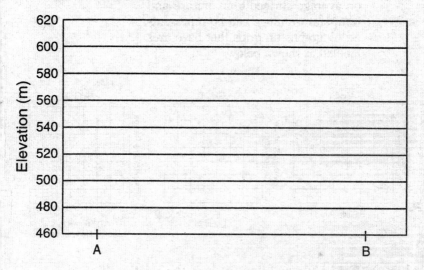

104 In the space provided, calculate the gradient of the slope between points *X* and *Y* on the topographic map, following the directions below.

 a Write the equation for gradient. [1]

 b Substitute data from the map into the equation. [1]

 c Calculate the gradient and label it with the proper units. [1]

Base your answers to questions 105 through 108 on the information below and on your knowledge of Earth science.

> The climate of an area is affected by many variables such as elevation, latitude, and distance to a large body of water. The effect of these variables on average surface temperature and temperature range can be represented by graphs on grids that have axes labeled as shown below.

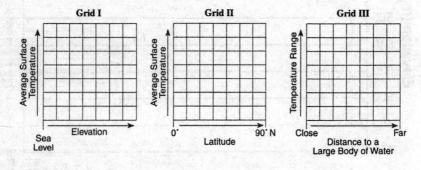

105 On Grid I, draw a line to show the relationship between elevation and average surface temperature. [1]

106 On Grid II, draw a line to show the relationship between latitude and average surface temperature. [1]

107 On Grid III, draw a line to show the relationship between distance to a large body of water and temperature range. [1]

108 The climate of most locations near the Equator is warm and moist.

  a Explain why the climate is warm. [1]

_____

_____

_____

  b Explain why the climate is moist. [1]

_____

_____

_____

  Base your answers to questions 109 through 111 on the diagram below and on your knowledge of Earth science. The diagram represents the apparent path of the Sun on the dates indicated for an observer in New York State. The diagram also shows the angle of Polaris above the horizon.

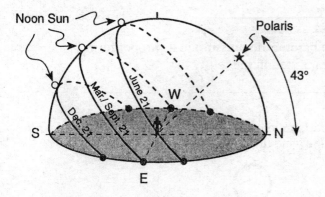

109 State the latitude of the location represented by the diagram to the *nearest degree*. Include the latitude direction in your answer.   [2]

_____

110 Label the zenith on the diagram. [1]

111 On the diagram, draw the apparent path of the Sun on May 21. Mark the position of sunrise on May 21 and label it *Sunrise*. [2]

Base your answers to questions 112 through 114 on the weather station data shown in the table below.

Air Temperature	21°C
Barometric Pressure	993.1 mb
Wind Direction	From the east
Windspeed	25 knots

112 State the air temperature in degrees Fahrenheit. [1]

_____ °F

113 State the barometric pressure in its proper form, as used on a station model.   [1]

_____

114 On the station model, draw a line with feathers to indicate the wind direction and speed. [2]

N
↑

Base your answer to question 115 on the newspaper article below and your knowledge of Earth science.

Legislation Protects Ozone

The governor of New York signed environmental legislation that restricted the use of ozone-depleting chemicals employed in refrigeration systems, air-conditioners, and fire extinguishers.

The law restricts, and in some cases bans, the sale of chlorofluorocarbons and halons. Both have been found to contribute to the destruction of the Earth's ozone layer, which protects the Earth from dangerous ultraviolet rays of the Sun.

115 Using one or more complete sentences, state one reason that ultraviolet rays are dangerous. [2]

_____

_____

_____

Answers
June 1996

Earth Science —
Program Modification Edition

Answer Key

PART I

1. 1	8. 2	15. 3	22. 1	29. 4	36. 1
2. 1	9. 3	16. 1	23. 1	30. 3	37. 1
3. 3	10. 3	17. 4	24. 2	31. 4	38. 2
4. 4	11. 1	18. 2	25. 3	32. 4	39. 3
5. 2	12. 4	19. 2	26. 2	33. 3	40. 1
6. 4	13. 3	20. 4	27. 1	34. 1	
7. 1	14. 2	21. 3	28. 3	35. 2	

PART II

Group A	Group B	Group C	Group D	Group E	Group F
41. 2	51. 4	61. 2	71. 2	81. 3	91. 4
42. 1	52. 4	62. 1	72. 3	82. 2	92. 3
43. 4	53. 3	63. 3	73. 4	83. 4	93. 1
44. 4	54. 1	64. 4	74. 1	84. 2	94. 3
45. 3	55. 4	65. 1	75. 3	85. 3	95. 4
46. 3	56. 4	66. 2	76. 2	86. 4	96. 1
47. 2	57. 2	67. 2	77. 4	87. 1	97. 2
48. 1	58. 1	68. 1	78. 1	88. 4	98. 2
49. 1	59. 2	69. 2	79. 3	89. 2	99. 4
50. 4	60. 3	70. 4	80. 2	90. 2	100. 4

PART III See answers explained.

Answers Explained

PART I

1. **1** Suppose the altitude of the Sun is measured at the same time at two different locations along the same line of longitude and that the angular difference between the two altitudes is 36°. Since the circumference of the Earth is 360°, the distance between the two locations represents 1/10 of the distance around the Earth (i.e., 36°/360°). The distance between the two locations can now be multiplied by 10 to determine the circumference of the Earth. The equation for calculating the circumference of the Earth by Eratosthenes' method appears in the Equations and Proportions chart in the *Earth Science Reference Tables*.

WRONG CHOICES EXPLAINED:
(2) The period of the Earth's revolution, which is 1 year or about 365 days, can be determined by noting the Earth's position in space in relation to the stars.

(3) The length of the major axis of the Earth's orbit can be determined from the shape of the orbit and measurements of the distance between the Earth and the Sun.

(4) The eccentricity of the Earth's orbit can also be determined from the shape of the orbit and the varying distances between the Earth and the Sun.

2. **1** Moh's scale is useful for identifying a mineral sample because each mineral has a distinct hardness. For example, there are a number of different varieties of quartz, which differ in color: citrine is yellow, amethyst is purple, and there are smoky quartz, milky quartz, rose (pink) quartz and so on. At first glance these might appear to be different minerals. Even before testing to show that these varieties have the same chemical composition (the color differences are due to impurities present), a hardness test using Moh's scale would reveal that all samples have the same hardness.

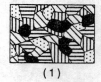

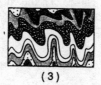

(1)   (2)   (3)   (4)

3. **3** Metamorphic rocks form when heat and pressure causes partial melting of the minerals in an existing rock. Sometimes this melting and the ensuing recrystallization cause the minerals that melted to migrate into bands

or layers. Gneiss is an example of a metamorphic rock that commonly contains bands or layers, as illustrated in diagram (3).

4. **4** Find in the *Earth Science Reference Tables* the chart entitled Scheme for Igneous Rock Identification. The fact that gabbro and basalt are in the same column means that they are composed of the same minerals, although they differ in grain size.

5. **2** Find in the *Earth Science Reference Tables* the map entitled Generalized Landscape Regions of New York State, and note the Adirondack Mountains in the upper right side of the map. Now find the corresponding map entitled Generalized Bedrock Geology of New York State. The symbols at the bottom of this map show that the area in the upper right, where the Adirondack Mountains are located, has surface bedrock composed of gneiss and quartz.

6. **4** According to the theory of plate tectonics, at some time in the past the current continents of South America and Africa were part of one large landmass. This landmass later split and the pieces drifted apart, being carried by convection cells in the mantle. By studying the shapes of present-day continents, scientists have inferred that South America and Africa once fit together like puzzle parts.

WRONG CHOICES EXPLAINED:
(1) Changes in the orbital velocity of the Earth would not move continents.
(2) If the continent's movements were caused by the Earth's rotation, all of the continents would be moving in the same direction. Observations have shown that continents are moving in many different directions, including the direction opposite that of the Earth's rotation.
(3) Find in the *Earth Science Reference Tables* the diagram entitled Tectonic Plates, and note that South America and Africa are shown to be moving apart.

7. **1** Find the chart entitled Tectonic Plates in the *Earth Science Reference Tables,* and locate the plate boundaries described in the answer choices. Note that the Tonga Trench is located along the boundary between the plates listed in choice (1), the Australian and Pacific plates. None of the other plate boundaries has a trench associated with it.

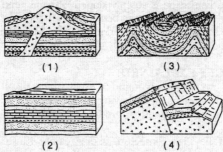

(1) (3)

(2) (4)

8. **2** In a plateau the rock layers are generally horizontal. Therefore the cross section in diagram (2) best represents the general bedrock structure of New York State's Allegheny Plateau. Although erosion may carve out steep valleys to create the appearance of mountains, the underlying rock layers remain horizontal. An example of such a landscape is the Grand Canyon, where the rock layers are horizontal despite the steep valley walls that have been carved by the Colorado River.

9. **3** The slope of a stream's bed is its gradient. A stream with a steep gradient is flowing down a steep slope, and water flows rapidly down a steep slope. The faster a stream flows, the more sediment it can carry and the faster it will cut downward into its bed. Rapid downcutting deepens a stream valley faster than it widens, resulting in a steep-sided V-shaped valley. Therefore, the best description of a stream with a steep gradient is that it flows rapidly, producing a V-shaped valley.

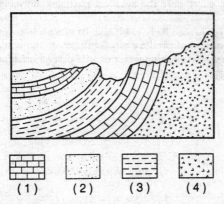

(1) (2) (3) (4)

10. **3** Note that the rock layer near the center of the diagram, represented by the symbol for choice (3), appears to be lower at the surface than the adjacent layers. It may be inferred, therefore, that the rock type (3) has weathered and eroded most, leaving the indentation at the surface of that layer.

11. **1** When a landscape is steeply sloped, streams tend to flow rapidly and follow a straight path. As the area becomes level, the velocity of the water decreases and the streams follow meandering paths. Meandering streams are characteristic of the level areas of gently sloping plains.

WRONG CHOICES EXPLAINED:

(2) Waterfalls are commonly found in areas where the elevation is dropping rapidly and the streams are following straighter paths.

(3) Steeply sloping hills tend to produce streams that flow rapidly over straight paths.

(4) V-shaped valleys are characteristic of streams with steep slopes where the water is flowing rapidly and is eroding the beds of the streams faster than the valley walls.

12. **4** As a pebble is being transported by a stream, it strikes other particles being carried by the stream and located along the bed of the stream. As a result, pieces of the pebble will break off, particularly the sharper points, so that the pebble becomes rounded and decreases in volume.

13. **3** Glaciers pick up and carry particles with a wide range of sizes. Along the base of the glacier, the particles being carried scratch the rock surface over which they pass. When the ice melts, this mixture is deposited in one place and often remains as a deposit that contains particles ranging in size from clay to boulders.

WRONG CHOICES EXPLAINED:

(1) Winds can carry only the smallest particles, normally sand-sized or even smaller.

(2) Ocean waves are unlikely to be able to move a boulder. Furthermore, the water action would not produce any scratches on the boulder.

(4) Even high-speed running water cannot carry a boulder.

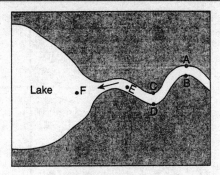

14. **2** The amount of deposition exceeds the amount of erosion in a river when the velocity of the water is decreasing. Along a curve in a river, the velocity slows down along the inside of the curve and speeds up along the outside. Net deposition will occur at locations B and C, therefore, because they are on the inside of curves in the river. Deposition will also exceed erosion at point F because the velocity of the water decreases as the river flows into the lake.

15. **3** As the velocity of the stream decreases, the largest particles are deposited first, with the smallest particles remaining in suspension the longest. Of the choices listed, the smallest particle is clay. The actual size of each of these particle types can be found in the graph entitled Relationship of Transported Particle Size to Water Velocity in the *Earth Science Reference Tables*.

16. **1** A fault is younger than all the rock layers it cuts across because the rock layers were in place before the fault occurred. If a continuous, undisturbed rock layer is found above the fault, this rock layer formed after the fault occurred and the fault is older than the rock layer.

17. **4** Index fossils are used to estimate the age of a rock layer. To be useful as an index fossil, an organism must have lived in many different locations so that rocks from widely separated areas can be correlated. It must also have been in existence for only a short period of geologic time. If an organism existed for millions of years, its presence as a fossil in a rock layer could not be used to infer the age of the layer.

WRONG CHOICES EXPLAINED:
(1) Most index fossils represent organisms that lived in water. The remains of an organism are more likely to be preserved when the organism falls to the bottom of a body of water and is quickly covered with sediments.
(2) The remains of organisms are more likely to be preserved in deeper water, in which the covering sediments at the bottom are less likely to be disturbed.
(3) The high temperatures of volcanic ash are more likely to destroy the remains of organisms than is the quiet water at the bottom of the ocean.

Early Pleistocene mermaids

18. **2** Find in the *Earth Science Reference Tables* the diagram entitled Geologic History of New York State at a Glance. In the column labeled "Epoch," find the Pleistocene Epoch near the top. In the column to the right, labeled "Life on Earth," note that one of the three life forms present during this epoch was mastodonts.

WRONG CHOICES EXPLAINED:
(1) Note in the geologic history diagram that armored fish became extinct during the Devonian Period, which occurred much before the Pleistocene epoch.

(3) According to the geologic history diagram, trilobites became extinct toward the end of the Permian Period.

(4) The geologic history diagram shows that dionsaurs become extinct during the Cretaceous Period.

19. **2** Find in the *Earth Science Reference Tables* the diagram entitled Geologic History of New York State at a Glance. Locate in the column headed "Period" the Jurassic Period. The diagram shows that the upper boundary or end of this period occurred 144 million years ago.

Half-Life	Mass of Original C-14 Remaining (grams)	Number of Years
0	1	0
1	$\frac{1}{2}$	5,700
2	$\frac{1}{4}$	11,400
3	$\frac{1}{8}$	17,100
4		
5		
6		

20. **4** Note from the diagram that the half-life of carbon-14 is 5,700 years. At the end of each half-life the amount of carbon-14 is reduced by half. Therefore, after 17,100 years only 1/8 of the original mass remains. After another 5,700 years only 1/16 would remain. Since a period of 34,200 years represents three half-lives beyond 17,100 years, the mass would be reduced by half three more times until only 1/64 of the original mass remained (i.e., there would be 1/16, then 1/32, finally 1/64). The original carbon-14 had a mass of 1 gram; hence, after 34,200 years, 1/64 gram remains.

21. **3** The half-life of a radioactive substance is the amount of time required for one-half of a sample of that substance to decay. Radioactive decay rates are not affected by changes in physical and environmental conditions. The rate of decay of a substance remains the same if the sample is cut in half, crushed, heated, and so on.

22. **1** Find in the *Earth Science Reference Tables* the diagram entitled Geologic History of New York State at a Glance. Locate the column headed "Important Geologic Events in New York." The Grenville Orogeny appears near the bottom of the column. The diagram also indicates that the ancestral Adirondack Mountains formed at this time, during the Precambrian Era.

23. **1** The troposphere is the layer of the Earth's atmosphere that is closest to the surface, and most of the gases of the atmosphere are found in this layer. Above the troposphere the air is very thin. Find in the *Earth Science Reference Tables* the graphs entitled Selected Properties of the Earth's Atmosphere, and note that the concentration of water vapor approaches zero at the top of the troposphere.

24. **2** When air is cooled to the dewpoint temperature, condensation can begin to occur. For vapor to change to droplets of water, there must also be nuclei upon which the droplets can form. In the atmosphere, the starting nuclei for condensation are usually dust particles. Therefore, for clouds to form, the air must be saturated and condensation nuclei must be present.

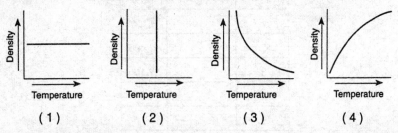

(1)   (2)   (3)   (4)

25. **3** When the temperature of the air increases, the air expands. This increase in volume causes the air to become less dense. Graph (3) shows this pattern in which increasing air temperature produces a decrease in air density.

26. **2** The evaporation of water from the surface of the oceans is the greatest source of atmospheric moisture because more than half of the Earth's surface is covered by the oceans. Evaporation of water from the land, lakes and streams, and ice sheets and glaciers provides a much smaller total proportion of the moisture entering the atmosphere.

27. **1** The higher the Sun is in the sky, the greater is the angle between the Sun's position and the horizon. When the angle is greatest, the intensity of sunlight per unit of area is also greatest. Of the choices given, 70° is the greatest angle.

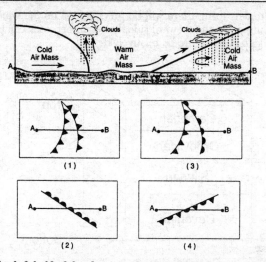

28. **3** The left half of the diagram shows a cold air mass on the left pushing a warm air mass on the right. The boundary between these two air masses is a cold front. The right half of the diagram shows the warm air on the left pushing the cold air on the right. The boundary here is a warm front. Find in the *Earth Science Reference Tables* the charts headed Weather Map Information, and note in the chart on the right the symbols for cold and warm fronts. Diagram (3) shows a cold front on the left and a warm front on the right, a pattern that corresponds to the frontal system in the diagram.

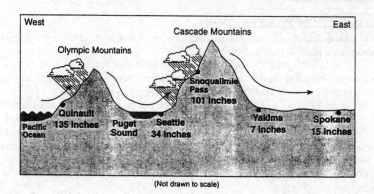

(Not drawn to scale)

29. **4** When air first reaches the base of the mountains, it begins to rise. As air rises, it expands and cools. Then, if the air is cooled to the dewpoint temperature, moisture begins to condense. If enough moisture condenses and

large water droplets form, there will be precipitation on the windward side. The air traveling down the leeward side, however, becomes compressed and warmer. This air will therefore be relatively dry.

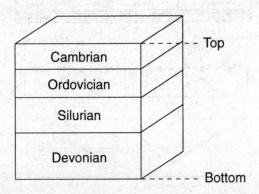

30. **3** Find the chart entitled Geologic History of New York State at a Glance in the *Earth Science Reference Tables,* and locate the column headed "Period." In this column, find the geologic ages of the rock layers in the diagram in the question, and note that the oldest layer is on the top and the youngest layer is on the bottom. According to the principle of superposition, the bottom layer of a sedimentary series, which was deposited first, is the oldest unless it has been overturned or had older rock thrust over it. The best explanation for the order of the rock layers in the diagram is that the layers have been overturned.

WRONG CHOICES EXPLAINED:
(1) As described above, in the diagram the oldest layer is on the top and the youngest layer is on the bottom.
(2) The symbol used to indicate a buried erosional surface is a wavy line. There is no wavy line in the diagram. Also, there is no missing geologic age in the sequence of layers that would indicate missing layers.
(4) Find the chart entitled Geologic History of New York State at a Glance in the *Earth Science Reference Tables,* and locate Permian in the column headed "Period." Note that the Permian Period is more recent than the Devonian and falls outside the sequence of ages in the diagram. The question contains no additional evidence to support the notion that a Permian layer was eroded, nor does absence of a Permian layer explain the inverted time sequence of the layers.

31. **4** In New York State the Sun's path through the sky is at its lowest and shortest on December 21. Therefore, the period of daylight is shortest on this date.

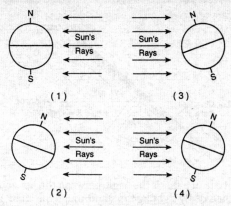

32. **4** On a winter day in the Northern Hemisphere the angle of the Earth's axis is tilted away from the direction of the Sun's rays, so that the solar rays strike the Earth's surface at a smaller angle. This situation is shown in diagram (4). The fact that the intensity of insolation is less at this time results in colder temperatures.

33. **3** Since the Earth's axis is tilted at an angle of $23^{1}/2°$, the vertical ray of the Sun passes to the north and south of the Equator as the Earth revolves around the Sun. The Sun is directly over the Equator only at the equinoxes. Of the choices given, only (3) is the date of an equinox.

34. **1** For an eclipse to occur, the umbra of the Moon's shadow must reach the Earth's surface. However, the umbra of the Moon's shadow barely reaches the Earth, and the small, circular shadow the Moon casts on the Earth's surface is never more than 269 kilometers in diameter. Solar eclipses are rare because the 5° tilt of the Moon's orbit, together with the small size of the shadow that can reach the Earth, makes it likely that the shadow will miss the Earth at the new-moon phase.

WRONG CHOICES EXPLAINED:
(2) During the new-moon phase, the Moon is on the side of the Earth facing the Sun. Night occurs on the side of the Earth *opposite* the Sun. Therefore, during the new-moon phase, the Moon is not visible during the night.
(3) During the new-moon phase, the night side of the Moon always faces toward the Earth; therefore this fact cannot explain why eclipses do not occur during every new-moon phase.
(4) The apparent diameter of the Moon depends on the distance of the Moon from the Earth when viewed; the diameter is greatest when the moon is at perigee and smallest when the Moon is at apogee. The new-moon phase depends on the relative positions of the Earth, Moon, and Sun and can occur when the Moon is at any point in its orbit, including both apogee and perigee.

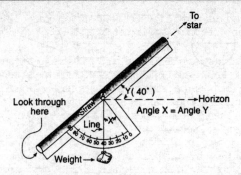

35. **2** The altitude of a star is the angle between the horizon and the position of the star in the sky. By sighting the star through the straw, a person using the instrument in the diagram can measure the angle between the star and the horizon.

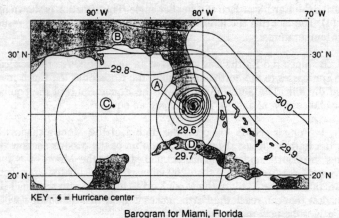

KEY - **⌇** = Hurricane center

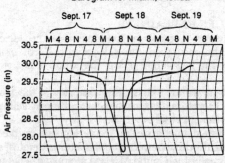

36. **1** Locate the center of the hurricane on the map, and note that it is currently near the southern tip of Florida. It is located just north of the 25° N

line of latitude and just west of the 80° W line of longitude. The current location of the center of the hurricane is therefore about 26° N 81° W.

37. **1** Winds blow from regions of high pressure to regions of low pressure. The greater the pressure gradient, the stronger the wind. On a weather map, the greater the pressure gradient, the more closely spaced the isobars. In the map in the question, note that the most closely spaced isobars are located around the hurricane center near point A. Therefore, the winds of this hurricane were strongest at point A.

38. **2** Winds blow from regions of high pressure to regions of low pressure. On the weather map in the question, note that the pressure decreases as you approach the hurricane center from any direction. Therefore, the surface winds will blow toward the center. However, they will not blow directly toward the center because of the Coriolis effect. In the Northern Hemisphere the Coriolis effect causes winds to be deflected toward the right, resulting in a counterclockwise pattern of movement. Therefore, the direction of movement of surface winds associated with this hurricane will be counterclockwise and toward the center.

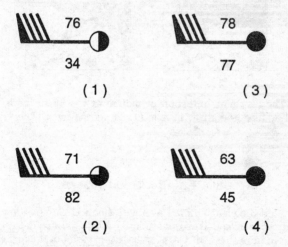

39. **3** Location A is close to the hurricane center, a region of low pressure, high winds, thick clouds, and torrential rains. Refer to the sample station model in the chart entitled Weather Map Information in the *Earth Science Reference Tables*. The station model for location A should show winds blowing from the west at high speed, completely overcast skies, and an air temperature at, or close to, the dewpoint. Only the station model depicted in choice (3) has all of these characteristics.

WRONG CHOICES EXPLAINED:

(1) This station model shows skies that are only partially overcast. Skies near a hurricane center would be completely overcast. Furthermore, the large difference between the temperature and dewpoint would indicate a low relative humidity, which is not the case in a region of heavy rainfall.

(2) This station model shows a dewpoint temperature higher than the air temperature, which is not possible. It also shows partly clear skies, which is highly unlikely near the hurricane center.

(4) This station model shows cool temperatures, which are unlikely in a hurricane. Furthermore, the large difference between the temperature and dewpoint would indicate a low relative humidity, which is not the case in a region of heavy rainfall.

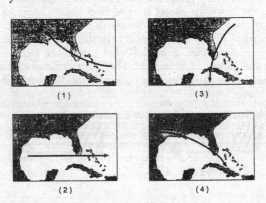

40. **1** The prevailing direction of motion of the hurricane is from east to west. Therefore, the arrow in map (1) shows the most likely track of the hurricane.

PART II

GROUP A—Rocks and Minerals

41. **2** Find the chart entitled Average Chemical Composition of Earth's Crust, Hydrosphere, and Troposphere in the *Earth Science Reference Tables*. Note in the columns headed "Crust" that oxygen has both the highest percent by mass and the highest percent by volume. Oxygen is the most abundant element in the Earth's crust.

42. **1** Find the chart entitled Scheme for Sedimentary Rock Identification in the *Earth Science Reference Tables*. Note that the chart is divided into an upper section labeled "Inorganic Land-Derived Sedimentary Rocks" and a lower section labeled "Chemically and/or Organically Formed Sedimentary Rocks." Since the sedimentary rock in the question is described as formed from land-derived sediments, look in the upper section of the chart

for the correct choice. Of the choices given, only (1), siltstone, is formed from land-derived sediments.

43. **4** The physical properties of minerals result from the internal arrangement of their atoms, which, in turn, is due to the nature of the chemical bonds holding the atoms together. Different types of bonds have different characteristics and will cause minerals to behave differently. For example, ionic bonds are strong, causing a mineral to resist breaking and have a high hardness. Since much energy is needed to break these strong bonds, the mineral will also have a high melting point. Furthermore, since the ions are stable, they do not gain or lose electrons easily and the mineral will, therefore, be a poor conductor of electricity. The electrical nature of an ionic bond makes it soluble in a polar solvent such as water, yielding a solution containing ions. Finally, since the bond is nondirected, atoms forming crystals may move into positions of high coordination and symmetry, yielding cleavage planes.

All of the other answer choices name specific physical properties of minerals, not a condition that could explain all of a mineral's physical properties.

44 **4** Find the chart entitled Scheme for Igneous Rock Identification in the *Earth Science Reference Tables,* and locate basalt on the chart. Follow this row to the left to the description of "Environment Of Formation," and note that the origin of basalt is extrusive or volcanic. Follow the row to the right and note that basalt has fine grains, which indicate rapid cooling. The basalt bedrock of the flat areas of the Moon most likely formed by rapid cooling of molten rock in lava flows.

WRONG CHOICES EXPLAINED:
(1) Cementing and compaction of sediments is one of the ways in which sedimentary rocks form. Basalt is an igneous rock formed by the crystallization of molten rock.
(2) Changes caused by heat and pressure on preexisting rocks are characteristic of metamorphic rocks, not extrusive igneous rocks such as basalt.
(3) Slow cooling of magma deep under the surface would result in large crystals. Basalt has fine crystals.

45. **3** Heat and pressure due to magma intrusions may cause rocks to change, or metamorphose, along the contact between the preexisting rock and the magma intrusion. This type of change is called contact metamorphism.

WRONG CHOICES EXPLAINED:
(1) Vertical sorting is the result of the deposition of sediments by water that is decreasing in velocity.
(2) Graded bedding is the result of sediments settling to the bottom of a standing body of water such as a lake or an ocean.
(4) Chemical evaporites are the result of water evaporating and leaving behind minerals that were in solution.

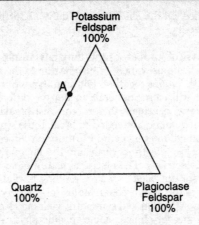

46. **3** The question states that rock A is coarse grained and igneous, and the position of rock A in the diagram indicates that the rock contains potassium feldspar and quartz. Find the chart entitled Scheme for Igneous Rock Identification in the *Earth Science Reference Tables*, and locate "coarse" in the column headed "Texture." Now, in the lower portion of the chart, find potassium feldspar and quartz. Follow the section in which they are found upward until it intersects the row of coarse-grained rocks. The only coarse-grained rock that contains quartz and potassium feldspar is granite.

47. **2** Find the chart entitled Scheme for Metamorphic Rock Identification in the *Earth Science Reference Tables*. Locate both quartzite and hornfels in the column headed "Rock Name," and compare the characteristics listed for these rocks in each of the columns to the left. One difference between hornfels and quartzite is found in the column headed "Composition"; hornfels contains plagioclase, but quartzite does *not* contain plagioclase.

48. **1** Find the chart entitled Scheme for Igneous Rock Identification in the *Earth Science Reference Tables*. As indicated by the arrow labeled "composition," igneous rocks toward the left side of the chart are considered felsic. This designation means that the minerals present have high concentrations of aluminum [noted as "FELSIC (Al)" on the chart] and low concentrations of iron and magnesium. Felsic rocks include granite, rhyolite, and pumice. The two major divisions of the Earth's crust are the oceanic crust and the continental crust. The oceanic crust is composed mostly of basaltic rocks, rocks with a composition similar to that of basalt. The continental crust, however, is composed mainly of granitic rocks, rocks with a composition similar to that of granite. Therefore, felsic rocks will most likely be found in the continental crust.

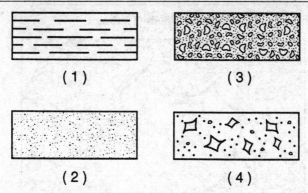

(1)

(3)

(2)

(4)

49. **1** Find the chart entitled Scheme for Sedimentary Rock Identification in the *Earth Science Reference Tables,* and locate each of the choices in the column headed "Map Symbol." Then, follow each symbol to the left to the column headed "Grain Size." The sedimentary rock represented by the symbol in choice (1), shale, has the smallest grain size—clay particles less than 0.0006 centimeter in diameter.

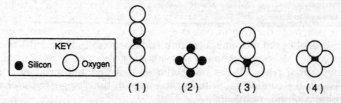

KEY

● Silicon ○ Oxygen

(1)   (2)   (3)   (4)

50. **4** The silicon-oxygen tetrahedron contains one atom of silicon surrounded by four atoms of oxygen. Diagram (4) is the only one that illustrates this pattern.

GROUP B—Plate Tectonics

51. **4** Part of the movement of convection currents in the Earth's mantle is horizontal. This slow movement of material carries with it sections of the overlying crustal plates. It has been inferred that this movement of crustal plates causes the continental drift that accounts for the changes in the shapes and positions of the continents.

52. **4** The oceanic crust is composed primarily of basaltic rock. Since it is crustal material that is being pushed upward at the mid-ocean ridges, the crust at these ridges is composed mainly of basalt.

53. **3** As material is pushed upward along mid-ocean ridges such as the East Pacific Rise and the Oceanic Ridge, the new material pushes some of the existing material outward. The result is that near the ridges the oceanic crust is younger because it has recently moved upward. The oceanic crust becomes progressively older as you travel outward in both directions.

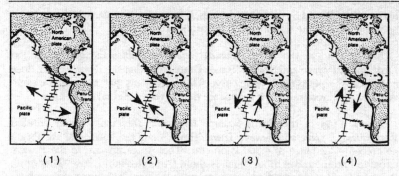

54. **1** Find in the *Earth Science Reference Tables* the map entitled Tectonic Plates, and locate on this map the corresponding area from the map appearing on the examination. The direction of the arrows in choice (1) corresponds to the direction of movement illustrated in the reference table map.

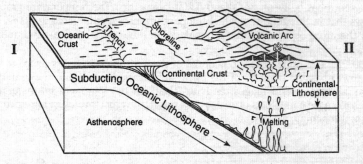

55. **4** Note that the cross section shows continental areas on the right and the ocean on the left. This description applies to areas *A* and *D*, but only area *D* shows the trench illustrated in the cross section near the continental shoreline.

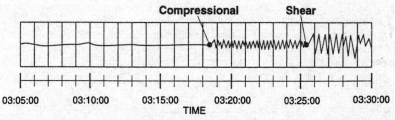

56. **4** Since *P*-waves (primary or compressional waves) travel faster than *S*-waves (secondary or shear waves), they arrive first at seismic recording stations, as illustrated in the diagram. The farther away the seismic station is from the epicenter, the greater the gap in arrival times. By measuring the difference

between the *P*- and *S*-wave arrival times, it is possible to calculate the distance between the seismic station and the earthquake epicenter.

57. **2** Find in the *Earth Science Reference Tables* the graph entitled Earthquake *P*-wave and *S*-wave Travel Time. Note that the *P*-wave curve does not rise as sharply as the *S*-wave curve. This fact means that, to travel the same distance, *P*-waves travel at a faster rate than *S*-waves and take less time.

58. **1** Find in the *Earth Science Reference Tables* the graph entitled Earthquake *P*-wave and *S*-wave Travel Time. Along the horizontal axis locate an epicenter distance of 4,800 kilometers Then trace upward to the *P*-wave curve to find the travel time, which is about 8 minutes in this case. Therefore, the earthquake occurred about 8 minutes before 5:10 p.m. or at about 5:02 p.m.

59. **2** Find in the *Earth Science Reference Tables* the diagram entitled Inferred Properties of Earth's Interior. According to the diagram, the iron-rich outer core lies between 2,900 and 5,200 kilometers. The temperature graph shows that in this zone the actual temperature exceeds the melting point. It may therefore be inferred that the material in this zone is an iron-rich liquid.

60. **3** The upper granitic material of the Earth's crust is less dense than the lower basaltic material of the crust and mantle. As a result, the granitic material in mountain ranges tends to "float" on the lower crustal material, much as an ice cube floats on water. As a result, the crust is thickest under continental mountain ranges.

GROUP C—**Oceanography**

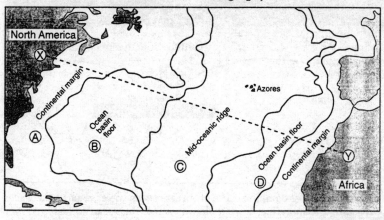

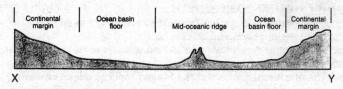

61. **2** In the profile, the ocean bottom is classified into the areas shown based on topography. For example, note that the two areas labeled "Continental margin" have similar slopes that are inclined more steeply than the "Ocean basin floors" in the adjacent areas. Also note that the boundary between "Ocean basin floor" and "Mid-oceanic ridge" occurs, in each case, where the slope changes direction, which is some distance from where the ridge suddenly rises from the ocean floor. The use of terms such as *floor, ridge,* and *margin* in the classifications also indicate that topography is the basis of the classification system.

62. **1** Land-derived sediments would most likely accumulate on the ocean bottom at a location close to land. Location *A* is closest to land.

63. **3** Mid-oceanic ridges, such as the one at location *C*, are places where mantle rock rises to the surface, forces older, preexisting oceanic crust aside, and solidifies to form new oceanic crust. Over time, this process forces older rocks farther and farther from the mid-oceanic ridges. Therefore, the mid-oceanic ridges contain the youngest oceanic crust, a fact that is consistent with the statement "Rocks at location *C* are generally younger than rocks at locations *A* and *B*."

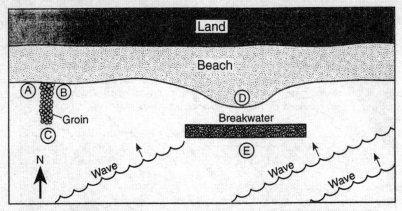

64. **4** The most common cause of surface ocean waves, such as the waves approaching the shoreline, is wind blowing across the ocean surface.

WRONG CHOICES EXPLAINED:

(1) Underwater earthquakes do cause waves known as tsunami. However, tsunami are rare, not common, occurrences.

(2) Variations in ocean water density cause water to rise or sink, forming a current rather than a wave.

(3) The gravitational effect of the Moon produces tides, not waves.

65. **1** Find the map entitled Surface Ocean Currents in the *Earth Science Reference Tables*. Note that the current that flows along the east coast of North America is the Florida Current, which is, therefore, the ocean current most likely to modify the climate of this shoreline.

66. **2** Waves approaching a shoreline at an angle create a longshore current parallel to the shore. The waves shown in the diagram would create a longshore current moving from right to left (from *D* toward *B*). The moving water in longshore currents can transport sand. When longshore currents slow down or stop, however, the water loses its carrying power, and the sand is deposited. When the motion of the water is stopped by the groin, sand transported by the longshore current in the diagram will be deposited at location *B*, where the beach will first begin to widen.

67. **2** The breakwater protects the beach at position *D* from wave action and produces a quiet zone behind it that serves as an accumulation place for sand. The accumulation of sand behind the breakwater will cause the bulge in the beach at position *D* to increase.

68. **1** Find in the *Earth Science Reference Tables* the diagram entitled Planetary Wind and Moisture Belts in the Troposphere and the map entitled Surface Ocean Currents. The planetary winds exert a stress on the ocean surface and produce the wind-driven circulation of the ocean shown in the Surface Ocean Currents map. The curved pattern of surface currents is due to the curved pattern of planetary winds, as indicated in the Planetary Wind diagram. The curved pattern of planetary winds is due to the Coriolis effect, which, in turn, is caused by the Earth's rotation.

69. **2** A large percentage of the minerals in seawater are carried there by rivers as weathering and erosion products from continental rocks. At present, rivers carry about 4 billion tons of dissolved materials to the oceans each year.

WRONG CHOICES EXPLAINED:

(1) Seafloor rocks are not in contact with the atmosphere and undergo almost no weathering.

(3) Minerals are inorganic compounds, whereas deep-ocean organic-matter sediments consist of organic compounds.

(4) The major components of the minerals dissolved in seawater are sodium, chlorine, magnesium, sulfate, calcium, and potassium—almost all solids. Although chlorine is a gas, it is not a major component of the gases given off during underwater volcanic eruptions.

70. **4** Tsunamis have amplitudes of only a few meters, but their wavelengths may exceed 200 kilometers. This kind of wave is so flat that it could pass unnoticed beneath a ship at sea. When a tsunami enters a bay or narrow channel, however, the water may be funneled into a huge wave up to 20 meters high. The kinetic energy of such a wave enables it to cause severe destruction in coastal areas.

WRONG CHOICES EXPLAINED:

(1) The small amplitude of a tsunami prevents it from reaching and transporting deep-ocean sediments.

(2) Destruction of some type occurs on the seafloor whenever an earthquake strong enough to produce a large tsunami occurs. At the very least, rocks are displaced.

(3) The magnitude of the earthquake determines the amplitude, not the direction, of the tsunami. A tsunami moves outward in all directions from the earthquake that caused it.

GROUP D—**Glacial Processes**

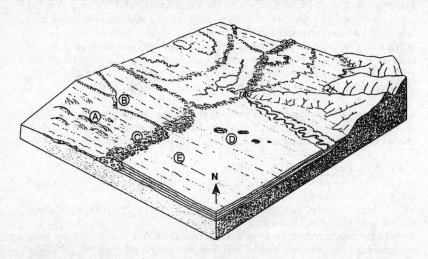

71. **2** The diagram shows that location *E* is part of a flat plain. Look at the edges of the block diagram, and note the horizontal layers shown in the cross section of this plain. Horizontal layers are deposited along glacial margins by meltwater streams. Since the deposition occurs in water, sorting of the sediments will occur. Therefore, the sediments deposited at location *E* will most likely be sorted and layered.

72. **3** A terminal moraine is a ridgelike accumulation of till deposited along the margin of a glacier. Till is poorly sorted, is generally unlayered, and contains angular particles. Location *C* is situated on a ridgelike structure, and the cross section of this structure at the edge of the block diagram shows it to be unlayered and poorly sorted. Therefore, location *C* is most likely a terminal moraine that marks the farthest advance of the glacier.

73. **4** The melting of isolated ice blocks forms rounded depressions in a body of drift called kettles. Only location *D* is associated with such depressions.

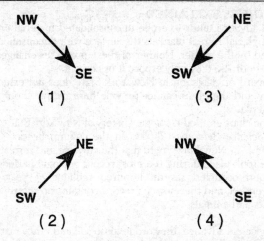

74. **1** Several features indicate the direction of ice movement. The major axis of drumlins aligns with the direction of ice movement, and the major axis of the drumlins at location *A* is aligned along a northwest-southeast line. The esker at location *B* also indicates a northwest-southeast orientation. The terminal moraine at location *C* marks the farthest extent of the glacier, and the outwash plain at locations *D* and *E* indicates that the water melting out of the glacier flowed southeast. Such movement would not have been possible if the glacier were moving from southeast to northwest—the ice would have blocked the flow of water. Therefore, the ice flowed from northwest to southeast as indicated by the arrow in choice (1).

75. **3** The many deposits of sand and gravel left by glaciers have had a positive effect on the economy of New York State. Eighty-five percent of the 2,000 or so mines in New York State produce sand and gravel used in the manufacture of concrete for building purposes and in road construction.

WRONG CHOICES EXPLAINED:

(1) Oil and natural gas are products of the decomposition of both plant and animal remains. Oil is almost always found in layers of marine sedimentary rock. This type of rock does not form as a result of glaciation.

(2) A reduction in the number of usable water reservoirs would represent a loss of a valuable resource and would not have a positive effect on New York State's economy.

(4) Fertile soil is a valuable resource, and a loss in such deposits would have a negative, not a positive, effect on New York State's economy.

76. **2** The advance and retreat of glaciers is triggered by climatic changes. Therefore, the record of glaciation in New York State provides geologists with information about changes in global climate.

WRONG CHOICES EXPLAINED:

(1) Glacial advance tends to erode unconsolidated material and exposed bedrock, while glacial retreat blankets the surface with unconsolidated debris that was carried by the glacier. Neither of these processes changes the composition of the bedrock; they only expose it or cover it.

(3) The record of glaciation in New York State does not extend across a plate boundary and therefore cannot provide information about the movement of crustal plates.

(4) Find the chart entitled Geologic History of New York State at a Glance in the *Earth Science Reference Tables.* In the column headed "Important Geological Events in New York," note that the advance and retreat of the last continental ice took place during the most recent geologic period. The most recent glaciation occurred several hundred millions of years after the Paleozoic Era ended, and therefore its record contains no information about Paleozoic plants and animals.

77. **4** As glaciers advance, they erode loose soil and other materials from the Earth's surface and expose bedrock. As glaciers retreat, rock material that was frozen into the ice drops out as it melts and blankets the surface with a thin layer of loose, unsorted rock fragments. The last glaciation in New York occurred so recently that this material has not weathered extensively, and the soils in New York State are thin and rocky.

78. **1** At the present time, glaciers occur mostly in areas with high latitudes or high altitudes where extreme conditions lead to year-round cold temperatures. Areas with high latitudes are near the poles. There, year-round cold exists because the tilt of the Earth's axis and the curvature of the Earth's surface cause solar radiation reaching the Earth's surface to be low in intensity and short in duration. Refer to the chart entitled Selected Properties of Earth's Atmosphere in the *Earth Science Reference Tables,* and note that the temperature in the troposphere drops rapidly with altitude. At high altitudes, the temperature may remain below freezing year-round.

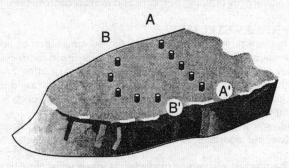

79. **3** In the diagram, at positions *B–B'*, the stakes in the center of the glacier have moved farther from their original positions in *A–A'* than the stakes

near the edge of the glacier. Since all the stakes started out in a straight line and were moving for the same length of time, this pattern of movement means that the glacier is moving faster in the center than on the sides.

80. **2** Find the map entitled Generalized Landscape Regions of New York State in the *Earth Science Reference Tables*. Of the four choices—the Adirondacks, Long Island, the Catskills, and the Taconics, only Long Island is located in a landscape region that is a plain—the Atlantic Coastal Plain. Now find the map entitled Generalized Bedrock Geology of New York State in the *Earth Science Reference Tables*. According to the codes at the bottom of the map, Long Island consists of unconsolidated gravels, sands, and clays.

GROUP E—Atmospheric Energy

81. **3** The major source of energy for most Earth processes is radiation received from the Sun. The 60% of solar rays that reach the Earth and are not reflected back into space are absorbed by the atmosphere, the land, and the sea. The energy absorbed by the land eventually warms the air and causes convection. Winds resulting from convection move clouds and, as they move over the sea, create waves. The energy absorbed by the sea warms the water, resulting in evaporation. The water vapor thus produced forms clouds and eventually rain, snow, and other types of precipitation. Thus, the driving force for almost all external processes that change the Earth's surface—wind, rain, streams, glaciers, and waves—is the radiation received from the Sun.

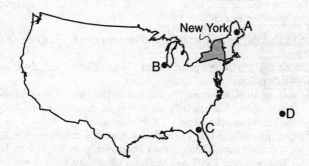

82. **2** Find the diagram entitled Planetary Wind and Moisture Belts in the Troposphere in the *Earth Science Reference Tables*. Note, in the latitudes found in the United States, a planetary wind belt that consists of southwesterly winds. These winds tend to move weather systems from southwest to northeast across the United States. Therefore, knowledge of the weather at location B, which is southwest of New York State, would be of most interest to a person predicting the next day's weather for New York State.

83. **4** Find the chart entitled Electromagnetic Spectrum in the *Earth Science Reference Tables,* and note that the wavelength of the radiation increases from left to right. Of the choices given, the only correct statement is that gamma rays, with a wavelength of 10^{-13} meter, have a shorter wavelength than visible light, with wavelengths between 10^{-6} and 10^{-7} meter.

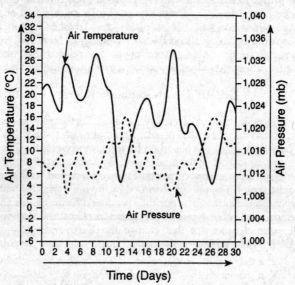

84. **2** To find the air temperature and air pressure readings on day 6, proceed as follows:

Locate 6 on the horizontal axis labeled "Time (Days)," and draw a vertical line through this point.

Follow this vertical line until it intersects the dotted line labeled "Air Pressure." Then move horizontally to the *right* to the scale labeled "Air Pressure (mb)" and read the value—1,016 millibars.

Now follow the vertical line through day 6 until it intersects the solid line labeled "Air Temperature." Then move horizontally to the *left* to the scale labeled "Air Temperature (°C)" and read the value—19°C.

The approximate air temperature and air pressure readings on day 6 were 19°C and 1,016 millibars.

85. **3** Find the diagram entitled Planetary Wind and Moisture Belts in the Troposphere in the *Earth Science Reference Tables,* and locate the areas labeled "DRY." Note that the arrows indicating wind patterns adjacent to these areas move apart, or diverge. The arrows indicating air movement in the troposphere surrounding the Earth show that the air is sinking in these areas.

86. **4** When the air pressure decreases because of an increase in the moisture content of the air, the air becomes less dense and rises. As the air rises, it cools and condensation forms clouds that may eventually lead to precipitation.

87. **1** Darker colors are better absorbers of electromagnetic energy than lighter colors. The more energy that is absorbed, the less is reflected. Dark colored sand will absorb more energy (and reflect less energy) than lighter colored water. Find the chart labeled "Specific Heats of Common Materials" in the Physical Constants section of the *Earth Science Reference Tables*. Dark colored sand probably consists of dark colored minerals such as those found in basalt. Note that liquid water has the highest specific heat of all the materials shown, including basalt.

88. **4** Find the chart labeled "Properties of Water" in the Physical Constants section of the *Earth Science Reference Tables*. Note that the latent heat of vaporization (540 cal/g) is greater than the latent heat of fusion (80 cal/g). The latent heat of vaporization is absorbed when water evaporates and given off when water condenses. Thus water *absorbs* the most heat during evaporation.

89. **2** Find the chart entitled Relative Humidity (%) in the *Earth Science Reference Tables*. Since the air temperature is the same as the dry-bulb temperature, find 16°C in the left column headed "Dry Bulb Temperature (°C)." Follow this row to the right until a value of 71 is reached. Then follow this column up until it intersects the "Difference Between Wet-Bulb and Dry-Bulb Temperatures (C°) at 3 C°. Finally, subtract this 3 C° temperature difference from the dry-bulb temperature of 16°C to obtain the wet-bulb temperature of 13°C.

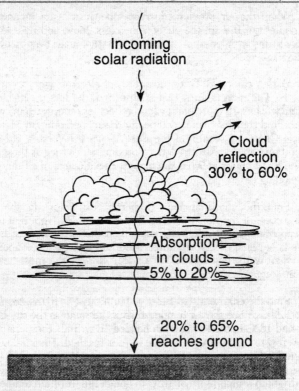

90. **2** According to the diagram, 30% to 60% of incoming solar radiation is reflected by clouds, and 5% to 20% of incoming solar radiation is absorbed by clouds. To find the incoming solar radiation reflected or absorbed on cloudy days, add the percentages reflected to the percentages absorbed as follows:

Reflected	30%	– 60%
Absorbed	5%	– 20%
Total	35%	– 80%

GROUP F—**Astronomy**

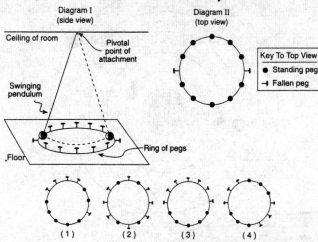

Diagram I
(side view)

Diagram II
(top view)

Ceiling of room

Pivotal
point of
attachment

Swinging
pendulum

Key To Top View
● Standing peg
⊣ Fallen peg

Ring of pegs

.Floor

(1)   (2)   (3)   (4)

91. **4** The pendulum will continue to swing in a straight line back and forth. Since the Earth is rotating on its axis, the path of the swinging pendulum will appear to follow a circular motion that is opposite to the direction in which the Earth is rotating. After several hours the pattern of standing and fallen pegs will appear as illustrated in diagram (4).

92. **3** Since the Earth rotates on its axis, a Foucault pendulum appears to swing along a circular path. If the Earth did not rotate, the pendulum would continue to swing back and forth along the same path.

WRONG CHOICES EXPLAINED:
   (1) The fact that the Sun rotates on its axis has no effect on the direction of swing of a Foucault pendulum.
   (2), (4) The Earth revolves around the Sun, but this movement does not affect the direction of swing of a Foucault pendulum.

93. **1** All of the planets move in elliptical orbits. One way to express the shape of an ellipse is by its eccentricity. Find the chart entitled Solar System Data in the *Earth Science Reference Tables*. In the column headed "Eccentricity of Orbit," note that Jupiter's eccentricity is 0.048. The orbit of Uranus, with an eccentricity of 0.047, is most similar to that of Jupiter in shape.

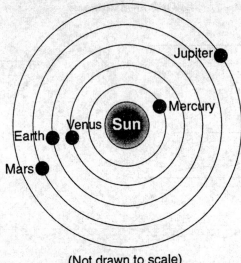

(Not drawn to scale)

94. **3** Night occurs on the side of the Earth opposite the Sun. Only objects on this side of the Earth are visible during the night. The only other planet on the side of the Earth opposite the Sun is Mars, so only Mars will be visible from the Earth at night.

95. **4** Although water has been found on several other planets, most notably Mars, according to current data Earth is the only planet in our solar system with *liquid* water on its surface.

96. **1** The word *geocentric* is derived from two word forms: *geo*, meaning Earth and *centric*, meaning in the center. Therefore, this model of our solar system is best described by the statement "The Earth is located at the center of the model."

97. **2** A day is defined by a planet's period of rotation; a year, by a planet's period of revolution. Find the chart entitled Solar System Data in the *Earth Science Reference Tables,* and locate the columns labeled "Period of Revolution" and "Period of Rotation." Note that Venus's period of rotation, or day, is 243 Earth days long and that its period of revolution, or year, is 224.7 Earth days long. Thus, Venus's day is longer than its year.

98. **2** According to the chart entitled Solar System Data, the greatest distance between two planets is the distance between Mercury and Pluto—5,842 million kilometers. The distance from our star, the sun, to the next closest star, Proxima Centauri, is 4.2 light-years, which is approximately 39,900,000 million

kilometers. The distances between stars are much greater than the distances between planets.

99. **4** All of the planets in our solar system, including Mercury, Venus, and Earth, have elliptical orbits with the Sun at one focus. According to the chart entitled Solar System Data in the *Earth Science Reference Tables*, the only other similarity between these three planets is their densities and density is not listed as an answer choice.

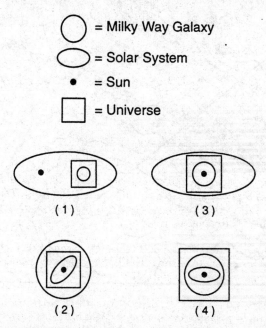

100. **4** The universe contains all galaxies, including our own—the Milky Way galaxy. The Milky Way galaxy contains many star systems, including our own solar system. Our solar system contains one star, the Sun. Therefore, the most accurate arrangement of symbols is choice (4), with Sun inside solar system, solar system inside Milky Way galaxy, and Milky Way galaxy inside universe.

PART III

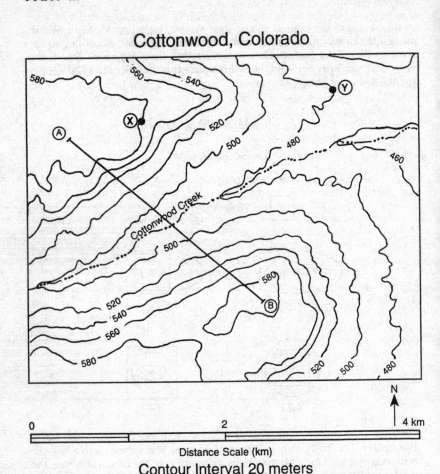

Cottonwood, Colorado

Distance Scale (km)

Contour Interval 20 meters

101. On the map, the line representing Cottonwood Creek is shown as a series of dashes separated by three dots, indicating that the creek is intermittent—it flows in times of abundant rainfall and runs dry during droughts. Streams flow downslope. To determine the downslope direction for Cottonwood Creek, find two places where it crosses contour lines; downslope will be from the higher contour toward the lower contour. For example, just

below the point labeled Y, the creek crosses the 460-meter contour. Just to the right of the label "Cottonwood Creek," it crosses the 480-meter contour. Since north is toward the top of the map, the downslope direction is toward the east-northeast. Since the question asks for the *general* direction, northeast and east are also acceptable answers.

102. On this map, the highest contour lines are labeled 580 meters. Since the contour interval is given as 20 meters, the next highest contour line would be 600 meters. However, since no 600-meter contour line appears on the map, no place on the map has that elevation. The highest possible elevation, to the *nearest meter*, of any location, including point *B*, on this topographic map would be the highest whole number of meters that is greater than 580 but less than 600, that is, 599 meters.

103. To construct a profile of the topography along line *AB* shown on the map, proceed as follows: Place the straight edge of a piece of scrap paper along line *AB*, and mark the edge of the paper at points *A* and *B* and wherever the paper intersects a contour line. Label each mark with the elevation of the intersecting contour line. Then place this paper against the lower edge of the grid provided on your answer sheet so that points *A* and *B* on the paper align with points *A* and *B* on the grid. Next, at each point where a contour line crosses the edge of the paper, draw a vertical line of the appropriate height on the grid. Finally, connect the endpoints of the lines in a smooth curve to form the finished profile. See the diagrams below.

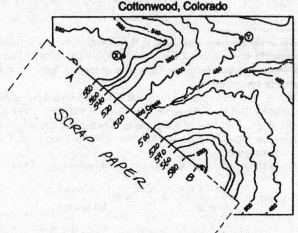

Cottonwood, Colorado

Marking the Edge of the Scrap Paper.

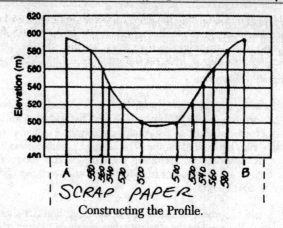

SCRAP PAPER

Constructing the Profile.

104. *a*. Find in the Equations and Proportions section in the *Earth Science Reference Tables* the equation for gradient:

$$\text{gradient} = \frac{\text{change in field value}}{\text{change in distance}}$$

This equation can also be expressed as

$$\text{gradient} = \frac{\Delta \text{ elevation}}{\Delta \text{ distance}}$$

b. The field value on a topographic map is the elevation, which is shown as isolines called contour lines. Point *X* lies on the 580-meter contour line, and point *Y* on the 480-meter contour line. Thus, the change in field value from point *X* to point *Y* is 100 meters.

The change in distance from point *X* to point *Y* can be found by using the distance scale beneath the map. Mark off the distance between point *X* and point *Y* along the edge of a piece of scrap paper. Then place the edge of the scrap paper along the distance scale and measure the distance from *X* to *Y*. The distance is 2 kilometers.

Finally, substitute these values in the equation for gradient as follows:

$$\text{gradient} = \frac{580 - 480}{2} = \frac{100}{2} = 50 \frac{\text{meters}}{\text{kilometer}}$$

c. The gradient is 50 meters/kilometer (50 m/km).

105. The Earth's surface lies entirely within the troposphere. Find the chart entitled Selected Properties of Earth's Atmosphere in the *Earth Science*

Reference Tables. The relationship between surface temperature and elevation (altitude) is shown on the graph at the right labeled "Temperature Zones." In the troposphere, the temperature decreases as the elevation (altitude) increases. This relationship is best represented by a line drawn as shown below.

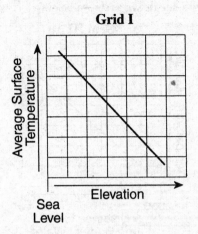

Grid I

106. The relationship between latitude and average surface temperature should show the highest temperatures near the Equator (0° latitude) decreasing to the lowest temperatures near the poles (90° latitude) as shown below.

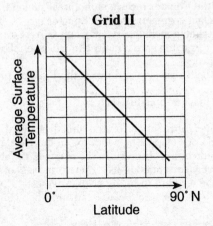

Grid II

107. Large bodies of water warm up and cool off slower than adjacent landmasses. Water has a very high specific heat, allowing it to gain or lose large quantities of heat with only a small change in temperature. Thus, large bodies

of water will be cooler in summer and warmer in winter than adjacent land-masses, thus moderating the temperatures of locations close to a large body of water. Thus, the closer a location is to a large body of water, the less extreme (smaller) its temperature range; the farther a location is from a large body of water, the more extreme (larger) its temperature range. This relationship is best represented by a line drawn as shown below.

Grid III

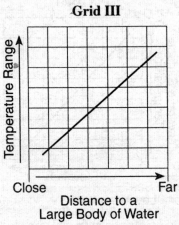

108. *a.* Your answer should include a scientifically correct explanation why most locations near the Equator have a warm climate. The following are examples of acceptable answers:

Locations near the Equator receive more direct sunlight.

Intensity of sunlight is greater near the Equator.

b. Your answer should include a scientifically correct explanation why most locations near the Equator have a moist climate. The following are examples of acceptable answers:

Planetary wind belts converge and rise at the Equator, producing precipitation.

The Equator is in a low-pressure belt.

109. Polaris, the North Star, is located directly above the North Pole. As a result, the altitude of Polaris at any location in the Northern Hemisphere is equal to the latitude of the observer. In the diagram, the altitude of Polaris to the observer in New York State is 43°. Therefore, the latitude of the location represented by the diagram is, to the *nearest degree,* 43°N. The direction of the latitude is North because Polaris can only be seen in the Northern Hemisphere.

110. By definition, the zenith is a point directly above an observer. The point directly above the observer in the diagram is shown correctly labeled in the diagram below the answer to question 111.

111. The month of May falls between March and June. The Sun's path in May will lie between its paths on March 21 and on June 21. Draw a line parallel to and between the paths for Mar/Sept 21 and June 21 as shown by the heavy line labeled "May 21" in the diagram below.

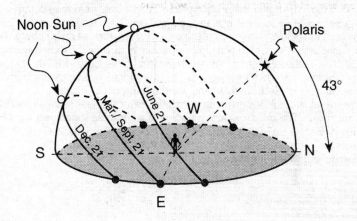

112. The table shows the air temperature in degrees Celsius. To convert this value to degrees Fahrenheit, find the scales labeled "Temperature" in the *Earth Science Reference Tables*. Locate 21° on the center, Celsius scale. Directly to the left is the Fahrenheit scale; 21° Celsius corresponds most closely to 70° Fahrenheit. (Since the scales don't correspond exactly, your answer may vary from the accepted value of 70°F by ± 1°F.)

113. The table shows the barometric pressure to be 993.1 millibars (mb). On a station model this reading is shortened by dropping the decimal point and the hundreds and thousands digits. Thus, 993.1 becomes 931 on a station model.

114. The table indicates that the wind is blowing from the east at a speed of 25 knots. Find the Weather Map Information diagram in the *Earth Science Reference Tables*. In the station model at the left, wind direction is indicated by drawing a line toward the direction from which the wind is blowing. Wind speed is indicated by drawing feathers on this line—a whole feather for every 10 knots and a half feather for 5 knots. Thus, 25 knots would be represented by two whole feathers and a half feather. The station model for the given data should have a line pointing to the east with two whole feathers and one half feather as shown below.

115. The article indicates that substances that damage the ozone layer and allow ultraviolet radiation to reach the Earth's surface are being banned. This extreme action is being taken because ultraviolet radiation is dangerous. It contains high levels of energy in the form of waves that have a short wavelength. Because of this short wavelength, these rays are able to penetrate living cells, where the high energy damages cell structures. In human beings this damage occurs most noticeably in the skin and over time may lead to skin cancer. In smaller organisms such as plankton and algae, the energy of ultraviolet rays may be lethal. A loss of phytoplankton would disrupt the oxygen-carbon dioxide cycle and could lead to the death of most life on Earth. Finally, if more energy reaches the Earth's surface, the Earth's radiative balance might shift, causing changes in global climate and weather patterns.

Any one of the reasons given above, stated as a complete sentence, would earn full credit. The following are examples of acceptable statements.

Ultraviolet rays are a major cause of skin cancer in humans.

Excessive ultraviolet radiation might kill most life on Earth.

Climate and weather patterns could change.

Unit (1–9) or Optional/Extended Topic (A–F)	Question Numbers (total)	Wrong Answers (x)	Grade
1. Earth Dimensions	1, 35, 36, 101–104, 109 (8)		$\dfrac{100(8-x)}{8} = \%$
2. Minerals and Rocks	4–8, 40 (6)		$\dfrac{100(6-x)}{6} = \%$
A. Rocks, Minerals, and Resources	41–50 (10)		$\dfrac{100(10-x)}{10} = \%$
3. The Dynamic Crust	6, 7 (2)		$\dfrac{100(2-x)}{2} = \%$
B. Earthquakes and the Earth's Interior	51–60 (10)		$\dfrac{100(10-x)}{10} = \%$
4. Surface Processes and Landscapes	4, 8, 10–15 (8)		$\dfrac{100(8-x)}{8} = \%$
C. Oceanography	61–70 (10)		$\dfrac{100(10-x)}{10} = \%$
D. Glacial Geology	71–80 (10)		$\dfrac{100(10-x)}{10} = \%$
5. Earth's History	16–22, 30 (8)		$\dfrac{100(8-x)}{8} = \%$
6. Meteorology	23–25, 28, 37–40 (8)		$\dfrac{100(8-x)}{8} = \%$
E. Latent Heat and Atmospheric Energy	81–90 (10)		$\dfrac{100(10-x)}{10} = \%$
7. Water Cycle and Climates	26, 27, 29, 105–108 (7)		$\dfrac{100(7-x)}{7} = \%$
8. The Earth in Space	31–34, 94 (5)		$\dfrac{100(5-x)}{5} = \%$
F. Astronomy Extensions	91–100 (10)		$\dfrac{100(10-x)}{10} = \%$
9. Environmental Awareness	115 (1)		$\dfrac{100(1-x)}{1} = \%$

To further pinpoint your weak areas, use the Topic Outline in the front of the book.

Examination June 1997

Earth Science — Program Modification Edition

PART I

Answer all 40 questions in this part.

Directions (1–40): For *each* statement or question, select the word or expression that, of those given, best completes the statement or answers the question. Record your answers in the spaces provided. [40]

1 A large earthquake occurred at 45° N 75° W on September 5, 1994. Which location in New York State was closest to the epicenter of the earthquake?

1 Buffalo   3 Albany
2 Massena   4 New York City   1 _____

2 Isolines on the map below show elevations above sae level, measured in meters.

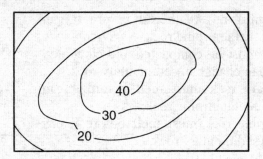

What is the highest possible elevation represented on this map?

(1) 39 m   (3) 49 m

(2) 41 m   (4) 51 m   2____

3 A conglomerate contains pebbles of limestone, sandstone, and granite. Based on this information, which inference about the pebbles in the conglomerate is most accurate?

(1) They had various origins.

(2) They came from other conglomerates.

(3) They are all the same age.

(4) They were eroded quickly.   3____

4 Which rock is usually composed of several different minerals?

1 rock gypsum   3 quartz

2 chemical limestone   4 gneiss   4____

5 Which statement about the formation of a rock is best supported by the rock cycle?

1 Magma must be weathered before it can change to metamorphic rock.
2 Sediment must be compacted and cemented before it can change to sedimentary rock.
3 Sedimentary rock must melt before it can change to metamorphic rock.
4 Metamorphic rock must melt before it can change to sedimentary rock.

5 _____

6 Which granite sample most likely formed from magma that cooled and solidified at the slowest rate?

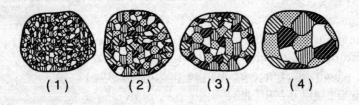

(1)   (2)   (3)   (4)   6 _____

7 The interpretation that the Earth's outer core is liquid was made primarily from

1 deep-sea drilling data   3 seismic data
2 magnetic data   4 satellite data

7 _____

8 What is the direction of crustal movement of the Australian plate?

1 northward   3 northwestward
2 southward   4 southeastward

8 _____

9 The cartoon below presents a humorous view of
Earth science.

The cartoon character on the right realizes that the
sand castle will eventually be

1 compacted into solid bedrock
2 removed by agents of erosion
3 preserved as fossil evidence
4 deformed during metamorphic change   9 _____

10 A deposit of rock particles that are scratched and
unsorted has most likely been transported and
deposited by

1 wind   3 running water
2 glacial ice   4 ocean waves   10 _____

11 The diagram below shows a process called frost
wedging.

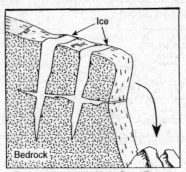

Frost wedging is an example of

1 weathering   3 metamorphism
2 cementing   4 deposition   11 ____

12 The diagram below shows a soil profile formed
in an area of granite bedrock. Four different
soil horizons, *A, B, C,* and *D,* are shown.

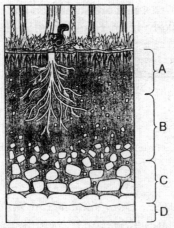

Which soil horizon contains the greatest amount
of material formed by biological activity?

(1) *A*   (3) *C*
(2) *B*   (4) *D*   12 ____

13 The map below shows the ancient location of evaporating seawater, which formed the Silurian-age deposits of rock salt and rock gypsum now found in some New York State crustal bedrock.

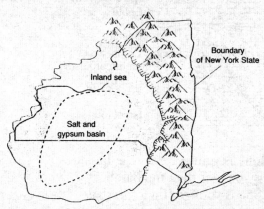

Within which two landscape regions are these large rock salt and rock gypsum deposits found?

1 Hudson Highlands and Taconic Mountains
2 Tug Hill Plateau and Adirondack Mountains
3 Erie-Ontario Lowlands and Allegheny Plateau
4 the Catskills and Hudson-Mohawk Lowlands   13 _____

14 The diagrams below represent geologic cross sections from two widely separated regions.

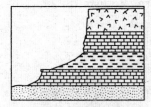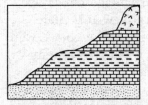

The layers of rock appear very similar, but the hillslopes and shapes are different. These differences are most likely the result of

1 volcanic eruptions   3 soil formation
2 earthquake activity   4 climate variations   14 _____

15 Which statement correctly describes an age relationship in the geologic cross section below?

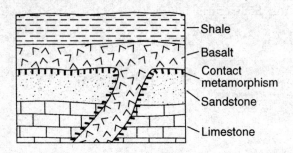

1 The sandstone is younger than the basalt.
2 The shale is younger than the basalt.
3 The limestone is younger than the shale.
4 The limestone is younger than the basalt.   15 _____

16 Present-day corals live in warm, tropical ocean water. Which inference is best supported by the discovery of Ordovician-age corals in the surface bedrock of western New York State?

1 Western New York State was covered by a warm, shallow sea during Ordovician time.
2 Ordovician-age corals lived in the forests of western New York State.
3 Ordovician-age corals were transported to western New York State by cold, freshwater streams.
4 Western New York was covered by a continental ice sheet that created coral fossils of Ordovician time.   16 _____

17 The diagram below represents the radioactive decay of uranium-238.

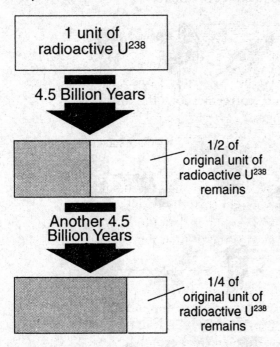

Shaded areas on the diagram represent the amount of

1 undecayed radioactive uranium-238 (U^{238})
2 undecayed radioactive rubidium-87 (Rb^{87})
3 stable carbon-14 (C^{14})
4 stable lead-206 (Pb^{206})   17 _____

18 By which process does water vapor change into clouds?

1 condensation	3 convection
2 evaporation	4 precipitation

18 _____

19 A geologist collected the fossils shown below from locations in New York State.

Which sequence correctly shows the fossils from oldest to youngest?

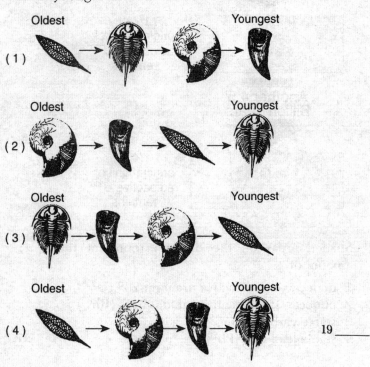

19 _____

20 Which list shows atmospheric layers in the correct order upward from the Earth's surface?

　1 thermosphere, mesosphere, stratosphere, troposphere

　2 troposphere, stratosphere, mesosphere, thermosphere

　3 stratosphere, mesosphere, troposphere, thermosphere

　4 thermosphere, troposphere, mesosphere, stratosphere

20 _____

21 Which statement best explains why precipitation occurs at frontal boundaries?

　1 Cold fronts move slower than warm fronts.

　2 Cold fronts move faster than warm fronts.

　3 Warm, moist air sinks when it meets cold, dry air.

　4 Warm, moist air rises when it meets cold, dry air.

21 _____

　Base your answers to questions 22 and 23 on the weather instrument shown in the diagram below.

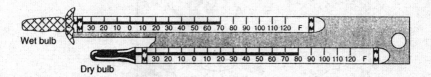

22 What are the equivalent Celsius temperature readings for the Fahrenheit readings shown?

　1 wet 21°C, dry 27°C

　2 wet 26°C, dry 37°C

　3 wet 70°C, dry 80°C

　4 wet 150°C, dry 176°C

22 _____

23 Which weather variables are most easily determined by using this weather instrument?

   1   air temperature and windspeed
   2   visibility and wind direction
   3   relative humidity and dewpoint
   4   air pressure and cloud type   23 _____

24 How do clouds affect the temperature at the Earth's surface?

   1   Clouds block sunlight during the day and prevent heat from escaping at night.
   2   Clouds block sunlight during the day and allow heat to escape at night.
   3   Clouds allow sunlight to reach the Earth during the day and prevent heat from escaping at night.
   4   Clouds allow sunlight to reach the Earth during the day and allow heat to escape at night.   24 _____

25 A low-pressure system is shown on the weather map below.

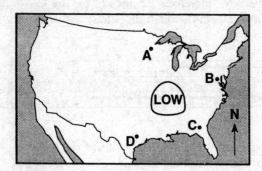

Toward which point will the low-pressure system move if it follows a normal storm track?

(1) *A*   (2) *B*   (3) *C*   (4) *D*   25 _____

26 Compared to an inland location of the same elevation and latitude, a coastal location is likely to have

1 warmer summers and cooler winters
2 warmer summers and warmer winters
3 cooler summers and cooler winters
4 cooler summers and warmer winters   26 _____

27 Which graph best represents the relationship between average yearly temperature and latitude?

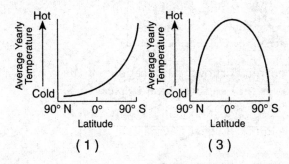

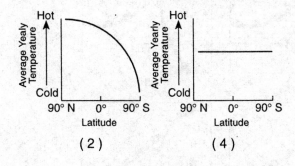

27 _____

Base your answers to questions 28 through 30 on the *Earth Science Reference Tables* and the weather map below. The map shows a low-pressure system. Weather data is given for cities *A* through *D*.

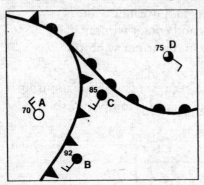

28 Which map correctly shows the locations of the continental polar (cP) and maritime tropical (mT) air masses?

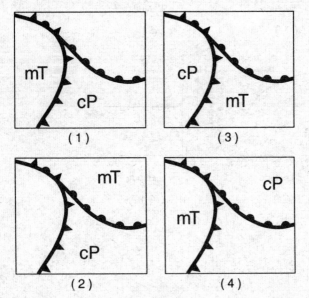

28 _____

29 Which city is *least* likely to have precipitation starting in the next few hours?

  (1) *A*
  (2) *B*
  (3) *C*
  (4) *D*

29 _____

30 Which map correctly shows arrows indicating the probable surface wind pattern?

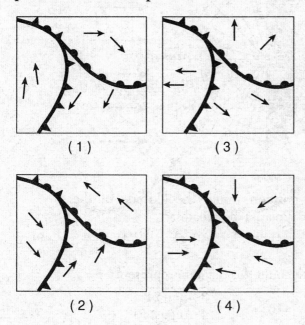

    (1)   (3)

    (2)   (4)

30 _____

Base your answers to questions 31 through 33 on the *Earth Science Reference Tables* and the map below. The map represents a view of the Earth looking down from above the North Pole, showing the Earth's 24 standard time zones. The Sun's rays are striking the Earth from the right. Points *A, B, C,* and *D* are locations on the Earth's surface.

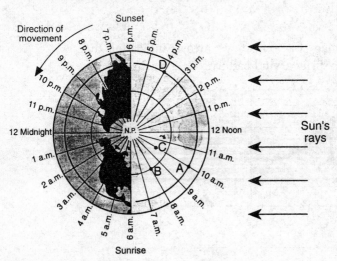

31 At which position would the altitude of the North Star (Polaris) be greatest?

(1) *A*   (2) *B*   (3) *C*   (4) *D*   31 _____

32 Which date could this diagram represent?

1 January 21   3 June 21
2 March 21   4 August 21   32 _____

33 Areas within a time zone generally keep the same standard clock time. In degrees of longitude, approximately how wide is one standard time zone?

(1) $7\frac{1}{2}°$ (2) 15°   (3) $23\frac{1}{2}°$ (4) 30°   33 _____

34 Why do the locations of sunrise and sunset vary in a cyclical pattern throughout the year?

　1 The Earth rotates on a tilted axis while revolving around the Sun.
　2 The Sun rotates on a tilted axis while revolving around the Earth.
　3 The Earth's orbit around the Sun is an ellipse.
　4 The Sun's orbit around the Earth is an ellipse.　　34 _____

35 Compared to Jupiter and Saturn, Venus and Mars have greater

　1 periods of revolution
　2 orbital velocities
　3 mean distances from the Sun
　4 equatorial diameters　　35 _____

36 Billions of stars in the same region of the universe are called

　1 solar systems　　　　3 constellations
　2 asteroid belts　　　　4 galaxies　　36 _____

　Base your answers to questions 37 and 38 on the diagram below, which represents the path of a planet in an elliptical orbit around a star. Points *A*, *B*, *C*, and *D* indicate four orbital positions of the planet.

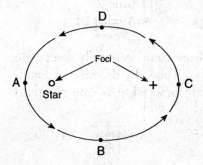

37 The eccentricity of the planet's orbit is approximately

    (1) 0.18   (2) 0.65   (3) 1.55   (4) 5.64   37 _____

38 Which graph best represents the gravitational attraction between the star and the planet?

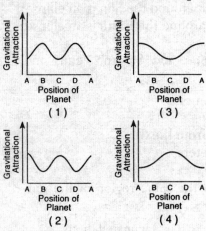

                                        38 _____

39 Which graph best represents the most common relationship between the amount of air pollution and the distance from an industrial city?

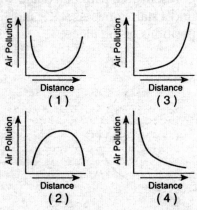

                                        39 _____

40 Which diagram best shows how air inside a greenhouse warms as a result of energy from the Sun?

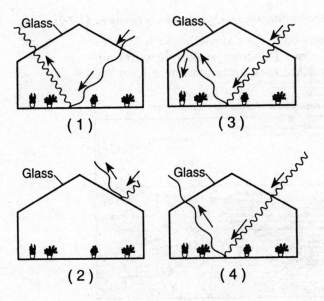

Key	
〰️	Visible light rays
〜	Infrared rays

(1)   (3)

(2)   (4)

40 _____

PART II

This part consists of six groups, each containing ten questions. Choose any one of these six groups. Be sure that you answer all ten questions in the single group chosen. Record the answers to these questions in the spaces provided. Some questions may require the use of the *Earth Science Reference Tables*. [10]

GROUP A — Rocks and Minerals

If you choose this group, be sure to answer questions 41–50.

Base your answers to questions 41 through 43 on the cross section below. The cross section shows the surface and subsurface rock formations near New York City.

Geologic Section Across the Hudson River

(Not drawn to scale)

41 The portion of the Palisades sill that contains large crystals of plagioclase feldspar and pyroxene is considered to be similar in texture and composition to

    1 obsidian   3 basalt glass

    2 granite   4 gabbro   41 ____

42 Which rock formation was originally limestone?

    1 Palisades sill   3 Inwood marble

    2 Fordham gneiss   4 Manhattan schist   42 ____

43 The rock types shown on the left side of this geologic cross section were mainly the result of

1 heat and pressure exerted on previously existing rock
2 melting and solidification of crustal rocks at great depths
3 tectonic plate boundaries diverging at the mid-ocean ridge
4 compaction and cementation of sediments under ocean waters

43 _____

Base your answers to questions 44 and 45 on the diagram below, which represents a cross section of an area of the Earth's crust. Letters *A* and *F* represent rock units.

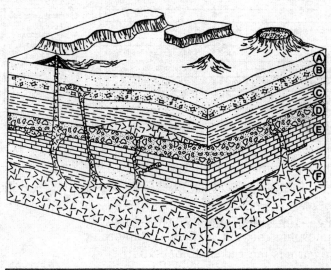

Key					
Sandstone		Shale		Limestone	
Breccia		Conglomerate		Igneous intrusion	

44 Which rock most likely had a chemical origin?

(1) *A*   (3) *C*

(2) *E*   (4) *F*   44 _____

45 Which statement best describes how rock layers *B* and *D* are different from each other?

1 One is intrusive and the other is extrusive.

2 One is clastic and the other is nonclastic.

3 One is foliated and the other is nonfoliated.

4 One has angular fragments and the other has rounded fragments.   45 _____

Base your answers to questions 46 through 48 on the table below, which gives the properties of four varieties of the mineral garnet.

Garnet						Chemical Symbols	
Variety	Composition	Density (g/cm³)	Hardness	Typical Color		Al — aluminum	
Pyrope	$Mg_3Al_2Si_3O_{12}$	3.6	7 to 7.5	Deep red to nearly black		Ca — calcium	
Almandine	$Fe_3Al_2Si_3O_{12}$	4.3	7 to 7.5	Brownish red		Fe — iron	
Spessartine	$Mn_3Al_2Si_3O_{12}$	4.2	7 to 7.5	Orange red		Mg — magnesium	
Grossular	$Ca_3Al_2Si_3O_{12}$	3.6	6.5 to 7	Yellowish green		Mn — manganese	
						O — oxygen	
						Si — silicon	

46 Which variety is correctly described as a calcium garnet?

1 pyrope   3 spessartine

2 almandine   4 grossular   46 _____

47 Garnets such as almandine are generally found in metamorphic rocks. In which metamorphic rocks are garnets most likely to be found?

1 schist and gneiss

2 gneiss and quartzite

3 quartzite and marble

4 marble and schist   47 _____

48 In which New York State landscape region are garnets most likely to be found in surface bedrock?

  1 Newark Lowlands
  2 Adirondack Mountains
  3 Erie-Ontario Lowlands
  4 Allegheny Plateau   48 _____

49 The cleavage or fracture of a mineral is normally determined by the mineral's

  1 density
  2 oxygen content
  3 internal arrangement of atoms
  4 position among surrounding minerals   49 _____

50 Oxygen is the most abundant element by volume in the Earth's

  1 inner core
  2 troposphere
  3 hydrosphere
  4 crust   50 _____

GROUP B — **Plate Tectonics**

If you choose this group, be sure to answer questions **51–60**.

Base your answers to questions 51 and 52 on the diagrams below of geologic cross sections of the upper mantle and crust at four different Earth locations, *A, B, C,* and *D*. Movement of the crustal sections (plates) is indicated by arrows, and the locations of frequent earthquakes are indicated by ✳. Diagrams are not drawn to scale.

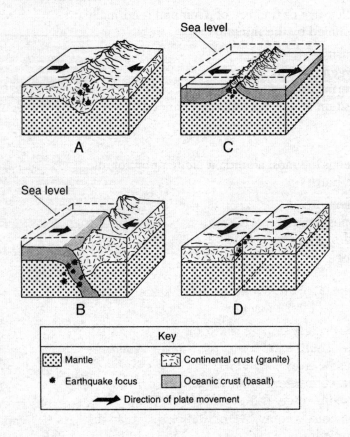

51 Which location best represents the boundary between the African plate and the South American plate?

(1) A   (2) B   (3) C   (4) D   51 _____

52 Which diagram represents plate movement associated with transform faults such as those causing California earthquakes?

(1) A   (2) B   (3) C   (4) D   52 _____

53 The diagram below represents a cross section of a portion of the Earth's crust.

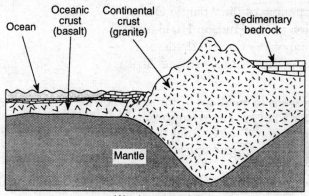

(Not drawn to scale)

Which statement about the Earth's crust is best supported by the diagram?

1 The oceanic crust is thicker than the mantle.

2 The continental crust is thicker than the oceanic crust.

3 The continental crust is composed primarily of sedimentary rock.

4 The crust is composed of denser rock than the mantle is.   53 _____

54 At a depth of 2,000 kilometers, the temperature of the stiffer mantle is inferred to be

(1) 6,500°C   (3) 3,500°C

(2) 4,200°C   (4) 1,500°C   54 _____

55 Hot springs on the ocean floor near the mid-ocean ridges provide evidence that

1 convection currents exist in the asthenosphere

2 meteor craters are found beneath the oceans

3 climate change has melted huge glaciers

4 marine fossils have been uplifted to high elevations   55 _____

56 Which geologic event occurred most recently?

1 initial opening of the Atlantic Ocean

2 formation of the Hudson Highlands

3 formation of the Catskill delta

4 collision of North America and Africa   56 _____

Base your answers to questions 57 through 60 on the four seismograms below. The seismograms show the arrival of *P*-waves and *S*-waves from the same earthquake at four different seismograph stations.

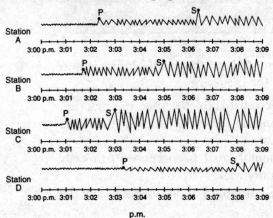

57 Which station is farthest from the epicenter of the earthquake?

(1) *A*
(2) *B*
(3) *C*
(4) *D*   57 _____

58 How many seismograms are needed to locate the epicenter of this earthquake?

1 Any one of the seismograms may be used.
2 Any two of the seismograms may be used.
3 Any three of the seismograms may be used.
4 All four seismograms must be used.   58 _____

59 What is the distance between station *A* and the epicenter of the earthquake?

(1) 1,000 km
(2) 2,000 km
(3) 2,600 km
(4) 3,200 km   59 _____

60 A fifth seismic station, located 5,600 kilometers from the earthquake epicenter, recorded the arrival of the *P*-wave at 3:06 p.m. What time did the earthquake occur?

(1) 2:05 p.m.
(2) 2:13 p.m.
(3) 2:34 p.m.
(4) 2:57 p.m.   60 _____

GROUP C — **Oceanography**

If you choose this group, be sure to answer questions **61–70**.

Base your answers to questions 61 and 62 on the diagram below. The diagram shows a coastal area with a mountain range and a portion of the ocean floor. A turbidity current through a submarine canyon has formed a fan-shaped sediment deposit on the ocean floor.

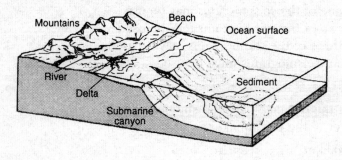

61 Which diagram best represents a cross section of the sediment in the fan-shaped deposit?

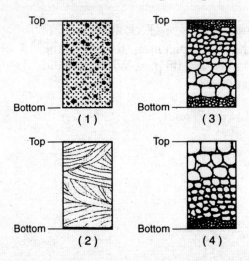

61 _____

62 In which region of the ocean is the submarine canyon located?

1 tidal zone
2 continental margin
3 deep ocean basin
4 mid-ocean ridge

62 _____

63 The scale below shows the age of rocks in relation to their distance from the Mid-Atlantic Ridge.

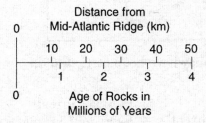

Distance from
Mid-Atlantic Ridge (km)

Age of Rocks in
Millions of Years

Some igneous rocks that originally formed at the Mid-Atlantic Ridge are now 37 kilometers from the ridge. Approximately how long ago did these rocks form?

(1) 1.8 million years ago
(2) 2.0 million years ago
(3) 3.0 million years ago
(4) 45.0 million years ago

63 _____

64 The shaded areas of the map below indicate concentrations of pollutants along the coastlines of North America.

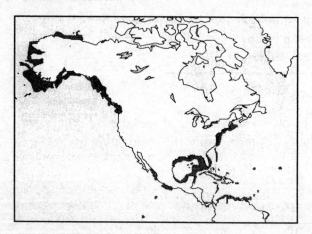

Polluting material may have been carried to the Alaska area by the

1 California Current   3 Florida Current
2 North Pacific Current   4 Labrador Current   64 _____

Base your answers to questions 65 and 66 on the diagram below. The diagram represents a shoreline with waves approaching at an angle. The exposed bedrock of the wavecut cliff is granite. Arrow *A* shows the direction of the longshore current and arrow *B* shows the general path of wave travel.

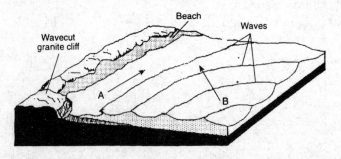

65 Which minerals are most likely to be found in the beach sand?

  1 olivine and hornblende
  2 pyroxene and plagioclase feldspar
  3 plagioclase feldspar and olivine
  4 quartz and potassium feldspar

65 _____

66 A large storm with high winds that develops out at sea is most likely to result in

  1 decreased erosion along the shoreline
  2 increased deposition along the shoreline
  3 increased wave height near the shore
  4 unchanged shoreline features

66 _____

Base your answers to questions 67 through 70 on the *Earth Science Reference Tables* and the graphs below. Graph I shows the yearly precipitation and evaporation at different latitudes, and graph II shows the salinity of the ocean at different latitudes. Salinity is a measure of the total amount of dissolved minerals in seawater, expressed as parts per thousand.

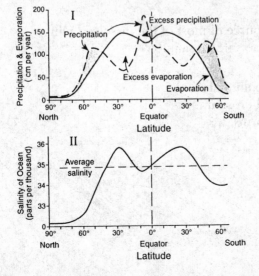

67 Compared to the amount of precipitation at the North Pole, the amount of precipitation at the Equator is

    1 half as much
    2 twice as much
    3 about the same
    4 more than ten times greater   67 _____

68 At which latitude is ocean salinity *least*?

    (1) 90° N
    (2) 30° N
    (3) 0°
    (4) 60° S   68 _____

69 The concentration of dissolved minerals in seawater would be increased by

    1 rapid evaporation
    2 heavy rainfall
    3 melting glaciers
    4 tsunamis   69 _____

70 What is the source of most of the dissolved minerals that cause surface salinity?

    1 ice sheets
    2 industrial pollutants
    3 continental erosion
    4 tropical storms   70 _____

GROUP D — Glacial Processes

If you choose this group, be sure to answer questions 71–80.

Base your answers to questions 71 through 74 on the diagrams below. Diagram I shows an imaginary present-day continent covered by an advancing glacial ice sheet. Isolines called isopachs are drawn, representing the thickness of the ice sheet in meters. Diagram II shows a cross section of the glacier with the land beneath it along reference line *XY*. Point *A* is a location on the glacier.

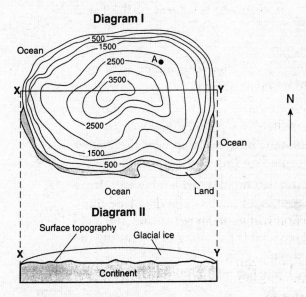

Diagram I

Diagram II

71 Chemical analysis of an ice sample taken from the central core of the glacier would most likely be used to study

1 subsurface bedrock
2 past atmospheric conditions
3 the glacier's rate of movement
4 the exact location of the ice sheet

71 _____

72 Which statement best describes the movement of this continental glacier?

  1 The glacier is advancing from north to south, only.
  2 The glacier is advancing from south to north, only.
  3 The glacier is moving outward in all directions from the central zone of accumulation.
  4 The glacier is moving inward from all directions toward the center of the continent.   72 _____

73 What is the approximate thickness of the ice at location A?

  (1) 1800 m
  (2) 2250 m
  (3) 2800 m
  (4) 3400 m   73 _____

74 Which statement best explains why the glacier originally formed on this continent?

  1 The accumulation of yearly rainwater froze every winter.
  2 The accumulation of snow during the cold season exceeded the melting of ice during the warm season.
  3 Icebergs from the surrounding sea accumulated.
  4 The continent has a low latitude and a low elevation.   74 _____

Base your answers to questions 75 through 78 on the diagrams below. Diagram I shows melting ice lobes of a continental glacier during the Pleistocene Epoch. Diagram II represents the landscape features of the same region at present, after the retreat of the continental ice sheet. Letters A through F indicate surface features in this region.

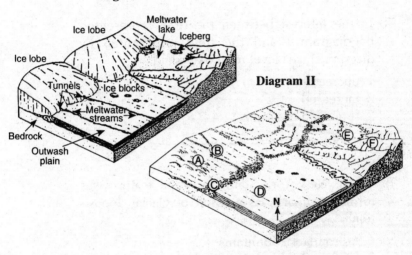

75 Which erosional feature most likely formed on the surface of the bedrock under the glacial ice?

1 sorted sands
2 sand dunes
3 parallel grooves
4 a V-shaped valley

75 _____

76 Which features in diagram II are composed of till directly deposited by the glacial ice?

(1) A and C   (3) C and E
(2) B and D   (4) E and F

76 _____

77 Which fossil has been found in glacial areas like those represented by *D* in diagram II?

1 placoderm fish
2 ammonoid
3 stromatolite
4 mastodont   77 _____

Note that question 78 has only three choices.

78 In the interval between the time represented by diagram I and the time represented by diagram II, sea level most likely had

1 decreased
2 increased
3 remained the same   78 _____

_____

79 Which New York State landscape feature was formed primarily as a result of glacial deposition?

1 Adirondack Mountains
2 Hudson-Mohawk Lowlands
3 Tug Hill Plateau
4 Long Island   79 _____

80 Glacial movement is caused primarily by

1 gravity
2 erosion
3 Earth's rotation
4 global winds   80 _____

GROUP E — **Atmospheric Energy**

If you choose this group, be sure to answer questions **81–90**.

81 Which source provides the most energy for atmospheric weather changes?

1 radiation from the Sun
2 radioactivity from the Earth's interior
3 heat stored in ocean water
4 heat stored in polar ice caps   81 _____

82 Which form of electromagnetic energy has a wavelength of 0.0001 meter?

(1) ultraviolet
(2) infrared
(3) FM and TV
(4) shortwave and AM radio   82 _____

83 Daily weather forecasts are based primarily on

1 ocean currents
2 seismic data
3 phases of the Moon
4 air-mass movements   83 _____

84 The diagram below shows air rising from the Earth's surface to form a thunderstorm cloud.

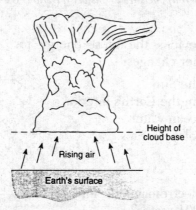

According to the Lapse Rate chart, what is the height of the base of the thunderstorm cloud when the air at the Earth's surface has a temperature of 20°C and a dewpoint of 12°C?

(1) 1.0 km   (3) 3.0 km
(2) 1.5 km   (4) 0.7 km   84 _____

85 Which cross section best shows the normal movement of the air over Oswego, New York, on a very hot summer afternoon?

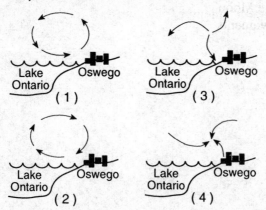

85 _____

86 The diagram below represents an activity in which an eye dropper was used to place a drop of water on a spinning globe. Instead of flowing due south toward the target point, the drop appeared to follow a curved path and missed the target.

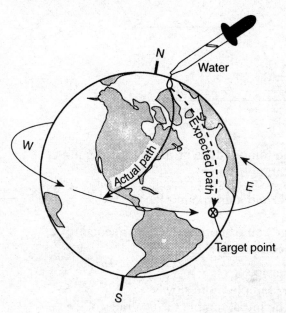

This curved-path phenomenon most directly affects the Earth's

1 tilt
2 Moon phases
3 wind belts
4 tectonic plates

86 _____

Base your answers to questions 87 through 90 on the graph below. The graph shows the results of a laboratory activity in which a sample of ice at –50°C was heated at a uniform rate for 80 minutes. The ice has a mass of 200 grams.

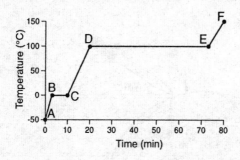

87 What was the temperature of the water 20 minutes after heating began?

(1) 70°C   (2) 100°C   (3) 110°C   (4) 150°C   87 _____

88 Which change could shorten the time needed to melt the ice completely?

1 using colder ice
2 stirring the sample more slowly
3 reducing the initial sample to 100 grams of ice
4 reducing the number of temperature readings taken   88 _____

89 What was the total amount of energy absorbed by the sample during the time between points B and C on the graph?

(1) 200 calories   (3) 10,800 calories
(2) 800 calories   (4) 16,000 calories   89 _____

90 During which interval of the graph is a phase change occurring?

(1) A to B   (2) E to F   (3) C to D   (4) D to E   90 _____

GROUP F — **Astronomy**

If you choose this group, be sure to answer questions **91–100**.

91 Which graph best illustrates the average temperatures of the planets in the solar system?

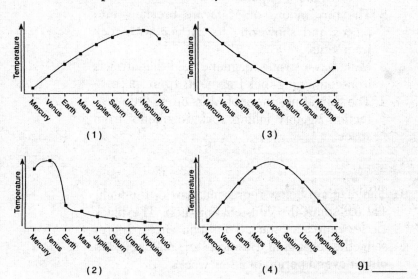

(1)

(2)

(3)

(4)

91 _____

92 A belt of asteroids is located an average distance of 503 million kilometers from the Sun. Between which two planets is this belt located?

1 Mars and Jupiter
2 Mars and Earth
3 Jupiter and Saturn
4 Saturn and Uranus

92 _____

93 Why are impact structures (craters) more common on the surface of Mars than on the surfaces of Venus, Earth, and Jupiter?

  1 Mars has the greatest surface area and receives more impacts.
  2 The tiny moons of Mars are breaking into pieces and showering its surface with rock fragments.
  3 Mars has a strong magnetic field that attracts iron-containing rock fragments from space.
  4 The thin atmosphere of Mars offers little protection against falling rock fragments from space.

93 _____

94 The diagram below represents part of the night sky including the constellation Leo. The black circles represent stars. The open circles represent the changing positions of one celestial object over a period of a few weeks.

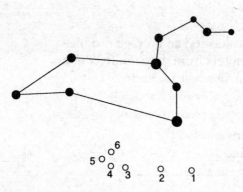

The celestial object represented by the open circles most likely is

  1 a galaxy   3 Earth's Moon
  2 a planet   4 another star

94 _____

Base your answers to questions 95 and 96 on the graphs below. The graphs show the composition of the atmospheres of Venus, Earth, Mars, and Jupiter.

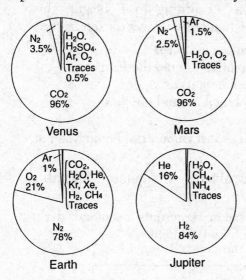

95 Which gas is present in the atmospheres of Venus, Earth, and Mars but is *not* present in the atmosphere of Jupiter?

1 argon (Ar)   3 hydrogen (H_2)
2 methane (CH_4)   4 water vapor (H_2O)   95 _____

96 Which planet has an atmosphere composed primarily of CO_2 and a period of rotation greater than its period of revolution?

1 Venus   2 Mercury   3 Earth   4 Mars   96 _____

97 In which type of model are the Sun, other stars, and the Moon in orbit around the Earth?

1 heliocentric model   3 concentric model
2 tetrahedral model   4 geocentric model   97 _____

98 In 1851, the French physicist Jean Foucault constructed a large pendulum that always changed
its direction of swing at the same rate in a clockwise direction. According to Foucault, this
change in direction of swing was caused by the

1 Moon's rotation on its axis
2 Moon's revolution around the Earth
3 Earth's rotation on its axis
4 Earth's revolution around the Sun   98 _____

99 Which planet has vast amounts of liquid water at
its surface?

1 Venus   2 Mars   3 Jupiter   4 Earth   99 _____

100 The diagram below represents a standard dark-
line spectrum for an element.

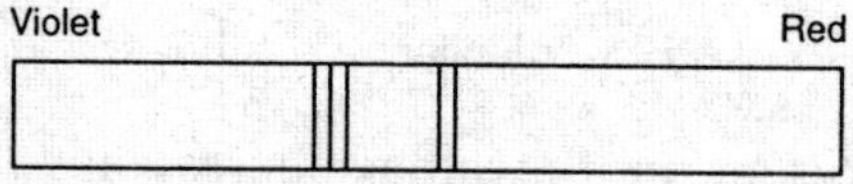

The spectral lines of this element are observed
in light from a distant galaxy. Which diagram
represents these spectral lines?

(1)

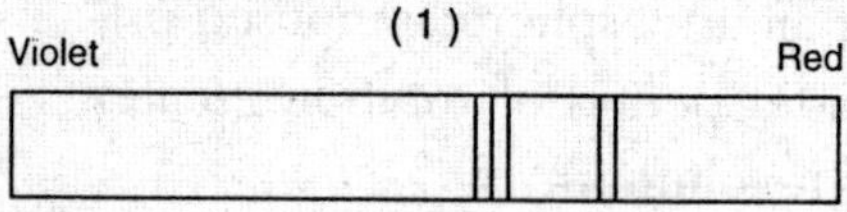

(2)

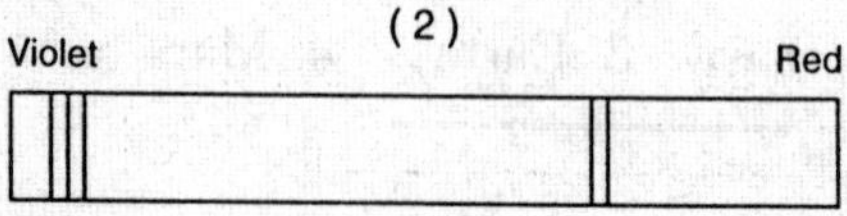

(3)

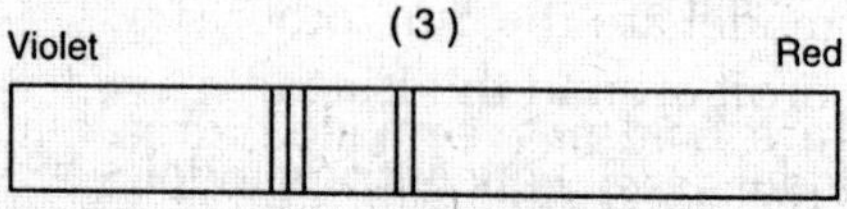

(4)   100 _____

PART III

This part consists of questions 101 through 120. Be sure that you answer all questions in this part. Record your answers in the spaces provided. Some questions may require the use of the *Earth Science Reference Tables*. [25]

Base your answers to questions 101 and 102 on the temperature field map below. The map shows 25 measurements (in °C) that were made in a temperature field and recorded as shown. The dots represent the exact locations of the measurements. *A* and *B* are locations within the field.

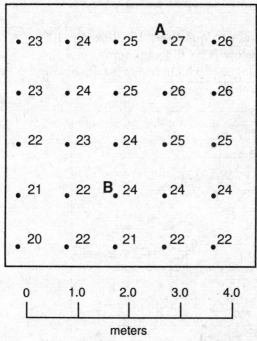

Temperature Field Map (°C)

101 On the temperature field map provided, draw three isotherms: the 23°C isotherm, the 24°C isotherm, and the 25°C isotherm.   [2]

102 In the space provided, calculate the temperature gradient between locations *A* and *B* on the temperature field map, following the directions below.

a Write the equation for gradient. [1]

b Substitute data from the map into the equation. [1]

c Calculate the gradient and label it with the proper units. [1]

Base your answers to questions 103 through 105 on the diagram below. The diagram represents the supercontinent Pangaea, which began to break up approximately 220 million years ago.

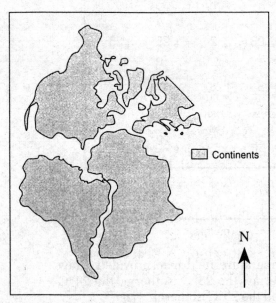

103 During which geologic period within the Mesozoic Era did the supercontinent Pangaea begin
to break apart?   [1]

_____

104 State one form of evidence that supports the
inference that Pangaea existed.   [1]

_____

_____

_____

105 State the compass direction toward which North
America has moved since Pangaea began to
break apart.   [1]

_____

_____

Base your answers to questions 106 through 108
on the information below.

A mountain is a landform with steeply
sloping sides whose peak is usually
thousands of feet higher than its base.
Mountains often contain a great deal of
nonsedimentary rock and have distorted
rock structures caused by faulting and folding of the crust.
A plateau is a broad, level area at a high
elevation. It usually has an undistorted,
horizontal rock structure. A plateau may
have steep slopes as a result of erosion.

106 State why marine fossils are not usually found in
the bedrock of the Adirondack Mountains.
[1]

_____

_____

_____

_____

107 State the agent of erosion that is most likely
responsible for shaping the Catskill Plateau so that
it physically resembles a mountainous region.
[1]

_____

_____

108 State the approximate age of the surface bedrock
of the Catskills.   [1]

_____

_____

Base your answers to questions 109 through 112 on the diagram and the stream data table below.

The diagram represents a stream flowing into a lake. Arrows show the direction of flow. Point *P* is a location in the stream. Line *XY* is a reference line across the stream. Points *X* and *Y* are locations on the banks. The data table gives the depth of water in the stream along line *XY*.

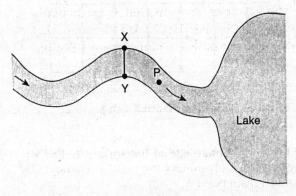

Stream Data Table

	Location X							Location Y
Distance from X (meters)	0	5	10	15	20	25	30	35
Depth of Water (meters)	0	5.0	5.5	4.5	3.5	2.0	0.5	0

Directions (109–110): Use the information in the data table to construct a profile of the depth of water. Use the grid provided, following the directions below.

109 On the vertical axis, mark an appropriate scale for the depth of water. Note that the zero (0) at the top of the axis represents the water surface. [1]

Stream Profile Graph

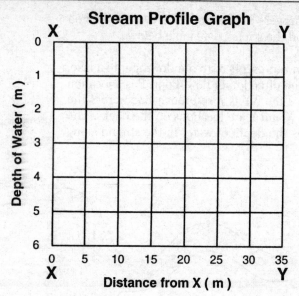

110 Plot the data for the depth of water in the stream along line *XY* and connect the points. (Distance is measured from point *X.*) [2]

Example:

111 State why the depth of water near the bank at point *X* is different from the depth of water near the bank at point *Y.* [1]

_____

_____

_____

112 At point *P,* the water velocity is 100 centimeters per second. State the name of the largest sediment that can be transported by the stream at point *P.* [1]

_____

Base your answers to questions 113 through 115 on the weather maps below. The weather maps show the positions of a tropical storm at 10 AM on July 2 and on July 3.

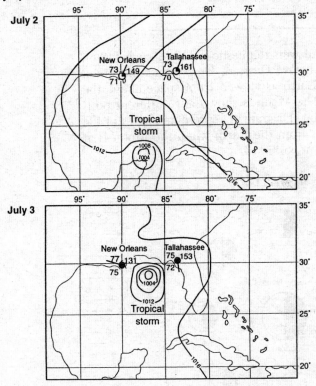

113 State the dewpoint temperature in Tallahassee on July 2. [1]

_____

114 Windspeed has been omitted from the station models. In one or more sentences, state how an increase in the storm's windspeed from July 2 to July 3 could be inferred from the maps. [2]

_____

_____

_____

115 The storm formed over warm tropical water. State what will most likely happen to the windspeed when the storm moves over land.   [1]

_____

_____

_____

Base your answers to questions 116 through 118 on the diagrams below. Diagram I represents the Moon orbiting the Earth as viewed from space above the North Pole. The Moon is shown at 8 different positions in its orbit. Diagram II represents phases of the Moon as seen from the Earth when the Moon is at position 2 and at position 4.

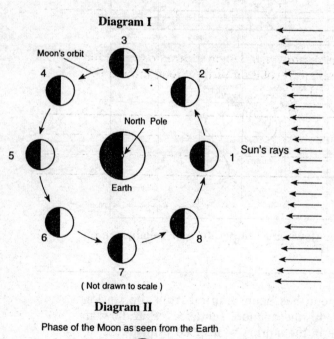

Diagram I

3

Moon's orbit

4

2

North Pole

5

Sun's rays

1

Earth

6

8

7

(Not drawn to scale)

Diagram II

Phase of the Moon as seen from the Earth

at position 2

at position 4

KEY:

☐ Lighted, visible part of Moon

■ Dark, invisible part of Moon

116 Shade the circle provided to illustrate the Moon's phase as seen from the Earth when the Moon is at position 7.   [1]

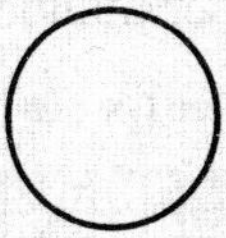

117 State the two positions of the Moon at which an eclipse could occur.   [1]

_____

_____

118 State the approximate length of time required for one complete revolution of the Moon around the Earth.   [1]

_____

_____

_____

_____

Base your answers to questions 119 through 120 on the graph below. The graph shows the average water temperature and the dissolved oxygen levels of water in a stream over a 12-month period. The level of dissolved oxygen is measured in parts per million (ppm).

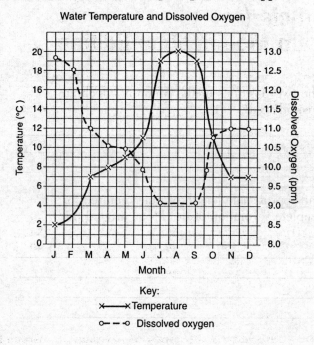

Water Temperature and Dissolved Oxygen

Key:
×——————× Temperature
o— — —o Dissolved oxygen

119 State the difference in average water temperature, in degrees Celsius, between January and August. [1]

_____

120 State the relationship between the temperature of the water and the level of dissolved oxygen in the water.   [1]

_____
_____
_____

Answers
June 1997

Earth Science —
Program Modification Edition

Answer Key

PART I

1. 2	8. 1	15. 2	22. 1	29. 1	36. 4
2. 3	9. 2	16. 1	23. 3	30. 2	37. 2
3. 1	10. 2	17. 4	24. 1	31. 3	38. 3
4. 4	11. 1	18. 1	25. 2	32. 2	39. 4
5. 2	12. 1	19. 3	26. 4	33. 2	40. 3
6. 4	13. 3	20. 2	27. 3	34. 1	
7. 3	14. 4	21. 4	28. 3	35. 2	

PART II

Group A	Group B	Group C	Group D	Group E	Group F
41. 4	51. 3	61. 3	71. 2	81. 1	91. 2
42. 3	52. 4	62. 2	72. 3	82. 2	92. 1
43. 1	53. 2	63. 3	73. 2	83. 4	93. 4
44. 2	54. 2	64. 2	74. 2	84. 1	94. 2
45. 4	55. 1	65. 4	75. 3	85. 1	95. 1
46. 4	56. 1	66. 3	76. 1	86. 3	96. 1
47. 1	57. 4	67. 4	77. 4	87. 2	97. 4
48. 2	58. 3	68. 1	78. 2	88. 3	98. 3
49. 3	59. 3	69. 1	79. 4	89. 4	99. 4
50. 4	60. 4	70. 3	80. 1	90. 4	100. 2

PART III See answers explained.

Answers Explained

PART I

1. **2** Find the map entitled Generalized Bedrock Geology of New York State in the *Earth Science Reference Tables*. Using the latitude and longitude markings along the perimeter of New York State, locate the intersection of 45° N and 75° W. The city closest to this point and hence to the epicenter of the earthquake is Massena, New York.

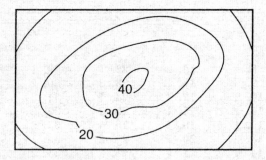

2. **3** From the three adjacent isolines in the diagram labeled 20, 30 and 40, it may be inferred that the contour interval is 10 meters. Since there are no additional lines above 40 meters, the highest possible elevation represented on this map must be greater than 40 meters but less than 50 meters. Of the choices given, 49 meters best meets this specification.

3. **1** Limestone is a nonclastic sedimentary rock, sandstone is a clastic sedimentary rock, and granite is an igneous rock. Therefore, an accurate inference is that the pebbles had various origins.

WRONG CHOICES EXPLAINED:
(2) The pebbles may have eroded out of older conglomerates, been re-deposited, and relithified into the conglomerate in the question, but there is no evidence of such events in the information given. Furthermore, since limestone is easily weathered, it is highly unlikely that the limestone pebbles would have survived erosion from a previous conglomerate.
(3) To accurately infer the relative or absolute ages of the pebbles, one would need information such as the relative positions of the rock structures from which the pebbles eroded, or the ratio of a radioisotope and its decay

product in the rock. Since no such information is given, an accurate inference about the ages of the pebbles cannot be made.

(4) To accurately infer the rate at which the pebbles eroded, one would need, at the very least, information such as the agent of erosion, or the climate in which the pebbles were eroded. Since no such information is given, an accurate inference about the rate at which the pebbles eroded cannot be made.

4. **4** Find in the *Earth Science Reference Tables* the chart entitled Scheme for Sedimentary Rock Identification. In the column labeled "Rock Name," find rock gypsum and chemical limestone. Then, in the column labeled "Composition," note that rock gypsum is composed of one mineral—gypsum, and chemical limestone is composed of one mineral—calcite. Find in the *Earth Science Reference Tables* the chart entitled Scheme for Metamorphic Rock Identification. In the column labeled "Rock Name," find quartzite and gneiss. Then, in the column labeled "Composition," note that quartzite is composed of one mineral—quartz, and gneiss is composed of several minerals—mica, quartz, feldspar, amphibole, garnet, and pyroxene. Choice (4), gneiss, is a rock that is usually composed of several different minerals.

5. **2** Find in the *Earth Science Reference Tables* the diagram entitled Rock Cycle in Earth's Crust. Note that the arrow leading to the box labeled "Sedimentary Rock" is labeled "Cementation" and "Dewatering/Compaction," indicating that these processes occur as sediments are changing to sedimentary rock.

WRONG CHOICES EXPLAINED:
(1) Magma is molten rock. Before it can be weathered, it must solidify into igneous rock and be exposed at the surface. This is indicated in the diagram entitled Rock Cycle in Earth's Crust by the arrow labeled "Solidification" and leading from the box labeled "Magma" to the box labeled "Igneous Rock."

(3) Sedimentary rock can be metamorphosed by heat and pressure directly into a metamorphic rock. This is indicated in the diagram entitled Rock Cycle in Earth's Crust by the arrow labeled "Heat and/or Pressure/Metamorphism," and leading from the box labeled "Sedimentary Rock" to the box labeled "Metamorphic Rock."

(4) Metamorphic rock can weather and erode directly into sediments that may then be compacted and cemented into sedimentary rock. This is indicated in the diagram entitled Rock Cycle in Earth's Crust by the arrow labeled "Weathering & Erosion/(Uplift)" and leading from the box labeled "Metamorphic Rock" directly to the box labeled "Sedimentary Rock." Note that if metamorphic rock "melts" it becomes igneous rock.

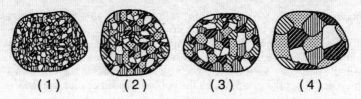

(1)   (2)   (3)   (4)

6. **4** The more slowly a magma cools, the larger the mineral crystals that result. The mineral crystals in choice (4) are the largest.

7. **3** The Earth's outer core is believed to be liquid because *S*-waves do not penetrate it and *P*-waves slow down in it. The behavior of *S*-waves and *P*-waves is seismic data derived from records of earthquake tremors collected at sites all over the Earth.

8. **1** Find the map entitled Tectonic Plates in the *Earth Science Reference Tables*. Locate the Australian plate at the right and left of the map, and note the three arrows on it. One arrow points north near the boundary with the Antarctic Plate. Another arrow, just below India, also points north. A third arrow near the boundary with the Philippine plate points northeastward. These arrows indicate that the direction of crustal movement of the Australian plate is primarily northward.

9. **2** The cartoon character on the right is looking at an approaching cloud with rain falling from it. If the rain falls on the sand castle, it will eventually be washed away by running water, or removed by an agent of erosion.

10. **2** Rock particles can become scratched by rubbing against each other or against bedrock while being transported by any of the agents of erosion listed as choices. However, sorting is typically associated with deposition by a *fluid* medium, such as wind, running water, or ocean waves. In deposition by solid glacial ice, there is no fluid medium through which sediments settle and can become sorted as they are deposited. Therefore, a deposit of unsorted rock particles is likely to have been transported and deposited by glacial ice.

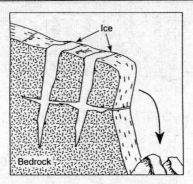

11. **1** The diagram shows ice freezing in cracks in the bedrock, causing the solid bedrock to break up into blocks. The breakdown of rock into smaller particles by natural processes is called weathering.

WRONG CHOICES EXPLAINED:

(2) Cementing is the binding together of particles by substances such as clay, carbonates, or hydrates of iron. In the diagram the ice is not binding the bedrock together; rather, it is clearly causing the bedrock to break apart, as shown by the arrow pointing to the blocks downslope.

(3) Metamorphism involves the changing of rock from one type to another, typically through heat, pressure, and chemical activity. Frost wedging breaks a solid mass of bedrock into loose fragments varying in size, but identical in composition to the original rock.

(4) Deposition is the process whereby transported sediment is dropped in a new place. Frost wedging only breaks the bedrock apart; it is gravity that causes the pieces to move downslope and be deposited in a new location.

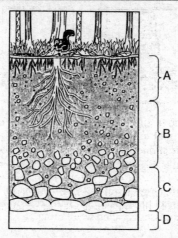

12. **1** Biological activity means the carrying out of life processes by living organisms. The air, water, and food required by the plants and animals shown in the diagram are most readily available near the surface. Fallen leaves and animal wastes are examples of materials formed by biological activities that are added at the surface, thus to horizon A. Therefore, soil horizon A, which is nearest the surface, is usually richest in material formed by biological activity.

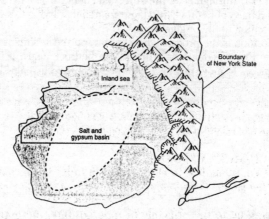

13. **3** Find the map entitled Generalized Landscape Regions of New York State in the *Earth Science Reference Tables*. Note that the area of New York State labeled "Salt and gypsum basin" on the map in the question falls within the landscape regions labeled "Erie-Ontario Lowlands" and "Allegheny Plateau."

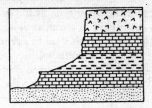 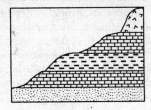

14. **4** The hillslopes and shapes of the rock layers in a region are the result of weathering and erosion of the exposed rock. For almost identical rock layers to weather and erode into very different hillslopes and shapes, the type and rate of weathering and erosion must have been significantly different. Climate is a key factor in determining both the type of weathering and erosion that occurs in a region and the rate at which it occurs. Thus, climate variations are most likely responsible for the differences in the hillslopes and shapes in the two regions.

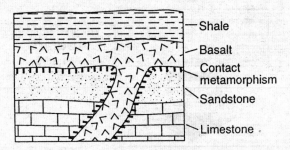

15. **2** Since the shale is on top of the basalt, but no contact metamorphism has occurred along the shale/basalt boundary, it can be inferred that the shale was deposited on top of the basalt after the basalt had already cooled. Thus, the statement "The shale is younger than the basalt." correctly describes an age relationship in the geologic cross section.

WRONG CHOICES EXPLAINED:
(1) The sandstone had to already exist in order to be contact metamorphosed by the basalt. Therefore, the sandstone is older, not younger, than the basalt.

(3) According to the principle of superposition, the bottom layer in a series of sedimentary layers is oldest unless it has been overturned or older rock has been thrust upon it. The cross section shows no evidence of overturn or thrust faulting. Thus, the limestone is probably the oldest layer since it is on the bottom, and the shale the youngest layer since it is on the top.

(4) The limestone had to already exist in order to be contact metamorphosed by the basalt. Therefore, the limestone is older, not younger, than the basalt.

16. **1** Fossils are generally deposited at the same time as the bedrock in which they are found. Since present-day corals live in warm, shallow seas, it is reasonable to infer that, during Ordovician time, corals lived in warm, shallow seas. If the bedrock in western New York State contains fossil corals of Ordovician age, then it is likely that a warm, shallow sea covered western New York State during Ordovician time and that both the coral and the bedrock were deposited in this sea.

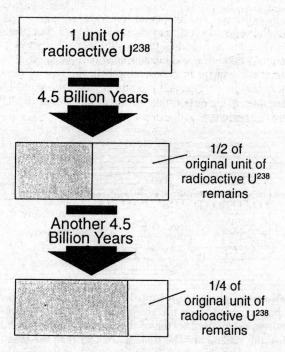

17. **4** Radioactive decay is the process by which unstable isotopes disintegrate spontaneously, giving off energy, subatomic particles, or both. The end product of radioactive decay is a stable isotope. Find the chart entitled Radioactive Decay Data in the Physical Constants section of the *Earth Science Reference Tables*. Note that uranium-238 has a half life of 4.5×10^9 years and disintegrates to form lead-206 (Pb^{206}). The diagram shows that after 4.5 billion years one half of the unstable uranium-238 remains unchanged, while the other half, shown as a shaded area, has decayed into stable lead-206.

18. **1** Clouds consist of droplets of liquid water or solid ice crystals. In order for water vapor to change into a cloud, it must change phase from a gas to a liquid or a solid. Condensation is the process by which water vapor changes into droplets of liquid water.

WRONG CHOICES EXPLAINED:

(2) Evaporation is the process by which liquid water changes to a gas; this process would cause a cloud to change into water vapor.

(3) Convection is the circular pattern of motion in a fluid caused by differences in the density within the fluid, not by a change in phase.

(4) Precipitation is condensed moisture that falls to the ground, not a change in phase.

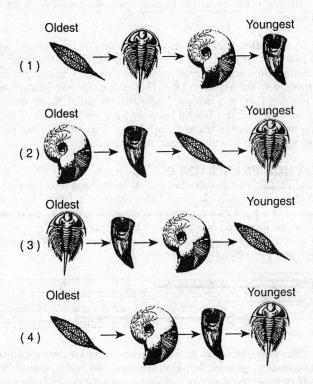

19. **3** Find the chart entitled Geologic History of New York State at a Glance in the *Earth Science Reference Tables*. In the column labeled "Important Fossils of New York," locate the fossils corresponding to the drawings in the question: the ammonoid, the figlike leaf, the trilobite, and the coral head. Next, to the left of each fossil, locate the column labeled "Period," and determine the time period during which the fossil organism lived: the ammonoid during the Devonian, the figlike leaf during the Cretaceous and Tertiary, the trilobite during the Cambrian, and the coral head during the Ordovician. Next, order the periods and their associated fossils from oldest to youngest: Cambrian—trilobite, Ordovician—coral head, Devonian—

ammonoid, and Cretaceous/Tertiary—figlike leaf. This order corresponds to the sequence of the fossils in choice (3).

20. **2** Locate the diagram entitled Selected Properties of Earth's Atmosphere in the *Earth Science Reference Tables*. In the section labeled "Temperature Zones" on the right-hand side of the diagram, note the altitude scale to the left. The correct order of the atmospheric layers from the Earth's surface (altitude = 0) upward is troposphere, stratosphere, mesosphere, thermosphere.

21. **4** Precipitation occurs when water droplets or ice crystals suspended in clouds join together and become heavy enough to fall. This is more likely to happen where the number of water droplets or ice crystals is increasing, thereby increasing the likelihood that they will come into contact with one another, join together, and become heavy enough to fall. When warm, moist air meets cold, dry air at a frontal boundary, it rises and cools. This cooling causes the moisture in the warm air to change from water vapor into water droplets or ice crystals and often results in precipitation.

WRONG CHOICES EXPLAINED:
(1) Cold fronts typically move faster, not slower, than warm fronts.
(2) Cold fronts typically do move faster than warm fronts. However, it is not the speed of a front that explains precipitation; rather, it is the processes that result in the formation of water droplets or ice crystals.
(3) Warm, moist air is less dense than cold, dry air. Therefore warm, moist air will rise, not sink, when it meets cold, dry air.

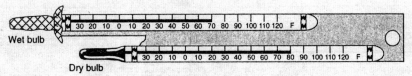

22. **1** The diagram shows a sling psychrometer consisting of two thermometers mounted on a holder. One thermometer is dry, and the other is kept wet by a small piece of cloth saturated with water. Both thermometers are labeled "F" for Fahrenheit at the right-hand end of their scales, so their readings are wet-bulb 70° F, dry-bulb 80° F. To convert to Celsius temperature readings, find the chart entitled Temperature in the *Earth Science Reference Tables*, and note that the left side of the diagram is the Fahrenheit scale and the center is the Celsius scale. Locate 70 on the Fahrenheit scale, and then look directly to the right to find that the corresponding temperature on the Celsius scale is 21. Repeat for 80 on the Fahrenheit scale; it corresponds to 27 on the Celsius scale. Thus, the equivalent Celsius temperatures are wet 21°C, dry 27°C.

23. **3** The idea behind the sling psychrometer is that water evaporating from the wet bulb cools that thermometer, while the dry-bulb temperature remains unchanged. The amount of evaporation, and the amount of cooling of the wet bulb, are directly related to the amount of moisture already in the air. Therefore, the sling psychrometer is used to determine moisture-related weather variables such as relative humidity and dewpoint.

WRONG CHOICES EXPLAINED:

(1) Air temperature could be determined from the dry-bulb reading, but to determine windspeed you would need an anemometer.

(2) To determine visibility, you would need to view objects at various distances; to determine wind direction, you would need a weather vane.

(4) Air pressure is measured with a barometer, and cloud type is determined by direct observation.

24. **1** The temperature at the Earth's surface is the result of a dynamic equilibrium between incoming sunlight (chiefly visible light) and outgoing terrestrial radiation (chiefly heat). Clouds are composed of water droplets, which reflect and absorb both visible light and heat. Therefore, clouds both block sunlight from reaching the Earth's surface during the day and prevent heat radiating from the Earth from escaping at night.

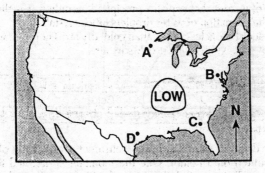

25. **2** The planetary winds that control the movement of weather systems in the continental United States are the prevailing southwesterlies. They normally move weather systems from southwest to northeast; therefore, if the low-pressure system on the weather map in the question follows a normal storm track, it will move toward point *B*.

26. **4** Water has a higher specific heat than land materials; therefore, when both water and land gain the same amount of heat energy, the land will increase in temperature to a greater degree than the water. Conversely, if

both land and water lose the same amount of heat energy, the land will decrease in temperature to a greater degree than the water. At a coastal location, the presence of a large body of water has a moderating effect on temperatures. In the summer, water does not increase in temperature as much as land, and the water has a cooling effect on coastal locations. Also, in the winter, water does not decrease in temperature as much as land, and the water has a warming effect on coastal locations. Thus, a coastal location has cooler summers and warmer winters than an inland location of the same elevation and latitude.

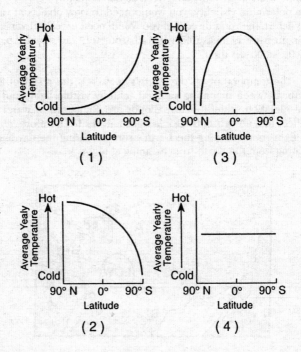

27. **3** Insolation does not strike all points on the Earth's surface at the same angle. Near the Equator insolation strikes the surface almost vertically, but near the poles insolation strikes at a more glancing angle. Therefore, the insolation near the Equator is more concentrated and results in a higher yearly average temperature at the Equator than at the poles. Choice (3) correctly shows cold temperatures at the poles (latitudes = 90° N and 90° S), and hot temperatures at the Equator (latitude = 0°).

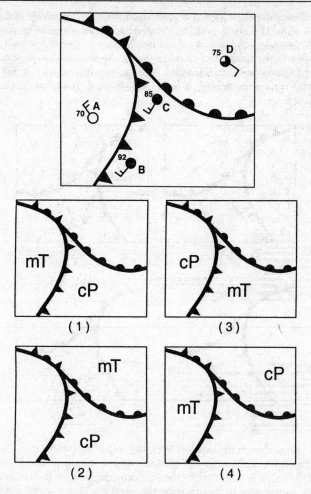

28. **3** Find the chart entitled Weather Map Information in the *Earth Science Reference Tables*, and note that the symbol for the front between city *A* and cities *B* and *C* in the map in the question represents a cold front, and that the terms "warm" and "cold" are relative: 70°F air is cold (cP) compared to 92°F or 85°F air.

Along a cold front, a cold air mass is advancing against a warm air mass in the direction in which the symbols are pointing. Thus, city *A* is in a cold air mass that is advancing toward cities *B* and *C*, which are in a warm air mass. Warm air masses are called tropical; cold air masses, polar. Only the diagram in choice (3) correctly shows a polar air mass to the left of the front and a tropical air mass to the right.

29. **1** Precipitation is typically associated with the approach and passing of a front. Find the chart entitled Weather Map Information in the *Earth Science Reference Tables*. Note that the shading of the circle in a station model indicates the amount of cloud cover. In the map in the question, cities *B*, *C*, and *D* have overcast skies and approaching fronts. City *A*, with clear skies and no front approaching, is *least* likely to have precipitation starting in the next few hours.

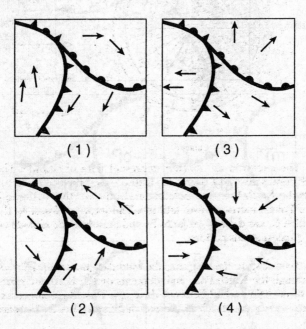

30. **2** Find the chart entitled Weather Map Information in the *Earth Science Reference Tables*. In the section labeled "Station Model," note that the wind direction is indicated by the line extending out of the station model. Now, note in the map in the question the wind directions at cities *A* through *D*. At city *A* the winds are from the northwest, at cities *B* and *C* the winds are from the southwest, and at city *D* the winds are from the southeast. The arrows in map (2) best match this pattern.

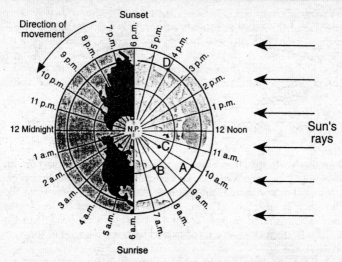

31. **3** The circles in the diagram represent lines of latitude. The outermost circle represents the Equator (0° latitude), and the center, labeled N.P., represents the North Pole (90° N latitude). In the Northern Hemisphere the altitude of Polaris increases with latitude. The location closest to the North Pole, position *C*, has the highest latitude, and therefore the altitude of Polaris would be greatest at this position.

32. **2** According to the diagram, the boundary between night and day is passing through the North Pole, and all points on the Earth are experiencing 12 hours of daylight and 12 hours of darkness. This situation occurs twice a year, approximately on March 21 (vernal) and September 21 (autumnal), the equinoxes.

33. **2** The 24 standard time zones span 360° of longitude (0° to 180° E plus 0° to 180° W). If 360° of longitude is divided by 24 time zones, the result is 15° of longitude per time zone.

34. **1** The Earth's axis of rotation is tilted at 23½° to the plane of its orbit around the Sun. Therefore, the position of the Earth in relation to the Sun varies in a cyclic pattern throughout the year.

When the Northern Hemisphere is tilted farthest toward the Sun, on June 21, the Sun will rise north of east and set north of west as shown in the diagram. When the Northern Hemisphere is tilted farthest away from the Sun, on December 21, the Sun will rise south of east and set south of west. The locations of sunrise and sunset vary in a cyclical pattern throughout the year because the Earth rotates on a tilted axis while revolving around the Sun.

35. **2** This question is most easily answered by a process of elimination (see Wrong Choices Explained below); however, the answer can also be worked out as detailed here.

Orbital velocity can be determined by dividing the circumference of a planet's orbit by its period of revolution. The four planets in question have almost circular orbits, so, for purposes of comparison, treat them as perfectly circular.

Find the chart entitled Equations and Proportions in the *Earth Science Reference Tables* and note that $C = 2\pi r$. Next, in the chart entitled Solar System Data, locate the column labeled "Mean Distance from Sun." Since the mean distance of a planet from the Sun is approximately the radius of its orbit, substitute the mean distance from the Sun for r in the equation $C = 2\pi r$

Jupiter: C = 778.3 million km × 2 × 3.14 = 4887.7 million km

Saturn: C = 1427 million km × 2 × 3.14 = 8961.6 million km

Venus: C = 108.2 million km × 2 × 3.14 = 679.5 million km

Mars: C = 227.9 million km × 2 × 3.14 = 1431.2 million km

Now divide each circumference of orbit by the period of revolution, in years, to get orbital velocity in millions of kilometers per year.

Jupiter: 4887.7 million km ÷ 11.86 yr = 412.1 million km/yr

Saturn: 8961.6 million km ÷ 29.46 yr = 304.2 million km/yr

Venus: 679.5 million km ÷ .62 yr   = 1095.5 million km/yr

Mars: 1431.2 million km ÷ 1.88 yr = 761.27 million km/yr

Thus, Mars and Venus have greater orbital velocities than Jupiter and Saturn.

WRONG CHOICES EXPLAINED:

(1) Locate the column labeled "Period of Revolution" in the Solar System Data chart, and note that Venus and Mars have shorter periods of revolution than Jupiter and Saturn.

(3) Locate the column labeled "Mean Distance from Sun" in the Solar System Data chart, and note that Venus and Mars have smaller mean distances from the Sun than Jupiter and Saturn.

(4) Locate the column labeled "Equatorial Diameter" in the Solar System Data chart, and note that Venus and Mars have smaller equatorial diameters than Jupiter and Saturn.

36. **4** A galaxy is a system consisting of hundreds of billions of stars in the same region of the universe.

WRONG CHOICES EXPLAINED:

(1) A solar system consists of only one star and any planets that orbit it.

(2) An asteroid belt consists of rocky debris orbiting a single star.

(3) A constellation is a configuration of stars, and usually consists of a dozen or so stars, not billions.

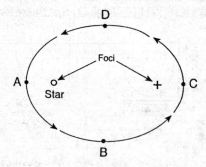

37. **2** Locate the chart entitled Equations and Proportions in the *Earth Science Reference Tables* and find the equation for the eccentricity of an ellipse:

$$\text{eccentricity} = \frac{\text{distance between foci}}{\text{length of major axis}}$$

Locate the centimeter ruler in the *Earth Science Reference Tables*, and use it to measure the distance in the diagram in the question between the foci of the ellipse and also the length of the major axis. The distance between the foci is 3.3 centimeters, and the length of the major axis (from points A to C) is 5.1 centimeters. Thus

$$\text{eccentricity of planet's orbit} = \frac{3.3 \text{ cm}}{5.1 \text{ cm}} = 0.65$$

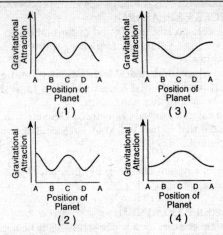

38. **3** Gravitational attraction between any two bodies is greatest when the distance between them is smallest. According to the diagram, the distance between the star and the planet is greatest at point C, smallest at point A. Thus, the gravitational attraction is greatest when the planet is at point A, and least when the planet is at point C. Graph (3) best shows this relationship.

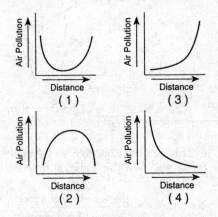

39. **4** Industrial cities are sources of pollution. The amount of pollution is greatest in a city and decreases with distance from the city because, as the pollution spreads out over a wider area, it becomes less concentrated. Graph (4) shows this relationship correctly, with pollution decreasing as distance increases.

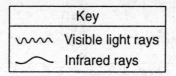

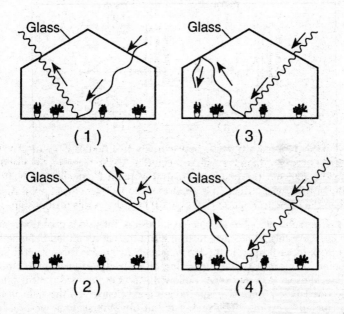

40. **3** The short-wave visible light rays from the Sun pass through the glass of the greenhouse and are absorbed by the plants and the soil. The energy that is radiated back is long-wave infrared rays. These long-wave rays cannot pass through the glass and are reflected back from the glass surface. In this manner, some of the Sun's energy is trapped inside the greenhouse, warming it. This pattern is illustrated in diagram (3).

PART II

GROUP A—**Rocks and Minerals**

Geologic Section Across the Hudson River

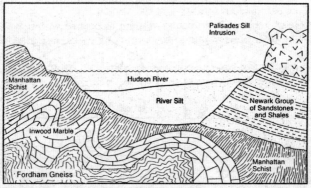

(Not drawn to scale)

41. **4** The Palisades sill is an igneous intrusion. Intrusions cool slowly and develop large crystals, or a coarse texture. Find the chart entitled Scheme for Igneous Rock Identification in the *Earth Science Reference Tables*. In the section labeled "Mineral Composition," locate the region that contains both plagioclase feldspar and pyroxene. Follow this section upward until it intersects the row labeled "Intrusive" on the left and "Coarse" on the right in the column labeled "Texture." The rock listed at this intersection, gabbro, has large crystals of plagioclase feldspar and pyroxene.

42. **3** Limestone consists of the mineral calcite. The words "was originally limestone" in the question indicate that the rock has changed. Rocks that have undergone changes are metamorphic. In the chart entitled Scheme for Metamorphic Rock Identification in the *Earth Science Reference Tables*, find the column labeled "Composition" and note that only one rock is composed of calcite. Follow this row right to the column labeled "Rock Name," and note that the name of the rock is marble. Also, note in the column labeled "Comments" that metamorphism of limestone or dolostone produces marble. Finally, refer to the column headed "Map Symbol" and then to the cross section in the question, and note that only one rock formation, the Inwood Marble, consists of a marble that was originally limestone.

43. **1** The left side of the geologic cross section consists of Manhattan Schist, Inwood Marble, and Fordham Gneiss. Find the Scheme for Metamorphic Rock Identification in the *Earth Science Reference Tables*. In the

column labeled "Rock Name" note that schist, marble, and gneiss are all metamorphic rocks. Metamorphic rocks form when heat and pressure are exerted on previously existing rock.

WRONG CHOICES EXPLAINED:

(2) Schist, marble, and gneiss are all metamorphic rocks. Melting and solidification yields igneous rocks.

(3) Divergent tectonic plate boundaries are associated with upwellings of magma, which cool underwater to form igneous rocks, not metamorphic rocks.

(4) Compaction and cementation of sediments are processes that result in the formation of sedimentary rocks, not metamorphic rocks.

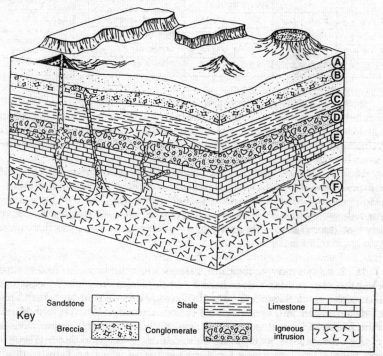

Key	Sandstone		Shale		Limestone	
	Breccia		Conglomerate		Igneous intrusion	

44. **2** Find the chart entitled Scheme for Sedimentary Rock Identification in the *Earth Science Reference Tables*, and locate the section labeled "Chemically and/or Organically Formed Sedimentary Rocks." Of the six rock types in the diagram—sandstone, breccia, shale, conglomerate, limestone, and the igneous intrusion—only limestone was chemically formed. Now, note, in the key for the diagram, the symbol for limestone. In the diagram, only layer

E corresponds to this symbol, and rock *E*, limestone, most likely had a chemical origin.

45. **4** Using the key to the diagram, note that rock layer *B* is breccia and rock layer *D* is a conglomerate. Now, find the chart entitled Scheme for Sedimentary Rock Identification in the *Earth Science Reference Tables*. In the column labeled "Rock Name" locate both conglomerate and breccia. In the column labeled "Comments" note that conglomerate has rounded fragments and breccia has angular fragments.

Garnet				
Variety	**Composition**	**Density** (g/cm³)	**Hardness**	**Typical Color**
Pyrope	$Mg_3Al_2Si_3O_{12}$	3.6	7 to 7.5	Deep red to nearly black
Almandine	$Fe_3Al_2Si_3O_{12}$	4.3	7 to 7.5	Brownish red
Spessartine	$Mn_3Al_2Si_3O_{12}$	4.2	7 to 7.5	Orange red
Grossular	$Ca_3Al_2Si_3O_{12}$	3.6	6.5 to 7	Yellowish green

Chemical Symbols	
Al	aluminum
Ca	calcium
Fe	iron
Mg	magnesium
Mn	manganese
O	oxygen
Si	silicon

46. **4** A calcium garnet contains calcium. In the chart entitled Chemical Symbols, note that the symbol for calcium is Ca. In the table, find the column labeled "Composition" and locate the symbol Ca. Note that only one variety of garnet, grossular, contains calcium.

47. **1** Find the chart entitled Scheme for Metamorphic Rock Identification in the *Earth Science Reference Tables*. In the column labeled "Composition" locate the name "garnet"; note that it is represented by a shaded column that is two rows high. Follow these rows right to the column labeled "Rock Name." The names of the two metamorphic rocks that contain garnet are schist and gneiss.

48. **2** As question 47 indicates, garnets are found in schists and gneisses. Find the map entitled Generalized Bedrock Geology of New York State. In the section at the lower left labeled "Geological Periods in New York," note that Cambrian and Ordovician bedrock includes schists, and Middle Proterozoic rocks contain gneisses. Also, to the left of their names, note the symbols for rocks formed during these periods, and locate these symbols on the Generalized Bedrock Geology map. You will find Cambrian and Ordovician rock in and just north of New York City, and Middle Proterozoic rock in the area surrounding Old Forge. Now, find the map entitled Generalized Landscape Regions of New York State. Note that one of these areas corresponds to the Adirondack Mountains, the landscape region where garnets are most likely to be found in surface bedrock.

49. **3** Cleavage and fracture are ways of describing how a mineral breaks. Cleavage is the tendency to break parallel to atomic planes of weakness in the crystalline structure, while fracture is the tendency to break unevenly because the atomic planes are equally strong in all directions and the breaking follows no particular direction. Thus, both cleavage and fracture are determined by the internal arrangement of atoms in a mineral.

50. **4** Find the chart entitled Average Chemical Composition of Earth's Crust, Hydrosphere, and Troposphere in the *Earth Science Reference Tables*. In the column labeled "Element" find oxygen. Follow this row right to the columns labeled "Percent by Volume" for the crust, hydrosphere, and troposphere, and compare the values listed: 94.04% in crust, 33% in hydrosphere, and 21% in troposphere. Oxygen is most abundant by volume in the Earth's crust.

GROUP B—Plate Tectonics

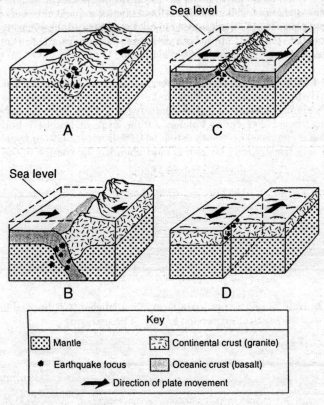

51. **3** Find the chart entitled Tectonic Plates in the *Earth Science Reference Tables*, and locate the boundary between the African plate and the South American plate. Note that the arrows illustrating plate movement on either side of the boundary point away from each another, indicating divergent plates. Note too that this plate boundary falls along the Mid-Atlantic Ridge, is a boundary between two plates of oceanic crust, and is below sea level. Now, refer to the key beneath the diagram in the question. Note that at location *C* in the diagram plates of oceanic crust diverge along a ridge beneath an ocean; therefore, location *C* best represents the boundary between the African plate and the South American plate.

52. **4** Transform boundaries are places where plates move past each other in strike-slip motions. In other words, the plates on either side of a transform boundary smash and grind past each other like the sides of two ships that came too close together and are grinding past each other. Model *D* best represents the type of plate motion associated with transform faults such as those causing California earthquakes.

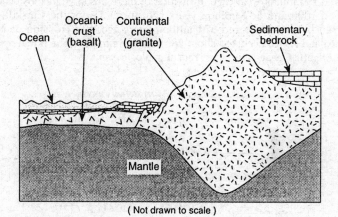

(Not drawn to scale)

53. **2** Note that in the diagram the section labeled "Continental crust" is many times thicker than the section labeled "Oceanic crust."

54. **2** Find the chart entitled Inferred Properties of the Earth's Interior in the *Earth Science Reference Tables*, and locate the graph labeled "Temperature (°C)" along the vertical axis and "Depth (km)" along the horizontal axis. Find the line labeled "2000" on the depth axis, and follow this line upward until it intersects the bold line labeled "Actual Temperature." (This intersection is a point in the stiffer mantle). Now, move horizontally to the left, and note on the vertical axis that the temperature at 2,000 kilometers is between 4,000°C and 5,000°C, a range that is consistent with choice (2).

55. **1** Convection is a method of heat transfer; thus, hot springs on the ocean floor near mid-ocean ridges indicate that the material *beneath* the ocean crust is hot. The ocean above these locations is colder in temperature and is therefore not the source of this heat. The most logical explanation is that material beneath the ocean crust is the heat source. The asthenosphere lies beneath the ocean crust, and convection currents in the asthenosphere would result in heat from deep within the Earth being transferred to the ocean crust. The hot springs thus produced provide evidence that convection currents exist in the asthenosphere.

56. **1** Find the chart entitled Geologic History of New York State at A Glance in the *Earth Science Reference Tables*. In the column labeled "Important Geologic Events in New York," locate the four events listed as choices. At each event, move horizontally to the left until you reach the vertical axis labeled "Millions of Years Ago" (the right-hand edge of the column labeled "Epoch"). Note that the initial opening of the Atlantic Ocean occurred 190 million years ago, formation of the Hudson Highlands occurred between 1300 and 540 million years ago, formation of the Catskill delta occurred between 360 and 374 million years ago, and collision of North America and Africa occurred about 286 million years ago. The initial opening of the Atlantic Ocean is the most recent geologic event.

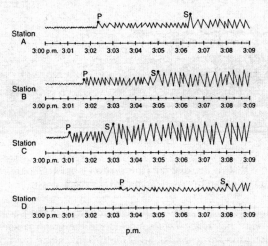

57. **4** *P*-waves travel faster than *S*-waves. Both waves originate simultaneously at the focus of an earthquake; but, as they travel away from the focus, and the epicenter directly above it, the *S*-waves lag farther and farther behind the *P*-waves. Thus, the farther a location from the epicenter, the greater is the difference between the arrival times of the *P*-waves and the *S*-waves. At station *D*, the *S*-waves arrived almost 5 minutes after the *P*-waves, the greatest difference between arrival times; therefore; station *D* is farthest from the epicenter of the earthquake.

58. **3** A minimum of three seismograms is required to determine the location of the epicenter. If circles are drawn on a map with their centers at the three stations, and with diameters equal to their respective distances from the epicenter, the three circles will intersect at one point. This point is the location of the epicenter of the earthquake.

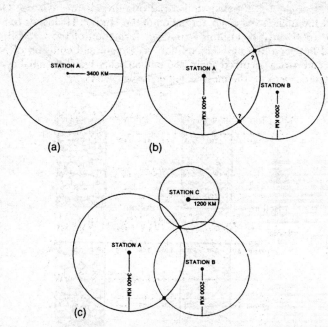

(a)   (b)

(c)

59. **3** Find the Earthquake P-wave and S-wave Travel Time graph in the *Earth Science Reference Tables*. According to the seismogram for station A in the question, the P-waves arrived at 3:02:20 and the S-waves arrived at 3:06:20, a difference in arrival times of 4 minutes. On the Travel Time graph, locate the point where the two curves are 4 minutes apart. Trace this position down to obtain an epicenter distance of 2,600 kilometers.

60. **4** Find the Earthquake P-wave and S-wave Travel Time graph in the *Earth Science Reference Tables*. For the fifth station, the distance to the epicenter is 5,600 kilometers. Locate this distance along the horizontal axis of the graph, and trace upward to the P-wave curve. Note that the P-wave travel time for that distance is 9 minutes. Since the P-wave arrived at this station at 3:06 p.m., the origin time, when the earthquake occurred, was 9 minutes earlier, or 2:57 p.m.

GROUP C—**Oceanography**

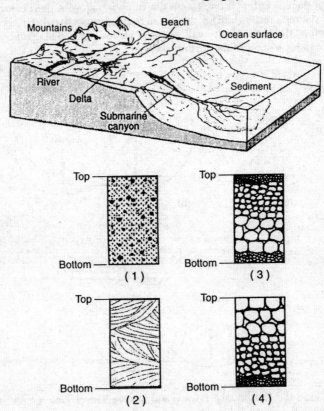

61. **3** Turbidity currents move downslope rapidly and can transport sediments ranging in size from fine to coarse. When these currents reach the end of the slope, they decrease in speed and deposit their sediment load. Sediments settle through water at different rates that depend on the characteristics of the sediments, resulting in sorting of the particles. Diagram (3) represents a cross section of sediment typical of turbidity-current deposits, in which larger particles have sunk to the bottom faster than smaller particles.

62. **2** As can be seen in the diagram, the submarine canyon originates in the shallow waters just off the coast and ends along the steep slope leading to a much deeper portion of the ocean floor. Such steep slopes can be found along the perimeter of every continent and mark the true margins of the continental landmass.

WRONG CHOICES EXPLAINED:

(1) The tidal zone is the portion of the shoreline that is submerged and exposed cyclically by high and low tides. The submarine canyon is too far from shore to fall within this zone.

(3) The deep ocean basin is located much farther offshore, well beyond even the fan-shaped sediment deposit shown in the diagram.

(4) The mid-ocean ridge is located in the center of the deep ocean basin. It is a region of submarine mountains split by a rift. It does not appear in the diagram.

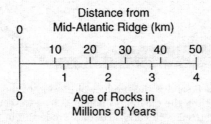

63. **3** Trace right along the upper scale labeled "Distance from Mid-Atlantic Ridge (km)," to a point corresponding to 37 kilometers. Now, look directly below that point at the lower scale, labeled "Age of Rocks in Millions of Years," to find the age of the igneous rocks—3 million years. The igneous rocks formed approximately 3 million years ago.

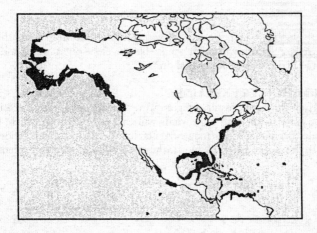

64. **2** Find the Surface Ocean Currents map in the *Earth Science Reference Tables*, and locate the area of the Alaskan coast corresponding to

the shaded areas on the map given in the question. Note that the ocean current off this section of the coast is labeled "N. Pacific C" and is flowing toward and along this area of the coast. This North Pacific Current may have carried polluting material to the Alaska area.

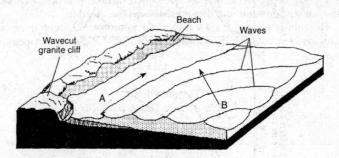

65. **4** The beach sand is the product of erosion of the surrounding land. Note that the beach in the diagram is bounded by a wavecut granite cliff and that a longshore current will carry materials eroded from the cliff along the beach. The beach sand will contain the same minerals as the granite from which it was derived. Find the Scheme for Igneous Rock Identification in the *Earth Science Reference Tables*. Locate the box labeled "Granite," and trace down to the mineral composition graph below it. Note that granite contains the minerals potassium feldspar, quartz, plagioclase feldspar, biotite and hornblende. Only the minerals in choice (4), quartz and potassium feldspar, are both present in granite and are therefore most likely to be found in the beach sand.

66. **3** One of the chief causes of waves is wind blowing across the ocean surface. A large storm with high winds that develops out at sea is most likely to result in higher waves near the shore.

WRONG CHOICES EXPLAINED:
   (1) High winds will result in higher waves, which will increase erosion.
   (2) A large storm with high winds will cause high waves, which, in turn, will create turbulence and strong currents that will decrease deposition.
   (4) High waves caused by the storm will erode the shoreline, changing its features.

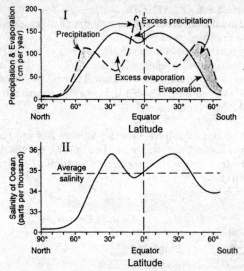

67. **4** Note that in Graph I, precipitation is represented by a dashed line. To compare the amounts of precipitation at the North Pole and the Equator, proceed as follows. Locate 0°—Equator along the horizontal axis labeled "Latitude," and trace upward to the intersection with the dashed line labeled "Precipitation." Then trace left to the intersection with the vertical axis labeled "Precipitation & Evaporation (cm per year)," and note the Equator has about 190 centimeters of precipitation per year. Repeat for 90°—North, and note that the North Pole has about 5 centimeters of precipitation per year. Precipitation is more than ten times greater at the Equator than at the North Pole.

68. **1** In Graph II, find the point along the solid line labeled "Average Salinity" that corresponds to the lowest value on the vertical scale labeled "Salinity of Oceans (parts per thousand)." Trace downward to the horizontal scale labeled "Latitude" to find that 90° N is the latitude at which ocean salinity is *least*.

69. **1** The concentration of dissolved minerals in seawater is the ratio of dissolved minerals to water in a given sample. Concentration can be increased by increasing the amount of dissolved minerals in the sample or by decreasing the amount of water. Rapid evaporation would decrease the amount of water, thereby increasing the concentration of dissolved minerals.

WRONG CHOICES EXPLAINED:

(2) Heavy rainfall would increase the amount of water, thus decreasing the concentration of dissolved minerals.

(3) Melting glaciers would increase the amount of water, thus decreasing the concentration of dissolved minerals.

(4) Tsunamis would neither increase nor decrease the amount of water or the amount of dissolved minerals and thus would have no effect on the concentration of dissolved minerals.

70. **3** Most of the dissolved minerals in seawater are carried into the ocean by streams flowing into the ocean basins. The dissolved minerals contained in these streams are the products of erosion of the continents over which the streams flow.

GROUP D—Glacial Processes

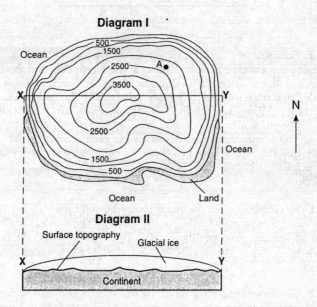

71. **2** Freshly fallen snow is a fluffy mass of delicate ice crystals and pockets of trapped air (atmospheric gases). As more and more snow builds up, melting and refreezing, together with the weight of overlying layers, compact the snow until the ice crystals grow together into a solid mass of glacial ice. Chemical analysis of an ice sample from the central core of the glacier would most likely be used to examine the dust and gases trapped in the glacial ice,

which could then be compared with those in the current atmosphere to determine what, if any, changes have occurred.

WRONG CHOICES EXPLAINED:
(1) Fragments of surface bedrock become frozen into glaciers as the glaciers move along, but the *sub*surface bedrock is untouched by a glacier and would not become part of the ice. Therefore, glacial ice would contain no evidence of the nature of the subsurface bedrock.

(3) A glacier's rate of movement is determined by comparing its position over time with a fixed reference point. Chemical analysis of the ice core would provide no information about position in relation to a reference point.

(4) An ice sheet's location is determined through the use of maps, celestial observations, or satellite data (global positioning system), not chemical analysis.

72. **3** Under pressure, ice behaves like a thick fluid. Just as thick syrup poured on a flat table builds up into a mound that then spreads out to form a circular puddle, snow accumulating in the central zone of an ice sheet exerts pressure on the ice beneath it, which then moves outward in all directions from the central zone of accumulation to form a roughly circular glacier, as shown in Diagram I. Note that glaciers will move down a slope; however, the cross section of Diagram II shows no gradient to produce such movement, in this question.

73. **2** Find the point marking location A in Diagram I. Note that it lies about halfway between the 2500-meter isopach and the next lowest isopach. Since there are two intervals between the 2500-meter line and the 1500-meter line, each interval is 500 meters (1000 meters/2 intervals = 500 meters/interval). Thus, point A lies halfway between the 2500-meter isopach and the 2000-meter isopach. The thickness of the ice at point A is halfway between 2500 and 2000 meters, or 2250 meters.

74. **2** Glaciers form when the snow accumulating during the cold season does not completely melt during the warm season. As a result, some snow is carried over into the next cold season, and the net amount of snow on the ground increases each year.

Diagram I

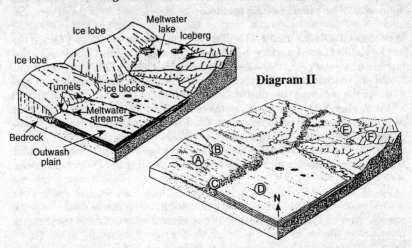

Diagram II

75. **3** As glaciers move across the surface bedrock, fragments of rock become frozen into the ice at the bottom of the glacier. These fragments scrape against the bedrock as the glacier slides over the bedrock, creating parallel grooves called glacial striations.

76. **1** Till is directly deposited from ice when rock particles are set down on the ground as the ice beneath them melts or they slide or fall off the melting ice. Till is found in places that were once covered by glacial ice. Locate in Diagram II the regions that correspond to ice lobes in Diagram I. Note that features *A*, *B* and *C* are located in a region that was once covered by glacial ice. Thus, features *A* and *C* are composed of till directly deposited by the glacial ice.

77. **4** The information in the question states that Diagram I represents the Pleistocene Epoch and Diagram II represents the present. Fossils found in areas such as *D* in Diagram II were buried in the sediments deposited by glacial meltwater streams during the Pleistocene. Find the chart entitled Geologic History of New York State at A Glance in the *Earth Science Reference Tables*. Locate the column labeled "Epoch," and trace down to the Pleistocene. Then trace right to the column labeled "Life on Earth," and note the organisms listed. Only choice (4), the mastodont, is an organism that lived during the Pleistocene. Fossils of mastodonts have been found in locations like the outwash plains of a Pleistocene glacier.

78. **2** The information in the question states that Diagram I represents the Pleistocene Epoch and Diagram II represents the present. Between the Pleistocene and the present, most of the glaciers that existed during the Pleistocene melted. This meltwater has run off into the oceans and caused the sea level to increase.

79. **4** Glaciers deposit unconsolidated sediments such as sand, gravel, and clay. Refer to the chart entitled Geologic History of New York State at A Glance in the *Earth Science Reference Tables*. The last glaciers covered New York State during the Pleistocene epoch. Find the map entitled Generalized Bedrock Geology of New York State in the *Earth Science Reference Tables*. Locate, in "Geological Periods in New York" at the lower left, the symbol for materials deposited during the Cretaceous, Tertiary, and Pleistocene epochs, and note that surface materials of this age are mostly unconsolidated gravels, sands, and clays.

Locate this symbol on the Generalized Bedrock Geology map and note that, of the four locations listed as choices, it is found only on Long Island.

80. **1** Glacial motion is caused primarily by gravity acting directly on the glacial ice. Since glacial ice behaves like a thick fluid, gravity causes it to spread outward and to flow downhill.

GROUP E—Atmospheric Energy

81. **1** The main source of energy in the atmosphere is radiation from the Sun, which is either directly absorbed by the atmosphere on its way down, or absorbed by the Earth's surface and reradiated as infrared radiation, which is almost totally absorbed by the atmosphere. This solar radiation provides the most energy for atmospheric weather changes.

WRONG CHOICES EXPLAINED:

(2) Radioactivity from the Earth's interior is only a minor source of atmospheric energy because most of it is absorbed by rocks in the Earth's crust and mantle. Rocks, which are poor conductors, release this heat at a very slow rate.

(3) The oceans act as both a heat source and a heat sink for the atmosphere and are in radiative balance with the atmosphere. Near the Equator, the air temperature is almost always higher than the water temperature, so the oceans act as a heat sink. Near the poles the converse is true. In temperate zones the oceans alternate from being a heat sink in the summer to being a heat source in the winter. Overall, the net gain equals the net loss.

(4) Heat stored in polar ice caps will flow into the atmosphere only if the atmosphere is colder than the ice caps. Currently, the polar ice caps are slowly melting, indicating that they are absorbing more heat from the atmosphere than they are reradiating back into it.

82. **2** Find the chart entitled Electromagnetic Spectrum in the *Earth Science Reference Tables*. Note that the scale labeled "Wavelength in meters" is in units of powers of 10, or scientific notation. Such units have the form $M \times 10^n$, where M is a number with a single nonzero digit to the left of the decimal point and n is a positive or negative exponent. To find M, shift the decimal point right or left in the original number until there is one nonzero digit to the left of it. To find n, count the number of places the decimal point has been shifted. If the decimal point has been shifted left, n is positive; if it has been shifted right, n is negative. Thus, a wavelength of 0.0001 meter becomes 1×10^{-4} meter, or simply 10^{-4} meter.

The Electromagnetic Spectrum chart indicates that the form of electromagnetic energy with a wavelength of 0.0001 (10^{-4}) meter is infrared rays.

83. **4** Synoptic forecasting is one of the methods of weather forecasting most commonly used today. Synoptic charts, or weather maps, depict the positions of air masses, the frontal boundaries between them, and storms. Studying a sequence of maps over time is the basis for projecting the future positions of air masses, their frontal boundaries, and associated storms and thus for forecasting the weather.

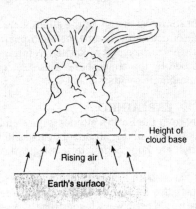

84. **1** As *moist* air rises, both the air temperature and the dewpoint temperature decrease due to adiabatic (without the gain or loss of heat) changes, but not at the same rate. When the air temperature reaches the dewpoint, condensation occurs. The cloud base in the diagram marks the point where

condensation first occurs. At the altitude of the cloud base, air temperature equals the dewpoint.

Find the chart entitled Lapse Rate in the *Earth Science Reference Tables*. First locate the intersection of 0 kilometers altitude and 20°C, and note the solid diagonal line labeled "Dry Adiabatic Lapse Rate 10°C/km" passing through it. Next, locate the intersection of 0 kilometers altitude and 12°C, and note the dashed diagonal labeled "Dewpoint Lapse Rate 2°C/km" passing through it. Trace the Dewpoint Lapse Rate 2°C/km dashed line (originating at 12°C) upward to the left until it intersects with the 20°C solid line. This intersection occurs at the point shown below, which represents the point where the air temperature and dewpoint temperature are equal and condensation occurs. Read to the left, and find that this occurs at an altitude of 1 km.

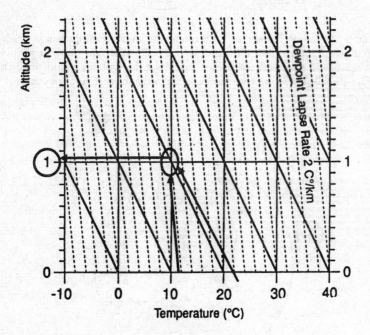

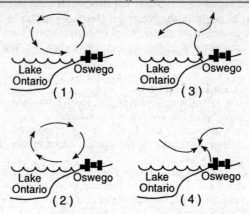

85. 1 On a very hot summer afternoon, the land at Oswego would reach a higher temperature than the water of Lake Ontario. The warmer air over Oswego would rise, and the cooler air over Lake Ontario would sink. The result would be a circular pattern of convection as shown in cross section (1).

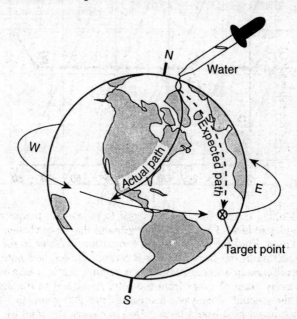

86. 3 The curved-path phenomenon, called the Coriolis effect, is the tendency of all matter in motion *over* the Earth's surface to be deflected to the right of its point of origin in the Northern Hemisphere and to the left in the Southern Hemisphere. Atmospheric wind belts are most directly affected by the curved-path phenomenon because they move *over* the Earth's surface as the Earth rotates.

WRONG CHOICES EXPLAINED:

(1) The Coriolis effect is an *effect* of the Earth's rotation, not a *cause* of either rotation or the tilt of the Earth's axis of rotation.

(2) The Moon's phases are due to the changing position of the Moon's illuminated half surface in relation to the Earth, caused by the Moon's revolution, not the Earth's rotation.

(4) The Earth's tectonic plates rotate along with the Earth and are bounded by other plates that limit their motion. Tectonic plate motion is most directly affected by convection currents beneath them in the upper mantle.

87. **2** Heating of the water began at time = 0 minute. On the horizontal axis, labeled "Time (min)," locate the value 20 minutes. Trace upward to the intersection with the solid line of the heating curve. Then trace left to the vertical axis, labeled "Temperature (°C)," and note that the temperature corresponding to 20 minutes of heating is 100°C.

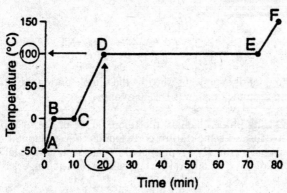

88. **3** Melting the ice involves heating it to its melting temperature and then adding latent heat of 80 calories per gram to the ice to change it from a solid to a liquid. Find the chart entitled Properties of Water in the Physical Constants section of the *Earth Science Reference Tables*, and note that the latent heat of fusion of water is 80 calories per gram; that is, it take 80 calories to change every gram of water from the solid state (ice) to the liquid state (water). If the amount of ice were decreased from 200 grams to 100 grams, half as much latent heat would be required to change the solid ice to liquid water because there would be half as many grams of ice. Since the ice is heated at a uniform rate, it would take half the time to supply the latent heat needed to melt 100 grams of ice.

WRONG CHOICES EXPLAINED:

(1) Using colder ice would require additional heat to raise the temperature of the ice to its melting point. Adding the additional heat would lengthen, not shorten, the time required to melt the ice.

(2) Stirring the sample more slowly would decrease the rate at which heat was brought into contact with the ice and would lengthen, not shorten, the time required to melt the ice.

(4) Reducing the number of temperature readings would simply result in fewer data; this change would have no effect on the amount of time required to melt the ice.

89. **4** The graph shows a horizontal line between points B and C. At this time the water was changing from a solid to a liquid. Find the chart entitled Properties of Water in the Physical Constants section of the *Earth Science Reference Tables*, and note that the latent heat of fusion of water is 80 calories per gram; that is, it takes 80 calories to change every gram of water from the solid state (ice) to the liquid state (water.) Now, find the latent heat equation in the chart entitled Equations and Proportions in the *Earth Science Reference Tables*. Substitute a mass of 200 grams and a heat of fusion of 80 calories per gram in the equation, and solve for heat:

$$Q = mHf$$
$$= 200 \, \cancel{g} \times 80 \frac{\text{cal}}{\cancel{g}} = 16{,}000 \text{ cal}$$

The total amount of energy absorbed by the sample during the time between points B and C on the graph was 16,000 calories.

90. **4** During phase changes, the temperature of the material remains constant. A constant temperature corresponds to a horizontal line on the graph. Choice (4), D to E, is represented by one such horizontal line on the graph and therefore is an interval during which a phase change is occurring.

GROUP F—Astronomy

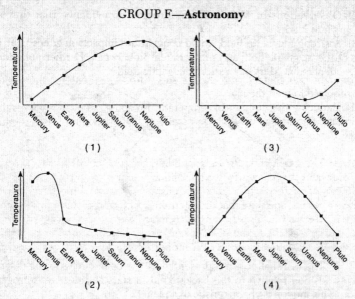

91. **2** The average temperatures of the planets are directly related to their mean distances from the Sun. The farther a planet is from the Sun, the less intense the solar radiation it receives, and the colder it will be. However, Venus has a higher than expected temperature because of the greenhouse effect associated with its very dense atmosphere. Process of elimination makes Graph (2) the best choice to show this relationship.

92. **1** Find the chart entitled Solar System Data in the *Earth Science Reference Tables*. In the column labeled "Mean Distance from Sun," note that a belt of asteroids 503 million kilometers from the Sun would be located between Mars and Jupiter.

93. **4** Impact structures are the result of rock fragments striking the surface of a planet. Where rock fragments must pass through an atmosphere before reaching the planet's surface, friction from interaction with the atmospheric gases heats the fragments and they often burn up before reaching the surface. Since Mars's atmosphere is very thin, there is less friction when rock fragments enter its atmosphere. Therefore, the fragments are more likely to reach the surface and create impact structures (craters).

WRONG CHOICES EXPLAINED:
(1) According to the chart entitled Solar System Data in the *Earth Science Reference Tables*, Mars has a relatively small diameter and would therefore have less, not more, surface area.

(2) Observations of Mars's moons provide no evidence that they are breaking into pieces.

(3) A strong magnetic field would concentrate impacts in zones near the poles of the magnetic field and protect areas in between. Furthermore, there is no evidence that Mars has a strong magnetic field.

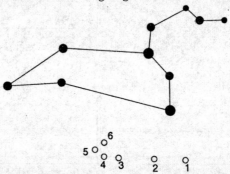

94. **2** The celestial object represented by the open circles in the diagram changes position relative to the background of stars and displays retrograde motion. This motion is characteristic of a planet.

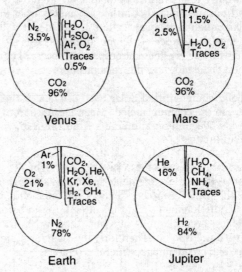

95. **1** Each gas in a planet's atmosphere is represented by the molecular formula for that gas within the circle. Comparison of the graphs indicates that, of the gases listed as choices, argon (Ar) is present in the atmosphere of Venus, Earth, and Mars but is absent from the atmosphere of Jupiter.

96. **1** In a pie graph, the area that a value occupies corresponds to its amount. Note that both Venus and Mars have atmospheres composed primarily of CO_2. Find the chart entitled Solar System Data in the *Earth Science Reference Tables*. In the columns labeled "Period of Revolution" and "Period of Rotation," note that Venus's period of rotation is greater than its period of revolution.

97. **4** A model in which the Sun, other stars, and the Moon are in orbit around the Earth has the Earth at its center. This model is called an Earth-centered or geocentric model.

98. **3** A pendulum swings back and forth in a straight line. If the pendulum's direction of swing is traced along the Earth's surface, it appears to be changing because the Earth is rotating on its axis beneath the pendulum while the pendulum is swinging. The change in direction of swing is caused by the Earth's rotation on its axis.

99. **4** The only planet in our solar system with large amounts of liquid water on its surface is Earth. Roughly two-thirds of the Earth's surface is covered with water. All of the other planets are either too cold or too hot for water at their surface to be in a liquid state.

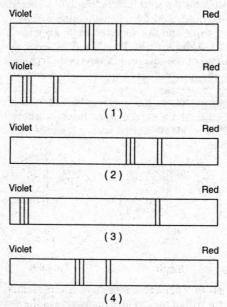

100. **2** Light consists of electromagnetic waves. If a source emits waves of a particular length, then a certain number of wavecrests will reach an

observer every second. If the source were emitting shorter waves, more wavecrests per second would reach the observer. If, however, the source is moving toward the observer, the observer would also perceive more crests per second, a phenomenon called the Doppler effect.

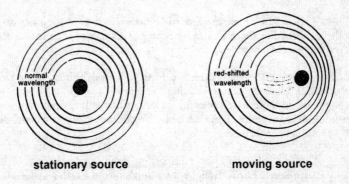

stationary source   **moving source**

The Doppler effect is the shift in wavelength of a spectrum line away from its normal wavelength, caused by relative motion between the observer and the source. If the source and the observer are moving toward each other, the light seems to have a shorter wavelength and its spectrum lines are shifted toward the violet end of the spectrum. Conversely, if the source and observer are moving apart, the spectrum lines appear shifted toward the red end of the spectrum.

Observations of objects in the universe indicate that the universe is expanding. Distant galaxies are moving away from us. Therefore, spectral lines from a distant galaxy would be shifted toward the red end of the spectrum, as shown in diagram (2).

Temperature Field Map (°C)

| 0 | 1.0 | 2.0 | 3.0 | 4.0 |

meters

PART III

101. Isotherms are lines connecting points of equal temperature. Begin by connecting the points labeled "23." Where adjacent points are labeled "22" and "24" it can be inferred that between these points lies a location whose temperature is 23°, so the isotherm you are drawing can pass between these points. Continue until you reach the edge of the map or the isotherm closes on itself. Repeat the process to draw the 24°C and 25°C isotherms. Your finished isoline map should match the one that follows.

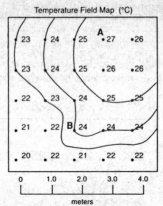

Temperature Field Map (°C)

The isotherms may be drawn as straight or as smoothly curved lines. The 25°C isotherm may be drawn in a circular pattern with closure around the location of the source or extended to the top and right sides of the map.

Two credits are allowed if all three lines are correctly drawn, and 1 credit if two are correctly drawn. If more than the three lines requested are drawn, 2 credits are awarded only if *all* lines are correctly drawn.

102. Find the chart entitled Equations and Proportions in the *Earth Science Reference Tables.*

a. Locate the equation for gradient and write it on the answer sheet.

$$\text{gradient} = \frac{\text{change in field value}}{\text{change in distance}}$$

or

$$\text{gradient} = \frac{\Delta \text{ temperature}}{\Delta \text{ distance}}$$

One credit is allowed for correctly writing the equation. The answer must be given in the form of an equation, which must include "gradient =" or "g =."

b. Note that point A is labeled 27 and point B is labeled 24. Using a piece of scrap paper, mark off the distance between points A and B. Compare this marked-off distance to the scale labeled "meters," and note that points A and B are 3 meters apart.

Substitute the data into the equation:

$$\text{gradient} = \frac{27°C - 24°C}{3\ m} = \frac{3°C}{3\ m}$$

or

$$\text{gradient} = \frac{3°C}{3\ m}$$

One credit is allowed for correctly substituting data into the equation.

c. Calculate the gradient, and state in simplest terms:

$$1°C/meter \quad or \quad 1°C/m$$

One credit is allowed for correctly calculating the gradient based upon your answer to part *b.* The answer must be labeled with the proper units to receive this credit.

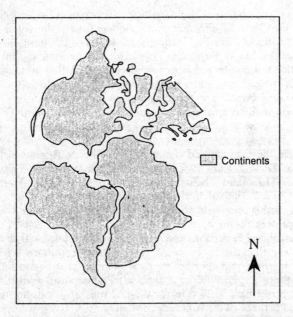

Continents

N

103. The directions include the information that Pangaea began to break up approximately 220 million years ago. Find the chart entitled Geologic History of New York State at A Glance in the *Earth Science Reference Tables*. Locate the column labeled "Period," and the scale labeled "Millions of Years Ago" along the right-hand edge of the column labeled "Epoch." Trace down this scale to 220 million years ago, then trace left to the "Period" column and note that 220 million years ago falls within the Triassic period.

On the far right of the same chart, the column labeled "Inferred Position of Earth's Landmasses" shows diagrams for the Triassic and Cretaceous periods. Note that Pangaea is together at the start of the Triassic period but apart by the Cretaceous period. Because the diagrams show only a single moment in time, in theory, Pangaea could have started to move apart during the Jurassic period, which is between the Triassic and Cretacious periods. Thus, Jurassic would also be an acceptable answer.

One credit is allowed for an answer of Triassic or Jurassic.

104. Any statement of scientifically correct evidence that supports the inference that Pangaea once existed is acceptable.

Examples of acceptable answers:
    The continents fit together like pieces of a jigsaw puzzle.
    Fossils correlate between continents.

One credit is allowed for any scientifically correct answer.

105. Find the chart entitled Geologic History of New York State at A Glance in the *Earth Science Reference Tables*. Locate the column labeled "Inferred Position of Earth's Landmasses," and note the positions of North America (shaded black) during the Triassic, the Cretaceous, and the Tertiary. Note also that this continent has moved successively farther north and west or northwest. (The compass positions can be inferred from the latitude and longitude lines superimposed on the diagrams.)

One credit is allowed for an answer of northwest (NW), north (N), or west (W).

106. Find the map entitled Generalized Landscape Regions of New York State in the *Earth Science Reference Tables*, and locate the region labeled "Adirondack Mountains." Next, find the map entitled Generalized Bedrock Geology of New York State in the *Earth Science Reference Tables*, and locate the region corresponding to the Adirondack Mountains. Note the symbols used to represent the rocks in this region, and find these symbols in the key labeled "Geological Periods in New York" in the lower left-hand corner of the map. You see that the rocks in the Adirondack Mountains are gneisses, quartzites, marbles, and anorthositic rocks, which, according to the key, are "Intensely Metamorphosed Rocks." Intense metamorphism would destroy any fossils that might have existed in the original rock, so fossils are *not* usually found in the bedrock of the Adirondack Mountains.

Examples of acceptable answers:
    Fossils are usually found in sedimentary rock.
    The Adirondacks are composed of metamorphic rock.

One credit is allowed for any scientifically correct answer.

107. Find the chart entitled Geologic History of New York State at A Glance in the *Earth Science Reference Tables*, and locate the column labeled "Important Geologic Events in New York." Trace down the column to the notation "Catskill Delta forms." Deltas are depositional features composed of sediments.

Next, find the map entitled Generalized Landscape Regions of New York State in the *Earth Science Reference Tables*, and locate the region labeled "The Catskills." Now, find the map entitled Generalized Bedrock Geology of New York State in the *Earth Science Reference Tables*, locate the region corresponding to the Catskills, and note the symbols used to represent the rocks in this region. Find these symbols in the key labeled "Geological Periods in New York" in the lower left-hand corner of the map. Note that the rocks in the Catskills are dominantly sedimentary.

For sedimentary rocks to physically resemble a mountainous region, the layers of sedimentary rock would need to be uplifted and eroded by an agent of erosion such as running water or glaciers.

One credit is allowed for an answer of running water, streams, rivers, or glaciers.

108. Find the map entitled Generalized Landscape Regions of New York State in the *Earth Science Reference Tables,* and locate the region labeled "The Catskills." Next, find the map entitled Generalized Bedrock Geology of New York State in the *Earth Science Reference Tables,* locate the region corresponding to the Catskills, and note the symbol used to represent the rocks in this region. Find this symbol in the key labeled "Geological Periods in New York" in the lower left-hand corner of the map. Note that the rocks in the Catskills are dominantly sedimentary rocks of Devonian age.

Find the chart entitled Geologic History of New York State at A Glance in the *Earth Science Reference Tables.* Locate the column labeled "Period," and the scale labeled "Millions of Years Ago" along the right-hand edge of the column labeled "Epoch." Trace down the "Period" column to the Devonian and then right to the "Millions of Years Ago" scale to find that the Devonian occurred 360–408 million years ago.

One credit is allowed for an answer of 360–408 million years or Devonian Age.

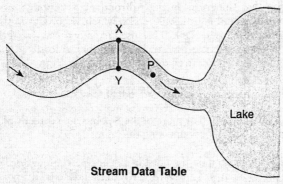

Stream Data Table

	Location X							Location Y
Distance from X (meters)	0	5	10	15	20	25	30	35
Depth of Water (meters)	0	5.0	5.5	4.5	3.5	2.0	0.5	0

109. According to the table in the question, the depth of the water ranges from 0.5 meter to 5.5 meters. The grid provided has six spaces along the vertical axis. Therefore, an appropriate scale would range from 0 to 6 meters, in 1-meter intervals. Since the top of the depth axis is labeled 0 meter to represent the water surface, number downward from this point as shown for question 110.

One credit is allowed for a scale that has numbers at equal intervals and accommodates all of the data.

110. Plot the data for the depths of the water, and connect the points as shown.

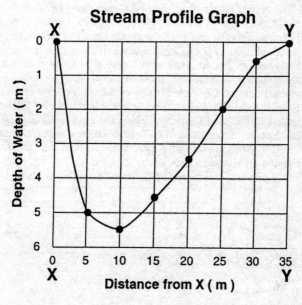

One credit is allowed if 6 or more of the 8 points are correctly plotted.
One credit is allowed for connecting the points.

111. Point *X* is located on the outside of a curve, where water flows faster, and erosion predominates. Therefore, the water tends to be deeper near point *X* than near point *Y*.

Examples of acceptable answers:
   Stream water moves faster on the outside of a curve.
   More deposition occurs on the inside of a bend.

One credit is allowed for any scientifically correct statement.

112. Find the graph entitled Relationship of Transported Particle Size to Water Velocity in the *Earth Science Reference Tables*. Since the water velocity at point *P* is 100 centimeters per second, locate the value 100 along the axis labeled "Stream Velocity (cm/sec)." Trace upward to the intersection with the bold line, and note that the particle diameter falls in the range labeled "Pebble."

One credit is allowed for the answer pebble.

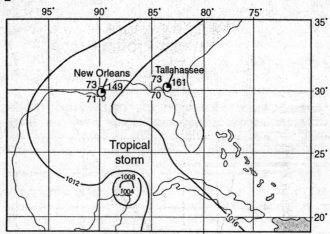

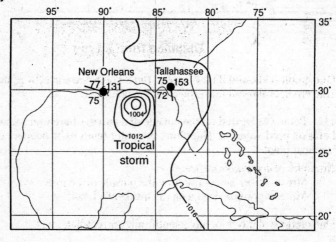

113. Find the chart entitled Weather Map Information in the *Earth Science Reference Tables*, and locate the key to the Station Model on the left side of the chart. Compare this key to the station model for Tallahassee on the July 2 weather map in the question. The dewpoint is 70°F.

One credit is allowed for the answer 70°F.

114. Windspeed is directly related to pressure gradient. The steeper the pressure gradient, the greater is the windspeed. A steep pressure gradient is indicated by closely spaced isobars on a weather map. Note that the isobars along the northern edge of the storm are more closely spaced on the map on July 3 than on the map on July 2. From this fact it can be inferred that there was an increase in windspeed from July 2 to July 3.

Examples of acceptable answers:
> Windspeed is higher where the isobars are spaced closer together.
> The storm has moved closer to these cities.

One credit is allowed for a scientifically correct answer.

One credit is allowed if the answer is stated in one or more complete sentences.

(No credit is allowed if the answer is scientifically incorrect, even if it is stated in a complete sentence.)

115. When the storm moves over land, the supply of moisture evaporating into the storm is cut off. As a result, the air becomes drier and more dense, and the air pressure begins to rise in the center of the storm. This, in turn, decreases the pressure gradient, and windspeed decreases.

One credit is allowed for an answer stating that windspeed will decrease.

Diagram I

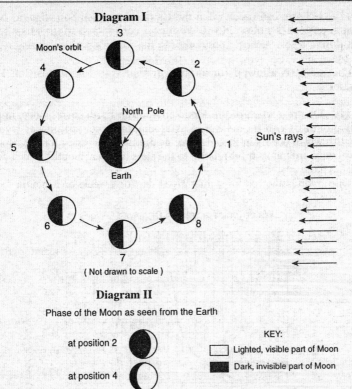

Moon's orbit

3

4

2

North Pole

Earth

5

1

Sun's rays

6

8

7

(Not drawn to scale)

Diagram II

Phase of the Moon as seen from the Earth

at position 2

at position 4

KEY:

☐ Lighted, visible part of Moon

■ Dark, invisible part of Moon

116. Position 7 in the diagram corresponds to a third-quarter phase Moon. In this phase, the left-hand side of the Moon appears illuminated to an observer on Earth. The illuminated half may appear tilted, depending on the relative position of the Moon and the plane of the ecliptic.

Examples of acceptable answers:

or   or

One credit is allowed if the right half of the Moon is shaded.

117. Eclipses can occur when the Earth, Moon, and Sun align, as occurs in both position 1 and position 5. Position 1 corresponds to the Moon's position during a solar eclipse; position 5, to the Moon's position during a lunar eclipse.

One point is allowed for an answer that states *both* position 1 and position 5.

118. One revolution of the Moon around the Earth corresponds, in time, to 1 month. It can be measured as a complete cycle of phases (synodic month), which occurs in 29.5 days, or as the time necessary for the Moon to travel once around its orbit relative to the stars (sidereal month), which occurs in 27. 3 days.

One credit is allowed for stating 27–31 days or 4 weeks or 1 month.

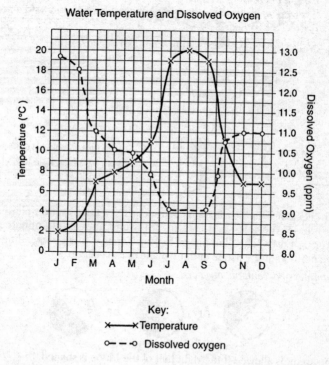

Water Temperature and Dissolved Oxygen

Key:
×———×Temperature
○— –○ Dissolved oxygen

119. The key to the diagram indicates that temperature is shown by a solid line; this line connects points marked by "X"s. On the horizontal axis, labeled "Month," locate J, representing January. Trace this line upward until it intersects with the temperature line. Then trace horizontally to the left to the

vertical axis, labeled "Temperature (°C)," and note that the temperature in January is 2°C. Repeat this process for August, and note that in August the temperature is 20°C. The difference in temperature between January and August is 20°C – 2°C, or 18°C.

One credit is allowed for an answer of 18°C.

120. Note that the line representing temperature and the line representing dissolved oxygen always slope in opposite directions. When temperature increases (from January to June), dissolved oxygen decreases. Conversely, when temperature decreases (from August to November), dissolved oxygen increases. The data in the graph show an indirect relationship.

Examples of acceptable answers:
> As the temperature of the water increases, the amount of dissolved oxygen decreases.
> There is an indirect relationship between temperature and amount of dissolved oxygen.

One credit is allowed for indicating an indirect relationship.

Unit (1–9) or Optional/Extended Topic (A–F)	Question Numbers (total)	Wrong Answers (x)	Grade
1. Earth Dimensions	1, 2, 20, 31, 101, 102, 109, 110 (8)		$\dfrac{100(8-x)}{8} = \%$
2. Minerals and Rocks	3–6 (4),		$\dfrac{100(4-x)}{4} = \%$
A. Rocks, Minerals and Resources	41–43, 45–47, 50, (8)		$\dfrac{100(8-x)}{8} = \%$
3. The Dynamic Crust	7, 8, 51, 103–106 (7)		$\dfrac{100(7-x)}{7} = \%$
B. Earthquakes and the Earth's Interior	52–55, 57–60 (8)		$\dfrac{100(8-x)}{8} = \%$
4. Surface Processes Landscapes	9–13, 107, 111, 112 (8)		$\dfrac{100(8-x)}{8} = \%$
C. Oceanography	61–70 (10)		$\dfrac{100(10-x)}{10} = \%$
D. Glacial Geology	71–80 (10)		$\dfrac{100(10-x)}{10} = \%$
5. Earth's History	14–17, 19, 56, 108 (7)		$\dfrac{100(7-x)}{7} = \%$
6. Meteorology	18, 21–23, 25, 28–30, 83, 85, 86, 113–115 (14)		$\dfrac{100(14-x)}{14} = \%$
E. Latent Heat and Atmospheric Energy	81–90 (10)		$\dfrac{100(10-x)}{10} = \%$
7. Water Cycle and Climates	24, 26, 27, 40 (4)		$\dfrac{100(4-x)}{4} = \%$
8. The Earth in Space	31–38, 116–118 (11)		$\dfrac{100(11-x)}{11} = \%$
F. Astronomy Extensions	91–100 (10)		$\dfrac{100(10-x)}{10} = \%$
9. Environmental Awareness	115 (1)		$\dfrac{100(1-x)}{1} = \%$

To further pinpoint your weak areas, use the Topic Outline in the front of the book.